STUDENT'S SOLUTIONS MANUAL

VOLUMES TWO AND THREE: CHAPTERS 21–44

SEARS & ZEMANSKY'S

UNIVERSITY PHYSICS

14TH EDITION

WAYNE ANDERSON
A. LEWIS FORD

PEARSON

Editor in Chief, Physical Sciences: Jeanne Zalesky
Executive Editor: Nancy Whilton
Project Manager: Beth Collins
Program Manager: Katie Conley
Development Manager: Cathy Murphy
Program and Project Management Team Lead: Kristen Flathman
Production Management, Composition, Illustration, and Proofreading: Lumina Datamatics
Marketing Manager: Will Moore
Manufacturing Buyer: Maura Zaldivar-Garcia
Cover Designers: Cadence Design Studio and Seventeenth Street Design
Cover and Interior Printer: Edwards Brothers Bar/Jackson Road
Cover Photo Credit: Knut Bry
About the Cover Image: www.leonardobridgeproject.org
The Leonardo Bridge Project is a project to build functional interpretations of Leonardo da Vinci's Golden Horn Bridge design, conceived and built first in Norway by artist Vebjørn Sand as a global public art project, linking people and cultures in communities in every continent.

Copyright ©2016, 2012, 2008 Pearson Education, Inc. All Rights Reserved. Printed in the United States of America. This publication is protected by copyright, and permission should be obtained from the publisher prior to any prohibited reproduction, storage in a retrieval system, or transmission in any form or by any means, electronic, mechanical, photocopying, recording, or otherwise. For information regarding permissions, request forms and the appropriate contacts within the Pearson Education Global Rights & Permissions department, please visit www.pearsoned.com/permissions/.

PEARSON, ALWAYS LEARNING and MasteringPhysics are exclusive trademarks in the U.S. and/or other countries owned by Pearson Education, Inc. or its affiliates.

Unless otherwise indicated herein, any third-party trademarks that may appear in this work are the property of their respective owners and any references to third-party trademarks, logos or other trade dress are for demonstrative or descriptive purposes only. Such references are not intended to imply any sponsorship, endorsement, authorization, or promotion of Pearson's products by the owners of such marks, or any relationship between the owner and Pearson Education, Inc. or its affiliates, authors, licensees or distributors.

www.pearsonhighered.com

ISBN 10: 0-13-396928-2
ISBN 13: 978-0-13-396928-3
1 2 3 4 5 6 7 8 9 10—**V031**—18 17 16 15

Contents

	Preface	v
Part I	**Mechanics**	
	Chapter 1 Units, Physical Quantities, and Vectors	1-1
	Chapter 2 Motion Along a Straight Line	2-1
	Chapter 3 Motion in Two or Three Dimensions	3-1
	Chapter 4 Newton's Laws of Motion	4-1
	Chapter 5 Applying Newton's Laws	5-1
	Chapter 6 Work and Kinetic Energy	6-1
	Chapter 7 Potential Energy and Energy Conservation	7-1
	Chapter 8 Momentum, Impulse, and Collisions	8-1
	Chapter 9 Rotation of Rigid Bodies	9-1
	Chapter 10 Dynamics of Rotational Motion	10-1
	Chapter 11 Equilibrium and Elasticity	11-1
	Chapter 12 Fluid Mechanics	12-1
	Chapter 13 Gravitation	13-1
	Chapter 14 Periodic Motion	14-1
Part II	**Waves/Acoustics**	
	Chapter 15 Mechanical Waves	15-1
	Chapter 16 Sound and Hearing	16-1
Part III	**Thermodynamics**	
	Chapter 17 Temperature and Heat	17-1
	Chapter 18 Thermal Properties of Matter	18-1
	Chapter 19 The First Law of Thermodynamics	19-1
	Chapter 20 The Second Law of Thermodynamics	20-1

Part IV Electromagnetism

- **Chapter 21** Electric Charge and Electric Field ... 21-1
- **Chapter 22** Gauss's Law ... 22-1
- **Chapter 23** Electric Potential ... 23-1
- **Chapter 24** Capacitance and Dielectrics ... 24-1
- **Chapter 25** Current, Resistance, and Electromotive Force ... 25-1
- **Chapter 26** Direct-Current Circuits ... 26-1
- **Chapter 27** Magnetic Field and Magnetic Forces ... 27-1
- **Chapter 28** Sources of Magnetic Field ... 28-1
- **Chapter 29** Electromagnetic Induction ... 29-1
- **Chapter 30** Inductance ... 30-1
- **Chapter 31** Alternating Current ... 31-1
- **Chapter 32** Electromagnetic Waves ... 32-1

Part V Optics

- **Chapter 33** The Nature and Propagation of Light ... 33-1
- **Chapter 34** Geometric Optics ... 34-1
- **Chapter 35** Interference ... 35-1
- **Chapter 36** Diffraction ... 36-1

Part VI Modern Physics

- **Chapter 37** Relativity ... 37-1
- **Chapter 38** Photons: Light Waves Behaving as Particles ... 38-1
- **Chapter 39** Particles Behaving as Waves ... 39-1
- **Chapter 40** Quantum Mechanics I: Wave Functions ... 40-1
- **Chapter 41** Quantum Mechanics II: Atomic Structure ... 41-1
- **Chapter 42** Molecules and Condensed Matter ... 42-1
- **Chapter 43** Nuclear Physics ... 43-1
- **Chapter 44** Particle Physics and Cosmology ... 44-1

PREFACE

This Student's Solutions Manual, Volumes 2 and 3, contains detailed solutions for approximately one-third of the Exercises and Problems in Chapters 21 through 44 of the Fourteenth Edition of *University Physics* by Roger Freedman and Hugh Young. The Exercises and Problems included in this manual are selected solely from the odd-numbered Exercises and Problems in the text (for which the answers are tabulated at the back of the textbook).

The Exercises and Problems included were not selected at random but rather were carefully chosen to include at least one representative example of each problem type. The remaining Exercises and Problems, for which solutions are not given here, constitute an ample set of problems for you to tackle on your own. In addition, there are the Challenge Problems in the text for which no solutions are given here.

This manual greatly expands the set of worked-out examples that accompanies the presentation of physics laws and concepts in the text. This manual was written to provide you with models to follow in working physics problems. The problems are worked out in the manner and style in which you should carry out your own problem solutions.

The Student's Solutions Manual Volume 1 companion volume is also available from your college bookstore.

Wayne Anderson
Lewis Ford
Sacramento, CA

21

ELECTRIC CHARGE AND ELECTRIC FIELD

21.1. **(a) IDENTIFY and SET UP:** Use the charge of one electron $(-1.602 \times 10^{-19}$ C) to find the number of electrons required to produce the net charge.
EXECUTE: The number of excess electrons needed to produce net charge q is

$$\frac{q}{-e} = \frac{-3.20 \times 10^{-9} \text{ C}}{-1.602 \times 10^{-19} \text{ C/electron}} = 2.00 \times 10^{10} \text{ electrons.}$$

(b) IDENTIFY and SET UP: Use the atomic mass of lead to find the number of lead atoms in 8.00×10^{-3} kg of lead. From this and the total number of excess electrons, find the number of excess electrons per lead atom.
EXECUTE: The atomic mass of lead is 207×10^{-3} kg/mol, so the number of moles in 8.00×10^{-3} kg is

$$n = \frac{m_{\text{tot}}}{M} = \frac{8.00 \times 10^{-3} \text{ kg}}{207 \times 10^{-3} \text{ kg/mol}} = 0.03865 \text{ mol.} \quad N_A \text{ (Avogadro's number) is the number of atoms in 1 mole,}$$

so the number of lead atoms is $N = nN_A = (0.03865 \text{ mol})(6.022 \times 10^{23} \text{ atoms/mol}) = 2.328 \times 10^{22}$ atoms.

The number of excess electrons per lead atom is $\frac{2.00 \times 10^{10} \text{ electrons}}{2.328 \times 10^{22} \text{ atoms}} = 8.59 \times 10^{-13}$.

EVALUATE: Even this small net charge corresponds to a large number of excess electrons. But the number of atoms in the sphere is much larger still, so the number of excess electrons per lead atom is very small.

21.3. **IDENTIFY and SET UP:** A proton has charge $+e$ and an electron has charge $-e$, with $e = 1.60 \times 10^{-19}$ C.

The force between them has magnitude $F = k\frac{|q_1 q_2|}{r^2} = k\frac{e^2}{r^2}$ and is attractive since the charges have opposite sign. A proton has mass $m_p = 1.67 \times 10^{-27}$ kg and an electron has mass 9.11×10^{-31} kg. The acceleration is related to the net force $\vec{F}$ by $\vec{F} = m\vec{a}$.

EXECUTE: $F = k\frac{e^2}{r^2} = (8.99 \times 10^9 \text{ N} \cdot \text{m}^2/\text{C}^2)\frac{(1.60 \times 10^{-19} \text{ C})^2}{(2.0 \times 10^{-10} \text{ m})^2} = 5.75 \times 10^{-9}$ N.

proton: $a_p = \frac{F}{m_p} = \frac{5.75 \times 10^{-9} \text{ N}}{1.67 \times 10^{-27} \text{ kg}} = 3.4 \times 10^{18}$ m/s^2.

electron: $a_e = \frac{F}{m_e} = \frac{5.75 \times 10^{-9} \text{ N}}{9.11 \times 10^{-31} \text{ kg}} = 6.3 \times 10^{21}$ m/s^2

The proton has an initial acceleration of 3.4×10^{18} m/s^2 toward the electron and the electron has an initial acceleration of 6.3×10^{21} m/s^2 toward the proton.
EVALUATE: The force the electron exerts on the proton is equal in magnitude to the force the proton exerts on the electron, but the accelerations of the two particles are very different because their masses are very different.

21.9. IDENTIFY: Apply Coulomb's law.
SET UP: Consider the force on one of the spheres.
EXECUTE: **(a)** $q_1 = q_2 = q$ and $F = \dfrac{1}{4\pi\epsilon_0}\dfrac{|q_1 q_2|}{r^2} = \dfrac{q^2}{4\pi\epsilon_0 r^2}$, so

$q = r\sqrt{\dfrac{F}{(1/4\pi\epsilon_0)}} = 0.150\text{ m}\sqrt{\dfrac{0.220\text{ N}}{8.988\times 10^9 \text{ N}\cdot\text{m}^2/\text{C}^2}} = 7.42\times 10^{-7}$ C (on each).

(b) $q_2 = 4q_1$

$F = \dfrac{1}{4\pi\epsilon_0}\dfrac{|q_1 q_2|}{r^2} = \dfrac{4q_1^2}{4\pi\epsilon_0 r^2}$ so $q_1 = r\sqrt{\dfrac{F}{4(1/4\pi\epsilon_0)}} = \dfrac{1}{2}r\sqrt{\dfrac{F}{(1/4\pi\epsilon_0)}} = \dfrac{1}{2}(7.42\times 10^{-7}\text{ C}) = 3.71\times 10^{-7}$ C.

And then $q_2 = 4q_1 = 1.48\times 10^{-6}$ C.

EVALUATE: The force on one sphere is the same magnitude as the force on the other sphere, whether the spheres have equal charges or not.

21.11. IDENTIFY: In a space satellite, the only force accelerating the free proton is the electrical repulsion of the other proton.
SET UP: Coulomb's law gives the force, and Newton's second law gives the acceleration:
$a = F/m = (1/4\pi\epsilon_0)(e^2/r^2)/m$.

EXECUTE:
(a) $a = (9.00\times 10^9\text{ N}\cdot\text{m}^2/\text{C}^2)(1.60\times 10^{-19}\text{ C})^2/[(0.00250\text{ m})^2(1.67\times 10^{-27}\text{ kg})] = 2.21\times 10^4$ m/s².
(b) The graphs are sketched in Figure 21.11.
EVALUATE: The electrical force of a single stationary proton gives the moving proton an initial acceleration about 20,000 times as great as the acceleration caused by the gravity of the entire earth. As the protons move farther apart, the electrical force gets weaker, so the acceleration decreases. Since the protons continue to repel, the velocity keeps increasing, but at a decreasing rate.

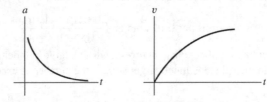

Figure 21.11

21.13. IDENTIFY: Apply Coulomb's law. The two forces on q_3 must have equal magnitudes and opposite directions.
SET UP: Like charges repel and unlike charges attract.
EXECUTE: The force $\vec{F}_2$ that q_2 exerts on q_3 has magnitude $F_2 = k\dfrac{|q_2 q_3|}{r_2^2}$ and is in the $+x$-direction.
$\vec{F}_1$ must be in the $-x$-direction, so q_1 must be positive. $F_1 = F_2$ gives $k\dfrac{|q_1||q_3|}{r_1^2} = k\dfrac{|q_2||q_3|}{r_2^2}$.

$|q_1| = |q_2|\left(\dfrac{r_1}{r_2}\right)^2 = (3.00\text{ nC})\left(\dfrac{2.00\text{ cm}}{4.00\text{ cm}}\right)^2 = 0.750$ nC.

EVALUATE: The result for the magnitude of q_1 doesn't depend on the magnitude of q_3.

21.17. IDENTIFY and SET UP: Apply Coulomb's law to calculate the force exerted by q_2 and q_3 on q_1. Add these forces as vectors to get the net force. The target variable is the x-coordinate of q_3.

EXECUTE: $\vec{F}_2$ is in the x-direction.

$F_2 = k\dfrac{|q_1 q_2|}{r_{12}^2} = 3.37$ N, so $F_{2x} = +3.37$ N

$F_x = F_{2x} + F_{3x}$ and $F_x = -7.00$ N

$F_{3x} = F_x - F_{2x} = -7.00$ N $- 3.37$ N $= -10.37$ N

For F_{3x} to be negative, q_3 must be on the $-x$-axis.

$F_3 = k\dfrac{|q_1 q_3|}{x^2}$, so $|x| = \sqrt{\dfrac{k|q_1 q_3|}{F_3}} = 0.144$ m, so $x = -0.144$ m

EVALUATE: q_2 attracts q_1 in the $+x$-direction so q_3 must attract q_1 in the $-x$-direction, and q_3 is at negative x.

21.21. IDENTIFY: We use Coulomb's law to find each electrical force and combine these forces to find the net force.

SET UP: In the O-H-N combination the O^- is 0.170 nm from the H^+ and 0.280 nm from the N^-. In the N-H-N combination the N^- is 0.190 nm from the H^+ and 0.300 nm from the other N^-. Like charges repel and unlike charges attract. The net force is the vector sum of the individual forces. The force due to each pair of charges is $F = k\dfrac{|q_1 q_2|}{r^2} = k\dfrac{e^2}{r^2}$.

EXECUTE: **(a)** $F = k\dfrac{|q_1 q_2|}{r^2} = k\dfrac{e^2}{r^2}$.

O-H-N:

$O^- - H^+$: $F = (8.99 \times 10^9 \text{ N} \cdot \text{m}^2/\text{C}^2)\dfrac{(1.60 \times 10^{-19} \text{ C})^2}{(0.170 \times 10^{-9} \text{ m})^2} = 7.96 \times 10^{-9}$ N, attractive

$O^- - N^-$: $F = (8.99 \times 10^9 \text{ N} \cdot \text{m}^2/\text{C}^2)\dfrac{(1.60 \times 10^{-19} \text{ C})^2}{(0.280 \times 10^{-9} \text{ m})^2} = 2.94 \times 10^{-9}$ N, repulsive

N-H-N:

$N^- - H^+$: $F = (8.99 \times 10^9 \text{ N} \cdot \text{m}^2/\text{C}^2)\dfrac{(1.60 \times 10^{-19} \text{ C})^2}{(0.190 \times 10^{-9} \text{ m})^2} = 6.38 \times 10^{-9}$ N, attractive

$N^- - N^-$: $F = (8.99 \times 10^9 \text{ N} \cdot \text{m}^2/\text{C}^2)\dfrac{(1.60 \times 10^{-19} \text{ C})^2}{(0.300 \times 10^{-9} \text{ m})^2} = 2.56 \times 10^{-9}$ N, repulsive

The total attractive force is 1.43×10^{-8} N and the total repulsive force is 5.50×10^{-9} N. The net force is attractive and has magnitude 1.43×10^{-8} N $- 5.50 \times 10^{-9}$ N $= 8.80 \times 10^{-9}$ N.

(b) $F = k\dfrac{e^2}{r^2} = (8.99 \times 10^9 \text{ N} \cdot \text{m}^2/\text{C}^2)\dfrac{(1.60 \times 10^{-19} \text{ C})^2}{(0.0529 \times 10^{-9} \text{ m})^2} = 8.22 \times 10^{-8}$ N.

EVALUATE: The bonding force of the electron in the hydrogen atom is a factor of 10 larger than the bonding force of the adenine-thymine molecules.

21.25. IDENTIFY: The acceleration that stops the charge is produced by the force that the electric field exerts on it. Since the field and the acceleration are constant, we can use the standard kinematics formulas to find acceleration and time.

(a) SET UP: First use kinematics to find the proton's acceleration. $v_x = 0$ when it stops. Then find the electric field needed to cause this acceleration using the fact that $F = qE$.

EXECUTE: $v_x^2 = v_{0x}^2 + 2a_x(x - x_0)$. $0 = (4.50 \times 10^6 \text{ m/s})^2 + 2a(0.0320 \text{ m})$ and $a = 3.16 \times 10^{14}$ m/s^2.
Now find the electric field, with $q = e$. $eE = ma$ and

$E = ma/e = (1.67 \times 10^{-27}$ kg$)(3.16 \times 10^{14}$ m/s$^2)/(1.60 \times 10^{-19}$ C$) = 3.30 \times 10^6$ N/C, to the left.

(b) SET UP: Kinematics gives $v = v_0 + at$, and $v = 0$ when the electron stops, so $t = v_0/a$.

EXECUTE: $t = v_0/a = (4.50 \times 10^6 \text{ m/s})/(3.16 \times 10^{14} \text{ m/s}^2) = 1.42 \times 10^{-8}$ s = 14.2 ns.

(c) SET UP: In part (a) we saw that the electric field is proportional to m, so we can use the ratio of the electric fields. $E_e/E_p = m_e/m_p$ and $E_e = (m_e/m_p)E_p$.

EXECUTE: $E_e = [(9.11 \times 10^{-31} \text{ kg})/(1.67 \times 10^{-27} \text{ kg})](3.30 \times 10^6 \text{ N/C}) = 1.80 \times 10^3$ N/C, to the right.

EVALUATE: Even a modest electric field, such as the ones in this situation, can produce enormous accelerations for electrons and protons.

21.29. IDENTIFY: The equation $\vec{F} = q\vec{E}$ gives the force on the particle in terms of its charge and the electric field between the plates. The force is constant and produces a constant acceleration. The motion is similar to projectile motion; use constant acceleration equations for the horizontal and vertical components of the motion.

SET UP: The motion is sketched in Figure 21.29a.

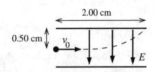

For an electron $q = -e$.

Figure 21.29a

$\vec{F} = q\vec{E}$ and q negative gives that $\vec{F}$ and $\vec{E}$ are in opposite directions, so $\vec{F}$ is upward. The free-body diagram for the electron is given in Figure 21.29b.

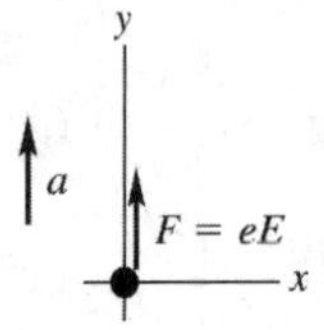

EXECUTE: (a) $\Sigma F_y = ma_y$

$eE = ma$

Figure 21.29b

Solve the kinematics to find the acceleration of the electron: Just misses upper plate says that $x - x_0 = 2.00$ cm when $y - y_0 = +0.500$ cm.

<u>x-component:</u>

$v_{0x} = v_0 = 1.60 \times 10^6$ m/s, $a_x = 0$, $x - x_0 = 0.0200$ m, $t = ?$

$x - x_0 = v_{0x}t + \frac{1}{2}a_x t^2$

$t = \dfrac{x - x_0}{v_{0x}} = \dfrac{0.0200 \text{ m}}{1.60 \times 10^6 \text{ m/s}} = 1.25 \times 10^{-8}$ s

In this same time t the electron travels 0.0050 m vertically.

<u>y-component:</u>

$t = 1.25 \times 10^{-8}$ s, $v_{0y} = 0$, $y - y_0 = +0.0050$ m, $a_y = ?$

$y - y_0 = v_{0y}t + \frac{1}{2}a_y t^2$

$a_y = \dfrac{2(y - y_0)}{t^2} = \dfrac{2(0.0050 \text{ m})}{(1.25 \times 10^{-8} \text{ s})^2} = 6.40 \times 10^{13}$ m/s^2.

(This analysis is very similar to that used in Chapter 3 for projectile motion, except that here the acceleration is upward rather than downward.) This acceleration must be produced by the electric-field force: $eE = ma$.

$$E = \dfrac{ma}{e} = \dfrac{(9.109 \times 10^{-31} \text{ kg})(6.40 \times 10^{13} \text{ m/s}^2)}{1.602 \times 10^{-19} \text{ C}} = 364 \text{ N/C}$$

Note that the acceleration produced by the electric field is much larger than g, the acceleration produced by gravity, so it is perfectly ok to neglect the gravity force on the electron in this problem.

(b) $a = \dfrac{eE}{m_p} = \dfrac{(1.602\times 10^{-19}\text{ C})(364\text{ N/C})}{1.673\times 10^{-27}\text{ kg}} = 3.49\times 10^{10}\text{ m/s}^2.$

This is much less than the acceleration of the electron in part (a) so the vertical deflection is less and the proton won't hit the plates. The proton has the same initial speed, so the proton takes the same time $t = 1.25\times 10^{-8}$ s to travel horizontally the length of the plates. The force on the proton is downward (in the same direction as $\vec{E}$, since q is positive), so the acceleration is downward and $a_y = -3.49\times 10^{10}$ m/s^2.

$y - y_0 = v_{0y}t + \tfrac{1}{2}a_y t^2 = \tfrac{1}{2}(-3.49\times 10^{10}\text{ m/s}^2)(1.25\times 10^{-8}\text{ s})^2 = -2.73\times 10^{-6}$ m. The displacement is 2.73×10^{-6} m, downward.

EVALUATE: **(c)** The displacements are in opposite directions because the electron has negative charge and the proton has positive charge. The electron and proton have the same magnitude of charge, so the force the electric field exerts has the same magnitude for each charge. But the proton has a mass larger by a factor of 1836 so its acceleration and its vertical displacement are smaller by this factor.
(d) In each case $a \gg g$ and it is reasonable to ignore the effects of gravity.

21.39. **IDENTIFY:** $E = \dfrac{1}{4\pi\varepsilon_0}\dfrac{|q|}{r^2}$ gives the electric field of each point charge. Use the principle of superposition and add the electric field vectors. In part (b) use $\vec{E} = \dfrac{\vec{F}_0}{q_0}$ to calculate the force, using the electric field calculated in part (a).

SET UP: The placement of charges is sketched in Figure 21.39a.

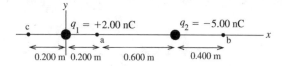

Figure 21.39a

The electric field of a point charge is directed away from the point charge if the charge is positive and toward the point charge if the charge is negative. The magnitude of the electric field is $E = \dfrac{1}{4\pi\varepsilon_0}\dfrac{|q|}{r^2}$, where r is the distance between the point where the field is calculated and the point charge.

(a) EXECUTE: (i) At point a the fields $\vec{E}_1$ of q_1 and $\vec{E}_2$ of q_2 are directed as shown in Figure 21.39b.

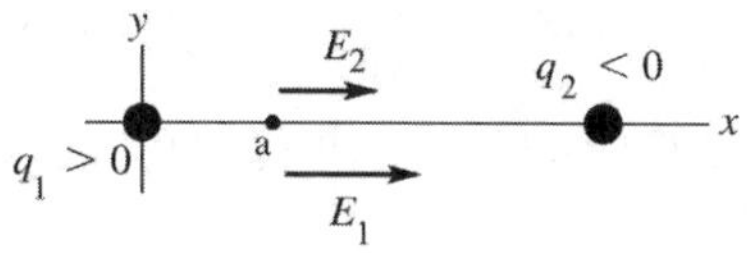

Figure 21.39b

$E_1 = \dfrac{1}{4\pi\varepsilon_0}\dfrac{|q_1|}{r_1^2} = (8.988\times 10^9\text{ N}\cdot\text{m}^2/\text{C}^2)\dfrac{2.00\times 10^{-9}\text{ C}}{(0.200\text{ m})^2} = 449.4$ N/C.

$E_2 = \dfrac{1}{4\pi\varepsilon_0}\dfrac{|q_2|}{r_2^2} = (8.988\times 10^9\text{ N}\cdot\text{m}^2/\text{C}^2)\dfrac{5.00\times 10^{-9}\text{ C}}{(0.600\text{ m})^2} = 124.8$ N/C.

$E_{1x} = 449.4$ N/C, $E_{1y} = 0$.

$E_{2x} = 124.8$ N/C, $E_{2y} = 0$.

$E_x = E_{1x} + E_{2x} = +449.4 \text{ N/C} + 124.8 \text{ N/C} = +574.2 \text{ N/C}.$

$E_y = E_{1y} + E_{2y} = 0.$

The resultant field at point a has magnitude 574 N/C and is in the +x-direction.

(ii) At point b the fields $\vec{E}_1$ of q_1 and $\vec{E}_2$ of q_2 are directed as shown in Figure 21.39c.

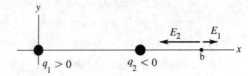

Figure 21.39c

$E_1 = \dfrac{1}{4\pi\varepsilon_0} \dfrac{|q_1|}{r_1^2} = (8.988 \times 10^9 \text{ N} \cdot \text{m}^2/\text{C}^2) \dfrac{2.00 \times 10^{-9} \text{ C}}{(1.20 \text{ m})^2} = 12.5 \text{ N/C}.$

$E_2 = \dfrac{1}{4\pi\varepsilon_0} \dfrac{|q_2|}{r_2^2} = (8.988 \times 10^9 \text{ N} \cdot \text{m}^2/\text{C}^2) \dfrac{5.00 \times 10^{-9} \text{ C}}{(0.400 \text{ m})^2} = 280.9 \text{ N/C}.$

$E_{1x} = 12.5 \text{ N/C}, E_{1y} = 0.$

$E_{2x} = -280.9 \text{ N/C}, E_{2y} = 0.$

$E_x = E_{1x} + E_{2x} = +12.5 \text{ N/C} - 280.9 \text{ N/C} = -268.4 \text{ N/C}.$

$E_y = E_{1y} + E_{2y} = 0.$

The resultant field at point b has magnitude 268 N/C and is in the $-x$-direction.

(iii) At point c the fields $\vec{E}_1$ of q_1 and $\vec{E}_2$ of q_2 are directed as shown in Figure 21.39d.

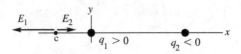

Figure 21.39d

$E_1 = \dfrac{1}{4\pi\varepsilon_0} \dfrac{|q_1|}{r_1^2} = (8.988 \times 10^9 \text{ N} \cdot \text{m}^2/\text{C}^2) \dfrac{2.00 \times 10^{-9} \text{ C}}{(0.200 \text{ m})^2} = 449.4 \text{ N/C}.$

$E_2 = \dfrac{1}{4\pi\varepsilon_0} \dfrac{|q_2|}{r_2^2} = (8.988 \times 10^9 \text{ N} \cdot \text{m}^2/\text{C}^2) \dfrac{5.00 \times 10^{-9} \text{ C}}{(1.00 \text{ m})^2} = 44.9 \text{ N/C}.$

$E_{1x} = -449.4 \text{ N/C}, E_{1y} = 0.$

$E_{2x} = +44.9 \text{ N/C}, E_{2y} = 0.$

$E_x = E_{1x} + E_{2x} = -449.4 \text{ N/C} + 44.9 \text{ N/C} = -404.5 \text{ N/C}.$

$E_y = E_{1y} + E_{2y} = 0.$

The resultant field at point b has magnitude 404 N/C and is in the $-x$-direction.

(b) SET UP: Since we have calculated $\vec{E}$ at each point the simplest way to get the force is to use $\vec{F} = -e\vec{E}$.

EXECUTE: (i) $F = (1.602 \times 10^{-19} \text{ C})(574.2 \text{ N/C}) = 9.20 \times 10^{-17}$ N, $-x$-direction.

(ii) $F = (1.602 \times 10^{-19} \text{ C})(268.4 \text{ N/C}) = 4.30 \times 10^{-17}$ N, $+x$-direction.

(iii) $F = (1.602 \times 10^{-19} \text{ C})(404.5 \text{ N/C}) = 6.48 \times 10^{-17}$ N, $+x$-direction.

EVALUATE: The general rule for electric field direction is away from positive charge and toward negative charge. Whether the field is in the +x- or −x-direction depends on where the field point is relative to the charge that produces the field. In part (a), for (i) the field magnitudes were added because the fields were in

21.41. **IDENTIFY:** $E = k\dfrac{|q|}{r^2}$. The net field is the vector sum of the fields due to each charge.

SET UP: The electric field of a negative charge is directed toward the charge. Label the charges $q_1, q_2,$ and q_3, as shown in Figure 21.41a. This figure also shows additional distances and angles. The electric fields at point P are shown in Figure 21.41b. This figure also shows the xy-coordinates we will use and the x- and y-components of the fields $\vec{E}_1$, $\vec{E}_2$, and $\vec{E}_3$.

EXECUTE: $E_1 = E_3 = (8.99 \times 10^9 \text{ N} \cdot \text{m}^2/\text{C}^2)\dfrac{5.00 \times 10^{-6} \text{ C}}{(0.100 \text{ m})^2} = 4.49 \times 10^6 \text{ N/C}.$

$E_2 = (8.99 \times 10^9 \text{ N} \cdot \text{m}^2/\text{C}^2)\dfrac{2.00 \times 10^{-6} \text{ C}}{(0.0600 \text{ m})^2} = 4.99 \times 10^6 \text{ N/C}.$

$E_y = E_{1y} + E_{2y} + E_{3y} = 0$ and $E_x = E_{1x} + E_{2x} + E_{3x} = E_2 + 2E_1 \cos 53.1° = 1.04 \times 10^7 \text{ N/C}.$

$E = 1.04 \times 10^7$ N/C, toward the $-2.00 \ \mu$C charge.

EVALUATE: The x-components of the fields of all three charges are in the same direction.

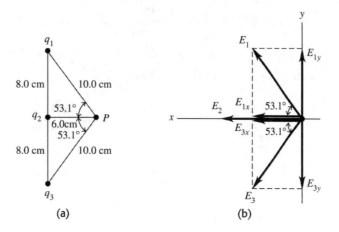

Figure 21.41

21.47. **IDENTIFY:** The electric field of a positive charge is directed radially outward from the charge and has magnitude $E = \dfrac{1}{4\pi\varepsilon_0}\dfrac{|q|}{r^2}$. The resultant electric field is the vector sum of the fields of the individual charges.

SET UP: The placement of the charges is shown in Figure 21.47a.

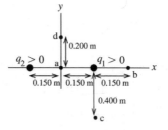

Figure 21.47a

EXECUTE: **(a)** The directions of the two fields are shown in Figure 21.47b.

$$E_1 = E_2 = \frac{1}{4\pi\varepsilon_0}\frac{|q|}{r^2} \text{ with } r = 0.150 \text{ m}.$$

$$E = E_2 - E_1 = 0; E_x = 0, E_y = 0.$$

Figure 21. 47b

(b) The two fields have the directions shown in Figure 21.47c.

$$E = E_1 + E_2, \text{ in the } +x\text{-direction}.$$

Figure 21. 47c

$$E_1 = \frac{1}{4\pi\varepsilon_0}\frac{|q_1|}{r_1^2} = (8.988\times10^9 \text{ N}\cdot\text{m}^2/\text{C}^2)\frac{6.00\times10^{-9} \text{ C}}{(0.150 \text{ m})^2} = 2396.8 \text{ N/C}.$$

$$E_2 = \frac{1}{4\pi\varepsilon_0}\frac{|q_2|}{r_2^2} = (8.988\times10^9 \text{ N}\cdot\text{m}^2/\text{C}^2)\frac{6.00\times10^{-9} \text{ C}}{(0.450 \text{ m})^2} = 266.3 \text{ N/C}.$$

$E = E_1 + E_2 = 2396.8 \text{ N/C} + 266.3 \text{ N/C} = 2660 \text{ N/C}; E_x = +2660 \text{ N/C}, E_y = 0.$

(c) The two fields have the directions shown in Figure 21.47d.

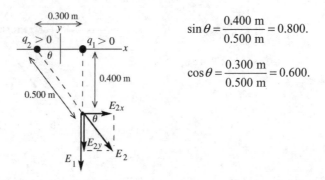

$$\sin\theta = \frac{0.400 \text{ m}}{0.500 \text{ m}} = 0.800.$$

$$\cos\theta = \frac{0.300 \text{ m}}{0.500 \text{ m}} = 0.600.$$

Figure 21. 47d

$$E_1 = \frac{1}{4\pi\varepsilon_0}\frac{|q_1|}{r_1^2} = (8.988\times10^9 \text{ N}\cdot\text{m}^2/\text{C}^2)\frac{6.00\times10^{-9} \text{ C}}{(0.400 \text{ m})^2} = 337.1 \text{ N/C}.$$

$$E_2 = \frac{1}{4\pi\varepsilon_0}\frac{|q_2|}{r_2^2} = (8.988\times10^9 \text{ N}\cdot\text{m}^2/\text{C}^2)\frac{6.00\times10^{-9} \text{ C}}{(0.500 \text{ m})^2} = 215.7 \text{ N/C}.$$

$E_{1x} = 0, E_{1y} = -E_1 = -337.1 \text{ N/C}.$

$E_{2x} = +E_2\cos\theta = +(215.7 \text{ N/C})(0.600) = +129.4 \text{ N/C}.$

$E_{2y} = -E_2\sin\theta = -(215.7 \text{ N/C})(0.800) = -172.6 \text{ N/C}.$

$E_x = E_{1x} + E_{2x} = +129 \text{ N/C}.$

$E_y = E_{1y} + E_{2y} = -337.1 \text{ N/C} - 172.6 \text{ N/C} = -510 \text{ N/C}.$

$E = \sqrt{E_x^2 + E_y^2} = \sqrt{(129 \text{ N/C})^2 + (-510 \text{ N/C})^2} = 526 \text{ N/C}.$

$\vec{E}$ and its components are shown in Figure 21.47e.

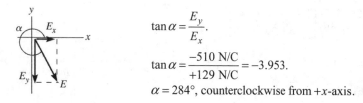

$\tan \alpha = \dfrac{E_y}{E_x}.$

$\tan \alpha = \dfrac{-510 \text{ N/C}}{+129 \text{ N/C}} = -3.953.$

$\alpha = 284°$, counterclockwise from $+x$-axis.

Figure 21.47e

(d) The two fields have the directions shown in Figure 21.47f.

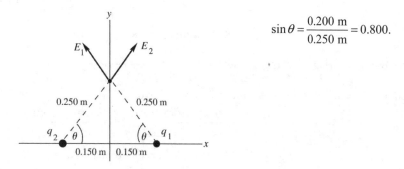

$\sin \theta = \dfrac{0.200 \text{ m}}{0.250 \text{ m}} = 0.800.$

Figure 21.47f

The components of the two fields are shown in Figure 21.47g.

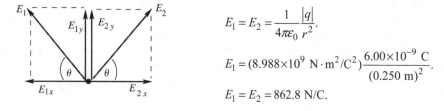

$E_1 = E_2 = \dfrac{1}{4\pi\varepsilon_0} \dfrac{|q|}{r^2}.$

$E_1 = (8.988 \times 10^9 \text{ N} \cdot \text{m}^2/\text{C}^2) \dfrac{6.00 \times 10^{-9} \text{ C}}{(0.250 \text{ m})^2}.$

$E_1 = E_2 = 862.8 \text{ N/C}.$

Figure 21.47g

$E_{1x} = -E_1 \cos\theta, \ E_{2x} = +E_2 \cos\theta.$
$E_x = E_{1x} + E_{2x} = 0.$
$E_{1y} = +E_1 \sin\theta, \ E_{2y} = +E_2 \sin\theta.$
$E_y = E_{1y} + E_{2y} = 2E_{1y} = 2E_1 \sin\theta = 2(862.8 \text{ N/C})(0.800) = 1380 \text{ N/C}.$
$E = 1380 \text{ N/C}$, in the $+y$-direction.

EVALUATE: Point a is symmetrically placed between identical charges, so symmetry tells us the electric field must be zero. Point b is to the right of both charges and both electric fields are in the $+x$-direction and the resultant field is in this direction. At point c both fields have a downward component and the field of q_2 has a component to the right, so the net $\vec{E}$ is in the fourth quadrant. At point d both fields have an upward component but by symmetry they have equal and opposite x-components so the net field is in the $+y$-direction. We can use this sort of reasoning to deduce the general direction of the net field before doing any calculations.

21.49. **IDENTIFY:** We must use the appropriate electric field formula: a uniform disk in (a), a ring in (b) because all the charge is along the rim of the disk, and a point-charge in (c).

(a) SET UP: First find the surface charge density (Q/A), then use the formula for the field due to a disk of charge, $E_x = \dfrac{\sigma}{2\varepsilon_0}\left[1 - \dfrac{1}{\sqrt{(R/x)^2 + 1}}\right]$.

EXECUTE: The surface charge density is $\sigma = \dfrac{Q}{A} = \dfrac{Q}{\pi r^2} = \dfrac{6.50\times 10^{-9}\text{ C}}{\pi(0.0125\text{ m})^2} = 1.324\times 10^{-5}\text{ C/m}^2$.

The electric field is

$$E_x = \dfrac{\sigma}{2\varepsilon_0}\left[1 - \dfrac{1}{\sqrt{(R/x)^2 + 1}}\right] = \dfrac{1.324\times 10^{-5}\text{ C/m}^2}{2(8.85\times 10^{-12}\text{ C}^2/\text{N}\cdot\text{m}^2)}\left[1 - \dfrac{1}{\sqrt{\left(\dfrac{1.25\text{ cm}}{2.00\text{ cm}}\right)^2 + 1}}\right]$$

$E_x = 1.14\times 10^5$ N/C, toward the center of the disk.

(b) SET UP: For a ring of charge, the field is $E = \dfrac{1}{4\pi\varepsilon_0}\dfrac{Qx}{(x^2 + a^2)^{3/2}}$.

EXECUTE: Substituting into the electric field formula gives

$$E = \dfrac{1}{4\pi\varepsilon_0}\dfrac{Qx}{(x^2 + a^2)^{3/2}} = \dfrac{(9.00\times 10^9\text{ N}\cdot\text{m}^2/\text{C}^2)(6.50\times 10^{-9}\text{ C})(0.0200\text{ m})}{[(0.0200\text{ m})^2 + (0.0125\text{ m})^2]^{3/2}}$$

$E = 8.92\times 10^4$ N/C, toward the center of the disk.

(c) SET UP: For a point charge, $E = (1/4\pi\varepsilon_0)q/r^2$.

EXECUTE: $E = (9.00\times 10^9\text{ N}\cdot\text{m}^2/\text{C}^2)(6.50\times 10^{-9}\text{ C})/(0.0200\text{ m})^2 = 1.46\times 10^5$ N/C.

(d) EVALUATE: With the ring, more of the charge is farther from P than with the disk. Also with the ring the component of the electric field parallel to the plane of the ring is greater than with the disk, and this component cancels. With the point charge in (c), all the field vectors add with no cancellation, and all the charge is closer to point P than in the other two cases.

21.53. (a) IDENTIFY and SET UP: Use $p = qd$ to relate the dipole moment to the charge magnitude and the separation d of the two charges. The direction is from the negative charge toward the positive charge.

EXECUTE: $p = qd = (4.5\times 10^{-9}\text{ C})(3.1\times 10^{-3}\text{ m}) = 1.4\times 10^{-11}$ C·m. The direction of $\vec{p}$ is from q_1 toward q_2.

(b) IDENTIFY and SET UP: Use $\tau = pE\sin\phi$ to relate the magnitudes of the torque and field.

EXECUTE: $\tau = pE\sin\phi$, with ϕ as defined in Figure 21.53, so

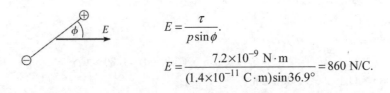

$E = \dfrac{\tau}{p\sin\phi}$.

$E = \dfrac{7.2\times 10^{-9}\text{ N}\cdot\text{m}}{(1.4\times 10^{-11}\text{ C}\cdot\text{m})\sin 36.9°} = 860$ N/C.

Figure 21.53

EVALUATE: The equation $\tau = pE\sin\phi$ gives the torque about an axis through the center of the dipole. But the forces on the two charges form a couple and the torque is the same for any axis parallel to this one. The force on each charge is $|q|E$ and the maximum moment arm for an axis at the center is $d/2$, so the maximum torque is $2(|q|E)(d/2) = 1.2\times 10^{-8}$ N·m. The torque for the orientation of the dipole in the problem is less than this maximum.

21.57. **(a) IDENTIFY:** Use Coulomb's law to calculate each force and then add them as vectors to obtain the net force. Torque is force times moment arm.
SET UP: The two forces on each charge in the dipole are shown in Figure 21.57a.

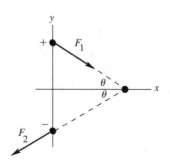

$\sin\theta = 1.50/2.00$ so $\theta = 48.6°$.

Opposite charges attract and like charges repel.

$F_x = F_{1x} + F_{2x} = 0.$

Figure 21.57a

EXECUTE: $F_1 = k\dfrac{|qq'|}{r^2} = k\dfrac{(5.00\times 10^{-6}\text{ C})(10.0\times 10^{-6}\text{ C})}{(0.0200\text{ m})^2} = 1.124\times 10^3$ N.

$F_{1y} = -F_1\sin\theta = -842.6$ N.

$F_{2y} = -842.6$ N so $F_y = F_{1y} + F_{2y} = -1680$ N (in the direction from the $+5.00\text{-}\mu$C charge toward the $-5.00\text{-}\mu$C charge).

EVALUATE: The x-components cancel and the y-components add.

(b) SET UP: Refer to Figure 21.57b.

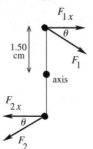

The y-components have zero moment arm and therefore zero torque.

F_{1x} and F_{2x} both produce clockwise torques.

Figure 21.57b

EXECUTE: $F_{1x} = F_1\cos\theta = 743.1$ N.

$\tau = 2(F_{1x})(0.0150\text{ m}) = 22.3$ N·m, clockwise.

EVALUATE: The electric field produced by the -10.00 μC charge is not uniform so $\tau = pE\sin\phi$ does not apply.

21.63. **IDENTIFY:** Use Coulomb's law for the force that one sphere exerts on the other and apply the first condition of equilibrium to one of the spheres.
SET UP: The placement of the spheres is sketched in Figure 21.63a.

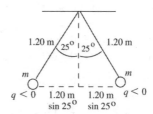

Figure 21.63a

EXECUTE: **(a)** The free-body diagrams for each sphere are given in Figure 21.63b.

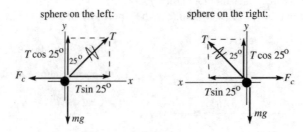

Figure 21.63b

F_c is the repulsive Coulomb force exerted by one sphere on the other.

(b) From either force diagram in part (a): $\Sigma F_y = ma_y$.

$T\cos 25.0° - mg = 0$ and $T = \dfrac{mg}{\cos 25.0°}$.

$\Sigma F_x = ma_x$.

$T\sin 25.0° - F_c = 0$ and $F_c = T\sin 25.0°$.

Use the first equation to eliminate T in the second: $F_c = (mg/\cos 25.0°)(\sin 25.0°) = mg\tan 25.0°$.

$F_c = \dfrac{1}{4\pi\varepsilon_0}\dfrac{|q_1 q_2|}{r^2} = \dfrac{1}{4\pi\varepsilon_0}\dfrac{q^2}{r^2} = \dfrac{1}{4\pi\varepsilon_0}\dfrac{q^2}{[2(1.20\text{ m})\sin 25.0°]^2}$.

Combine this with $F_c = mg\tan 25.0°$ and get $mg\tan 25.0° = \dfrac{1}{4\pi\varepsilon_0}\dfrac{q^2}{[2(1.20\text{ m})\sin 25.0°]^2}$.

$q = (2.40\text{ m})\sin 25.0°\sqrt{\dfrac{mg\tan 25.0°}{(1/4\pi\varepsilon_0)}}$.

$q = (2.40\text{ m})\sin 25.0°\sqrt{\dfrac{(15.0\times 10^{-3}\text{ kg})(9.80\text{ m/s}^2)\tan 25.0°}{8.988\times 10^9\text{ N}\cdot\text{m}^2/\text{C}^2}} = 2.80\times 10^{-6}$ C.

(c) The separation between the two spheres is given by $2L\sin\theta$. $q = 2.80\ \mu$C as found in part (b).

$F_c = (1/4\pi\varepsilon_0)q^2/(2L\sin\theta)^2$ and $F_c = mg\tan\theta$. Thus $(1/4\pi\varepsilon_0)q^2/(2L\sin\theta)^2 = mg\tan\theta$.

$(\sin\theta)^2 \tan\theta = \dfrac{1}{4\pi\varepsilon_0}\dfrac{q^2}{4L^2 mg} = (8.988\times 10^9\text{ N}\cdot\text{m}^2/\text{C}^2)\dfrac{(2.80\times 10^{-6}\text{ C})^2}{4(0.600\text{ m})^2(15.0\times 10^{-3}\text{ kg})(9.80\text{ m/s}^2)} = 0.3328$.

Solve this equation by trial and error. This will go quicker if we can make a good estimate of the value of θ that solves the equation. For θ small, $\tan\theta \approx \sin\theta$. With this approximation the equation becomes $\sin^3\theta = 0.3328$ and $\sin\theta = 0.6930$, so $\theta = 43.9°$. Now refine this guess:

θ	$\sin^2\theta\tan\theta$
45.0°	0.5000
40.0°	0.3467
39.6°	0.3361
39.5°	0.3335
39.4°	0.3309

so $\theta = 39.5°$.

EVALUATE: The expression in part (c) says $\theta \to 0$ as $L \to \infty$ and $\theta \to 90°$ as $L \to 0$. When L is decreased from the value in part (a), θ increases.

21.65. **IDENTIFY:** The electric field exerts a horizontal force away from the wall on the ball. When the ball hangs at rest, the forces on it (gravity, the tension in the string, and the electric force due to the field) add to zero.

SET UP: The ball is in equilibrium, so for it $\Sigma F_x = 0$ and $\Sigma F_y = 0$. The force diagram for the ball is given in Figure 21.65. F_E is the force exerted by the electric field. $\vec{F} = q\vec{E}$. Since the electric field is horizontal, $\vec{F}_E$ is horizontal. Use the coordinates shown in the figure. The tension in the string has been replaced by its x- and y-components.

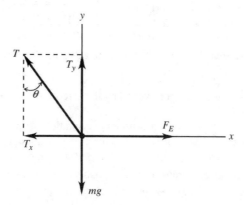

Figure 21.65

EXECUTE: $\Sigma F_y = 0$ gives $T_y - mg = 0$. $T\cos\theta - mg = 0$ and $T = \dfrac{mg}{\cos\theta}$. $\Sigma F_x = 0$ gives $F_E - T_x = 0$. $F_E - T\sin\theta = 0$. Combing the equations and solving for F_E gives

$F_E = \left(\dfrac{mg}{\cos\theta}\right)\sin\theta = mg\tan\theta = (12.3 \times 10^{-3}\text{ kg})(9.80\text{ m/s}^2)(\tan 17.4°) = 3.78 \times 10^{-2}$ N. $F_E = |q|E$ so

$E = \dfrac{F_E}{|q|} = \dfrac{3.78 \times 10^{-2}\text{ N}}{1.11 \times 10^{-6}\text{ C}} = 3.41 \times 10^4$ N/C. Since q is negative and $\vec{F}_E$ is to the right, $\vec{E}$ is to the left in the figure.

EVALUATE: The larger the electric field E the greater the angle the string makes with the wall.

21.67. IDENTIFY: For a point charge, $E = \dfrac{|q|}{r^2}$. For the net electric field to be zero, $\vec{E}_1$ and $\vec{E}_2$ must have equal magnitudes and opposite directions.

SET UP: Let $q_1 = +0.500$ nC and $q_2 = +8.00$ nC. $\vec{E}$ is toward a negative charge and away from a positive charge.

EXECUTE: The two charges and the directions of their electric fields in three regions are shown in Figure 21.67. Only in region II are the two electric fields in opposite directions. Consider a point a distance x from q_1 so a distance $1.20\text{ m} - x$ from q_2. $E_1 = E_2$ gives $k\dfrac{0.500\text{ nC}}{x^2} = k\dfrac{8.00\text{ nC}}{(1.20\text{ m} - x)^2}$. $16x^2 = (1.20\text{ m} - x)^2$.

$4x = \pm(1.20\text{ m} - x)$ and $x = 0.24$ m is the positive solution. The electric field is zero at a point between the two charges, 0.24 m from the 0.500 nC charge and 0.96 m from the 8.00 nC charge.

EVALUATE: There is only one point along the line connecting the two charges where the net electric field is zero. This point is closer to the charge that has the smaller magnitude.

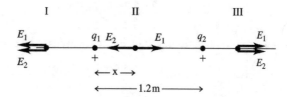

Figure 21.67

21.75. **IDENTIFY:** The only external force acting on the electron is the electrical attraction of the proton, and its acceleration is toward the center of its circular path (that is, toward the proton). Newton's second law applies to the electron and Coulomb's law gives the electrical force on it due to the proton.

SET UP: Newton's second law gives $F_C = m\dfrac{v^2}{r}$. Using the electrical force for F_C gives $k\dfrac{e^2}{r^2} = m\dfrac{v^2}{r}$.

EXECUTE: Solving for v gives $v = \sqrt{\dfrac{ke^2}{mr}} = \sqrt{\dfrac{(8.99 \times 10^9 \text{ N} \cdot \text{m}^2/\text{C}^2)(1.60 \times 10^{-19} \text{ C})^2}{(9.109 \times 10^{-31} \text{ kg})(5.29 \times 10^{-11} \text{ m})}} = 2.19 \times 10^6$ m/s.

EVALUATE: This speed is less than 1% the speed of light, so it is reasonably safe to use Newtonian physics.

21.77. **IDENTIFY:** $\vec{E} = \dfrac{\vec{F}_0}{q_0}$ gives the force exerted by the electric field. This force is constant since the electric field is uniform and gives the proton a constant acceleration. Apply the constant acceleration equations for the x- and y-components of the motion, just as for projectile motion.

SET UP: The electric field is upward so the electric force on the positively charged proton is upward and has magnitude $F = eE$. Use coordinates where positive y is downward. Then applying $\Sigma \vec{F} = m\vec{a}$ to the proton gives that $a_x = 0$ and $a_y = -eE/m$. In these coordinates the initial velocity has components $v_x = +v_0 \cos\alpha$ and $v_y = +v_0 \sin\alpha$, as shown in Figure 21.77a.

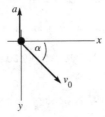

Figure 21.77a

EXECUTE: **(a)** Finding h_{max}: At $y = h_{max}$ the y-component of the velocity is zero.
$v_y = 0, v_{0y} = v_0 \sin\alpha, a_y = -eE/m, y - y_0 = h_{max} = ?$
$v_y^2 = v_{0y}^2 + 2a_y(y - y_0)$.
$y - y_0 = \dfrac{v_y^2 - v_{0y}^2}{2a_y}$.
$h_{max} = \dfrac{-v_0^2 \sin^2\alpha}{2(-eE/m)} = \dfrac{mv_0^2 \sin^2\alpha}{2eE}$.

(b) Use the vertical motion to find the time t: $y - y_0 = 0, v_{0y} = v_0 \sin\alpha, a_y = -eE/m, t = ?$
$y - y_0 = v_{0y}t + \dfrac{1}{2}a_y t^2$.

With $y - y_0 = 0$ this gives $t = -\dfrac{2v_{0y}}{a_y} = -\dfrac{2(v_0 \sin\alpha)}{-eE/m} = \dfrac{2mv_0 \sin\alpha}{eE}$.

Then use the x-component motion to find d: $a_x = 0, v_{0x} = v_0 \cos\alpha, t = 2mv_0 \sin\alpha/eE, x - x_0 = d = ?$
$x - x_0 = v_{0x}t + \dfrac{1}{2}a_x t^2$ gives $d = v_0 \cos\alpha \left(\dfrac{2mv_0 \sin\alpha}{eE}\right) = \dfrac{mv_0^2 2\sin\alpha\cos\alpha}{eE} = \dfrac{mv_0^2 \sin 2\alpha}{eE}$.

(c) The trajectory of the proton is sketched in Figure 21.77b.

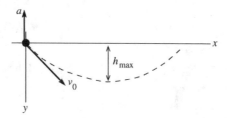

Figure 21.77b

(d) Use the expression in part (a): $h_{max} = \dfrac{[(4.00 \times 10^5 \text{ m/s})(\sin 30.0°)]^2 (1.673 \times 10^{-27} \text{ kg})}{2(1.602 \times 10^{-19} \text{ C})(500 \text{ N/C})} = 0.418 \text{ m}$.

Use the expression in part (b): $d = \dfrac{(1.673 \times 10^{-27} \text{ kg})(4.00 \times 10^5 \text{ m/s})^2 \sin 60.0°}{(1.602 \times 10^{-19} \text{ C})(500 \text{ N/C})} = 2.89 \text{ m}$.

EVALUATE: In part (a), $a_y = -eE/m = -4.8 \times 10^{10} \text{ m/s}^2$. This is much larger in magnitude than g, the acceleration due to gravity, so it is reasonable to ignore gravity. The motion is just like projectile motion, except that the acceleration is upward rather than downward and has a much different magnitude. h_{max} and d increase when α or v_0 increase and decrease when E increases.

21.79. IDENTIFY: Divide the charge distribution into infinitesimal segments of length dx'. Calculate E_x and E_y due to a segment and integrate to find the total field.

SET UP: The charge dQ of a segment of length dx' is $dQ = (Q/a)dx'$. The distance between a segment at x' and a point at x on the x-axis is $x - x'$ since $x > a$.

EXECUTE: (a) $dE_x = \dfrac{1}{4\pi\varepsilon_0} \dfrac{dQ}{(x-x')^2} = \dfrac{1}{4\pi\varepsilon_0} \dfrac{(Q/a)dx'}{(x-x')^2}$. Integrating with respect to x' over the length of the charge distribution gives

$E_x = \dfrac{1}{4\pi\varepsilon_0} \displaystyle\int_0^a \dfrac{(Q/a)dx'}{(x-x')^2} = \dfrac{1}{4\pi\varepsilon_0} \dfrac{Q}{a}\left(\dfrac{1}{x-a} - \dfrac{1}{x}\right) = \dfrac{1}{4\pi\varepsilon_0} \dfrac{Q}{a} \dfrac{a}{x(x-a)} = \dfrac{1}{4\pi\varepsilon_0} \dfrac{Q}{x(x-a)}$. $E_y = 0$.

(b) At the location of the charge, $x = r + a$, so $E_x = \dfrac{1}{4\pi\varepsilon_0} \dfrac{Q}{(r+a)(r+a-a)} = \dfrac{1}{4\pi\varepsilon_0} \dfrac{Q}{r(r+a)}$.

Using $\vec{F} = q\vec{E}$, we have $\vec{F} = q\vec{E} = \dfrac{1}{4\pi\varepsilon_0} \dfrac{qQ}{r(r+a)} \hat{i}$.

EVALUATE: (c) For $r \gg a$, $r + a \to r$, so the magnitude of the force becomes $F = \dfrac{1}{4\pi\varepsilon_0} \dfrac{qQ}{r^2}$. The charge distribution looks like a point charge from far away, so the force takes the form of the force between a pair of point charges.

21.85. IDENTIFY: Divide the charge distribution into small segments, use the point charge formula for the electric field due to each small segment and integrate over the charge distribution to find the x- and y-components of the total field.

SET UP: Consider the small segment shown in Figure 21.85a.

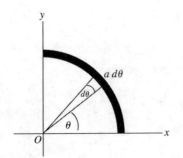

EXECUTE: A small segment that subtends angle $d\theta$ has length $a\,d\theta$ and contains charge $dQ = \left(\dfrac{a\,d\theta}{\frac{1}{2}\pi a}\right)Q = \dfrac{2Q}{\pi}d\theta.$ ($\frac{1}{2}\pi a$ is the total length of the charge distribution.)

Figure 21.85a

The charge is negative, so the field at the origin is directed toward the small segment. The small segment is located at angle θ as shown in the sketch. The electric field due to dQ is shown in Figure 21.85b, along with its components.

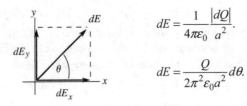

$$dE = \dfrac{1}{4\pi\varepsilon_0}\dfrac{|dQ|}{a^2}.$$

$$dE = \dfrac{Q}{2\pi^2\varepsilon_0 a^2}d\theta.$$

Figure 21.85b

$dE_x = dE\cos\theta = (Q/2\pi^2\varepsilon_0 a^2)\cos\theta\,d\theta.$

$E_x = \int dE_x = \dfrac{Q}{2\pi^2\varepsilon_0 a^2}\int_0^{\pi/2}\cos\theta\,d\theta = \dfrac{Q}{2\pi^2\varepsilon_0 a^2}(\sin\theta\big|_0^{\pi/2}) = \dfrac{Q}{2\pi^2\varepsilon_0 a^2}.$

$dE_y = dE\sin\theta = (Q/2\pi^2\varepsilon_0 a^2)\sin\theta\,d\theta.$

$E_y = \int dE_y = \dfrac{Q}{2\pi^2\varepsilon_0 a^2}\int_0^{\pi/2}\sin\theta\,d\theta = \dfrac{Q}{2\pi^2\varepsilon_0 a^2}(-\cos\theta\big|_0^{\pi/2}) = \dfrac{Q}{2\pi^2\varepsilon_0 a^2}.$

EVALUATE: Note that $E_x = E_y$, as expected from symmetry.

21.87. IDENTIFY: Each wire produces an electric field at P due to a finite wire. These fields add by vector addition.

SET UP: Each field has magnitude $\dfrac{1}{4\pi\varepsilon_0}\dfrac{Q}{x\sqrt{x^2+a^2}}$. The field due to the negative wire points to the left, while the field due to the positive wire points downward, making the two fields perpendicular to each other and of equal magnitude. The net field is the vector sum of these two, which is

$E_{net} = 2E_1\cos 45° = 2\dfrac{1}{4\pi\varepsilon_0}\dfrac{Q}{x\sqrt{x^2+a^2}}\cos 45°.$ In part (b), the electrical force on an electron at P is eE.

EXECUTE: (a) The net field is $E_{net} = 2\dfrac{1}{4\pi\varepsilon_0}\dfrac{Q}{x\sqrt{x^2+a^2}}\cos 45°.$

$E_{net} = \dfrac{2(9.00\times 10^9\ \text{N}\cdot\text{m}^2/\text{C}^2)(2.50\times 10^{-6}\ \text{C})\cos 45°}{(0.600\ \text{m})\sqrt{(0.600\ \text{m})^2+(0.600\ \text{m})^2}} = 6.25\times 10^4\ \text{N/C}.$

The direction is 225° counterclockwise from an axis pointing to the right at point P.

(b) $F = eE = (1.60\times 10^{-19}\ \text{C})(6.25\times 10^4\ \text{N/C}) = 1.00\times 10^{-14}\ \text{N}$, opposite to the direction of the electric field, since the electron has negative charge.

EVALUATE: Since the electric fields due to the two wires have equal magnitudes and are perpendicular to each other, we only have to calculate one of them in the solution.

21.89. **IDENTIFY:** Each sheet produces an electric field that is independent of the distance from the sheet. The net field is the vector sum of the two fields.

SET UP: The formula for each field is $E = \sigma/2\varepsilon_0$, and the net field is the vector sum of these.

$$E_{net} = \frac{\sigma_B}{2\varepsilon_0} \pm \frac{\sigma_A}{2\varepsilon_0} = \frac{\sigma_B \pm \sigma_A}{2\varepsilon_0},$$ where we use the + or − sign depending on whether the fields are in the same or opposite directions and σ_B and σ_A are the magnitudes of the surface charges.

EXECUTE: **(a)** The fields add and point to the left, giving $E_{net} = 1.15 \times 10^6$ N/C.

(b) The fields oppose and point to the left, so $E_{net} = 1.58 \times 10^5$ N/C.

(c) The fields oppose but now point to the right, giving $E_{net} = 1.58 \times 10^5$ N/C.

EVALUATE: We can simplify the calculations by sketching the fields and doing an algebraic solution first.

21.93. **IDENTIFY:** The net force on the third sphere is the vector sum of the forces due to the other two charges. Coulomb's law gives the forces.

SET UP: $F = k\dfrac{|q_1 q_2|}{r^2}$.

EXECUTE: **(a)** Between the two fixed charges, the electric forces on the third sphere q_3 are in opposite directions and have magnitude 4.50 N in the +x-direction. Applying Coulomb's law gives
4.50 N = $k[q_1(4.00\ \mu C)/(0.200\text{ m})^2 - q_2(4.00\ \mu C)/(0.200\text{ m})^2]$.
Simplifying gives $q_1 - q_2 = 5.00\ \mu C$.
With q_3 at $x = +0.600$ m, the electric forces on q_3 are all in the +x-direction and add to 3.50 N. As before, Coulomb's law gives
3.50 N = $k[q_1(4.00\ \mu C)/(0.600\text{ m})^2 + q_2(4.00\ \mu C)/(0.200\text{ m})^2]$.
Simplifying gives $q_1 + 9q_2 = 35.0\ \mu C$.
Solving the two equations simultaneously gives $q_1 = 8.00\ \mu C$ and $q_2 = 3.00\ \mu C$.

(b) Both forces on q_3 are in the –x-direction, so their magnitudes add. Factoring out common factors and using the values for q_1 and q_2 we just found, Coulomb's law gives
$F_{net} = kq_3\ [q_1/(0.200\text{ m})^2 + q_2/(0.600\text{ m})^2]$.
$F_{net} = (8.99 \times 10^9\ \text{N} \cdot \text{m}^2/\text{C}^2)\ [(8.00\ \mu C)/(0.200\text{ m})^2 + (3.00\ \mu C)/(0.600\text{ m})^2] = 7.49$ N, and it is in the –x-direction.

(c) The forces are in opposite direction and add to zero, so
$0 = kq_1q_3/x^2 - kq_2q_3/(0.400\text{ m} - x)^2$.
$(0.400\text{ m} - x)^2 = (q_2/q_1)x^2$.
Taking square roots of both sides gives
$0.400\text{ m} - x = \pm x\sqrt{q_2/q_1} = \pm 0.6124x$.
Solving for x, we get two values: $x = 0.248$ m and $x = 1.03$ m. The charge q_3 must be between the other two charges for the forces on it to balance. Only the first value is between the two charges, so it is the correct one: $x = 0.248$ m.

EVALUATE: Check the answers in part (a) by substituting these values back into the original equations. 8.00 μC − 3.00 μC = 5.00 μC and 8.00 μC + 9(3.00 μC) =35.0 μC, so the answers check in both equations. In part (c), the second root, $x = 1.03$ m, has some meaning. The condition we imposed to solve the problem was that the magnitudes of the two forces were equal. This happens at $x = 0.248$ mn, but it also happens at $x = 1.03$ m. However at the second root the forces are both in the +x-direction and therefore cannot cancel.

21.101. **IDENTIFY and SET UP:** Assume that the charge remains at the end of the stem and that the bees approach to 15 cm from this end of the stem. The electric field is $E = k\dfrac{|q|}{r^2}$.

EXECUTE: Using the numbers given, we have
$E = k\dfrac{|q|}{r^2} = (8.99 \times 10^9\ \text{N} \cdot \text{m}^2/\text{C}^2)\ (40 \times 10^{-12}\ \text{C})/(0.15\text{ m})^2 = 16$ N/C, which is choice (b).

EVALUATE: Even if the charge spread out a bit over the stem, the result would be in the neighborhood of the value we calculated.

GAUSS'S LAW

22.3. IDENTIFY: The electric flux through an area is defined as the product of the component of the electric field perpendicular to the area times the area.

(a) SET UP: In this case, the electric field is perpendicular to the surface of the sphere, so $\Phi_E = EA = E(4\pi r^2)$.

EXECUTE: Substituting in the numbers gives
$$\Phi_E = (1.25 \times 10^6 \text{ N/C}) 4\pi (0.150 \text{ m})^2 = 3.53 \times 10^5 \text{ N} \cdot \text{m}^2/\text{C}.$$

(b) IDENTIFY: We use the electric field due to a point charge.

SET UP: $E = \dfrac{1}{4\pi\varepsilon_0} \dfrac{|q|}{r^2}$

EXECUTE: Solving for q and substituting the numbers gives
$$q = 4\pi\varepsilon_0 r^2 E = \dfrac{1}{9.00 \times 10^9 \text{ N} \cdot \text{m}^2/\text{C}^2} (0.150 \text{ m})^2 (1.25 \times 10^6 \text{ N/C}) = 3.13 \times 10^{-6} \text{ C}.$$

EVALUATE: The flux would be the same no matter how large the sphere, since the area is proportional to r^2 while the electric field is proportional to $1/r^2$.

22.5. IDENTIFY: The flux through the curved upper half of the hemisphere is the same as the flux through the flat circle defined by the bottom of the hemisphere because every electric field line that passes through the flat circle also must pass through the curved surface of the hemisphere.

SET UP: The electric field is perpendicular to the flat circle, so the flux is simply the product of E and the area of the flat circle of radius r.

EXECUTE: $\Phi_E = EA = E(\pi r^2) = \pi r^2 E$

EVALUATE: The flux would be the same if the hemisphere were replaced by any other surface bounded by the flat circle.

22.9. IDENTIFY: Apply the results in Example 22.5 for the field of a spherical shell of charge.

SET UP: Example 22.5 shows that $E = 0$ inside a uniform spherical shell and that $E = k\dfrac{|q|}{r^2}$ outside the shell.

EXECUTE: **(a)** $E = 0$.

(b) $r = 0.060$ m and $E = (8.99 \times 10^9 \text{ N} \cdot \text{m}^2/\text{C}^2) \dfrac{49.0 \times 10^{-6} \text{ C}}{(0.060 \text{ m})^2} = 1.22 \times 10^8$ N/C.

(c) $r = 0.110$ m and $E = (8.99 \times 10^9 \text{ N} \cdot \text{m}^2/\text{C}^2) \dfrac{49.0 \times 10^{-6} \text{ C}}{(0.110 \text{ m})^2} = 3.64 \times 10^7$ N/C.

EVALUATE: Outside the shell the electric field is the same as if all the charge were concentrated at the center of the shell. But inside the shell the field is not the same as for a point charge at the center of the shell, inside the shell the electric field is zero.

22.13. **IDENTIFY:** Each line lies in the electric field of the other line, and therefore each line experiences a force due to the other line.

SET UP: The field of one line at the location of the other is $E = \dfrac{\lambda}{2\pi\varepsilon_0 r}$. For charge $dq = \lambda dx$ on one line, the force on it due to the other line is $dF = E dq$. The total force is $F = \int E dq = E \int dq = Eq$.

EXECUTE: $E = \dfrac{\lambda}{2\pi\varepsilon_0 r} = \dfrac{5.20 \times 10^{-6} \text{ C/m}}{2\pi(8.854 \times 10^{-12} \text{ C}^2/(\text{N} \cdot \text{m}^2))(0.300 \text{ m})} = 3.116 \times 10^5$ N/C. The force on one line due to the other is $F = Eq$, where $q = \lambda(0.0500 \text{ m}) = 2.60 \times 10^{-7}$ C. The net force is
$F = Eq = (3.116 \times 10^5 \text{ N/C})(2.60 \times 10^{-7} \text{ C}) = 0.0810$ N.

EVALUATE: Since the electric field at each line due to the other line is uniform, each segment of line experiences the same force, so all we need to use is $F = Eq$, even though the line is *not* a point charge.

22.15. **IDENTIFY and SET UP:** Example 22.5 derived that the electric field just outside the surface of a spherical conductor that has net charge $|q|$ is $E = \dfrac{1}{4\pi\varepsilon_0} \dfrac{|q|}{R^2}$. Calculate $|q|$ and from this the number of excess electrons.

EXECUTE: $|q| = \dfrac{R^2 E}{(1/4\pi\varepsilon_0)} = \dfrac{(0.130 \text{ m})^2 (1150 \text{ N/C})}{8.988 \times 10^9 \text{ N} \cdot \text{m}^2/\text{C}^2} = 2.162 \times 10^{-9}$ C.

Each electron has a charge of magnitude $e = 1.602 \times 10^{-19}$ C, so the number of excess electrons needed is
$\dfrac{2.162 \times 10^{-9} \text{ C}}{1.602 \times 10^{-19} \text{ C}} = 1.35 \times 10^{10}$.

EVALUATE: The result we obtained for q is a typical value for the charge of an object. Such net charges correspond to a large number of excess electrons since the charge of each electron is very small.

22.17. **IDENTIFY:** Add the vector electric fields due to each line of charge. $E(r)$ for a line of charge is given by Example 22.6 and is directed toward a negative line of charge and away from a positive line.

SET UP: The two lines of charge are shown in Figure 22.17.

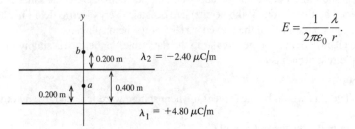

$E = \dfrac{1}{2\pi\varepsilon_0} \dfrac{\lambda}{r}$.

Figure 22.17

EXECUTE: (a) At point a, $\vec{E}_1$ and $\vec{E}_2$ are in the $+y$-direction (toward negative charge, away from positive charge).
$E_1 = (1/2\pi\varepsilon_0)[(4.80 \times 10^{-6} \text{ C/m})/(0.200 \text{ m})] = 4.314 \times 10^5$ N/C.
$E_2 = (1/2\pi\varepsilon_0)[(2.40 \times 10^{-6} \text{ C/m})/(0.200 \text{ m})] = 2.157 \times 10^5$ N/C.
$E = E_1 + E_2 = 6.47 \times 10^5$ N/C, in the y-direction.

(b) At point b, $\vec{E}_1$ is in the $+y$-direction and $\vec{E}_2$ is in the $-y$-direction.
$E_1 = (1/2\pi\varepsilon_0)[(4.80 \times 10^{-6} \text{ C/m})/(0.600 \text{ m})] = 1.438 \times 10^5$ N/C.
$E_2 = (1/2\pi\varepsilon_0)[(2.40 \times 10^{-6} \text{ C/m})/(0.200 \text{ m})] = 2.157 \times 10^5$ N/C.
$E = E_2 - E_1 = 7.2 \times 10^4$ N/C, in the $-y$-direction.

EVALUATE: At point a the two fields are in the same direction and the magnitudes add. At point b the two fields are in opposite directions and the magnitudes subtract.

22.19. IDENTIFY: The electric field inside the conductor is zero, and all of its initial charge lies on its outer surface. The introduction of charge into the cavity induces charge onto the surface of the cavity, which induces an equal but opposite charge on the outer surface of the conductor. The net charge on the outer surface of the conductor is the sum of the positive charge initially there and the additional negative charge due to the introduction of the negative charge into the cavity.

(a) SET UP: First find the initial positive charge on the outer surface of the conductor using $q_i = \sigma A$, where A is the area of its outer surface. Then find the net charge on the surface after the negative charge has been introduced into the cavity. Finally, use the definition of surface charge density.

EXECUTE: The original positive charge on the outer surface is

$$q_i = \sigma A = \sigma(4\pi r^2) = (6.37 \times 10^{-6} \text{ C/m}^2)4\pi(0.250 \text{ m})^2 = 5.00 \times 10^{-6} \text{ C}.$$

After the introduction of $-0.500 \ \mu\text{C}$ into the cavity, the outer charge is now

$$5.00 \ \mu\text{C} - 0.500 \ \mu\text{C} = 4.50 \ \mu\text{C}.$$

The surface charge density is now $\sigma = \dfrac{q}{A} = \dfrac{q}{4\pi r^2} = \dfrac{4.50 \times 10^{-6} \text{ C}}{4\pi(0.250 \text{ m})^2} = 5.73 \times 10^{-6} \text{ C/m}^2.$

(b) SET UP: Using Gauss's law, the electric field is $E = \dfrac{\Phi_E}{A} = \dfrac{q}{\varepsilon_0 A} = \dfrac{q}{\varepsilon_0 4\pi r^2}.$

EXECUTE: Substituting numbers gives

$$E = \dfrac{4.50 \times 10^{-6} \text{ C}}{(8.85 \times 10^{-12} \text{ C}^2/\text{N} \cdot \text{m}^2)(4\pi)(0.250 \text{ m})^2} = 6.47 \times 10^5 \text{ N/C}.$$

(c) SET UP: We use Gauss's law again to find the flux. $\Phi_E = \dfrac{Q_{\text{encl}}}{\varepsilon_0}.$

EXECUTE: Substituting numbers gives

$$\Phi_E = \dfrac{-0.500 \times 10^{-6} \text{ C}}{8.85 \times 10^{-12} \text{ C}^2/\text{N} \cdot \text{m}^2} = -5.65 \times 10^4 \text{ N} \cdot \text{m}^2/\text{C}.$$

EVALUATE: The excess charge on the conductor is still $+5.00 \ \mu\text{C}$, as it originally was. The introduction of the $-0.500 \ \mu\text{C}$ inside the cavity merely induced equal but opposite charges (for a net of zero) on the surfaces of the conductor.

22.21. IDENTIFY: The magnitude of the electric field is constant at any given distance from the center because the charge density is uniform inside the sphere. We can use Gauss's law to relate the field to the charge causing it.

(a) SET UP: Gauss's law tells us that $EA = \dfrac{q}{\varepsilon_0}$, and the charge density is given by $\rho = \dfrac{q}{V} = \dfrac{q}{(4/3)\pi R^3}.$

EXECUTE: Solving for q and substituting numbers gives

$q = EA\varepsilon_0 = E(4\pi r^2)\varepsilon_0 = (1750 \text{ N/C})(4\pi)(0.500 \text{ m})^2(8.85 \times 10^{-12} \text{ C}^2/\text{N} \cdot \text{m}^2) = 4.866 \times 10^{-8} \text{ C}.$ Using the

formula for charge density we get $\rho = \dfrac{q}{V} = \dfrac{q}{(4/3)\pi R^3} = \dfrac{4.866 \times 10^{-8} \text{ C}}{(4/3)\pi(0.355 \text{ m})^3} = 2.60 \times 10^{-7} \text{ C/m}^3.$

(b) SET UP: Take a Gaussian surface of radius $r = 0.200$ m, concentric with the insulating sphere. The

charge enclosed within this surface is $q_{\text{encl}} = \rho V = \rho\left(\dfrac{4}{3}\pi r^3\right)$, and we can treat this charge as a point-

charge, using Coulomb's law $E = \dfrac{1}{4\pi\varepsilon_0}\dfrac{q_{\text{encl}}}{r^2}.$ The charge beyond $r = 0.200$ m makes no contribution to the electric field.

EXECUTE: First find the enclosed charge:

$$q_{encl} = \rho\left(\frac{4}{3}\pi r^3\right) = (2.60\times 10^{-7} \text{ C/m}^3)\left[\frac{4}{3}\pi(0.200 \text{ m})^3\right] = 8.70\times 10^{-9} \text{ C}$$

Now treat this charge as a point-charge and use Coulomb's law to find the field:

$$E = (9.00\times 10^9 \text{ N}\cdot\text{m}^2/\text{C}^2)\frac{8.70\times 10^{-9} \text{ C}}{(0.200 \text{ m})^2} = 1.96\times 10^3 \text{ N/C}$$

EVALUATE: Outside this sphere, it behaves like a point-charge located at its center. Inside of it, at a distance r from the center, the field is due only to the charge between the center and r.

22.29. **IDENTIFY:** Apply Gauss's law to a Gaussian surface and calculate E.
(a) SET UP and EXECUTE: Consider the charge on a length l of the cylinder. This can be expressed as $q = \lambda l$. But since the surface area is $2\pi R l$ it can also be expressed as $q = \sigma 2\pi R l$. These two expressions must be equal, so $\lambda l = \sigma 2\pi R l$ and $\lambda = 2\pi R \sigma$.
(b) SET UP: Apply Gauss's law to a Gaussian surface that is a cylinder of length l, radius r, and whose axis coincides with the axis of the charge distribution, as shown in Figure 22.29.

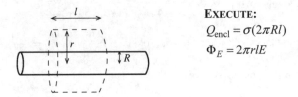

EXECUTE:
$Q_{encl} = \sigma(2\pi R l)$
$\Phi_E = 2\pi r l E$

Figure 22.29

$\Phi_E = \dfrac{Q_{encl}}{\varepsilon_0}$ gives $2\pi r l E = \dfrac{\sigma(2\pi R l)}{\varepsilon_0}$, so $E = \dfrac{\sigma R}{\varepsilon_0 r}$.

EVALUATE: (c) Example 22.6 shows that the electric field of an infinite line of charge is $E = \lambda/2\pi\varepsilon_0 r$.

$\sigma = \dfrac{\lambda}{2\pi R}$, so $E = \dfrac{\sigma R}{\varepsilon_0 r} = \dfrac{R}{\varepsilon_0 r}\left(\dfrac{\lambda}{2\pi R}\right) = \dfrac{\lambda}{2\pi\varepsilon_0 r}$, the same as for an infinite line of charge that is along the axis of the cylinder.

22.31. **IDENTIFY:** The uniform electric field of the sheet exerts a constant force on the proton perpendicular to the sheet, and therefore does not change the parallel component of its velocity. Newton's second law allows us to calculate the proton's acceleration perpendicular to the sheet, and uniform-acceleration kinematics allows us to determine its perpendicular velocity component.
SET UP: Let $+x$ be the direction of the initial velocity and let $+y$ be the direction perpendicular to the sheet and pointing away from it. $a_x = 0$ so $v_x = v_{0x} = 9.70\times 10^2$ m/s. The electric field due to the sheet is $E = \dfrac{\sigma}{2\varepsilon_0}$ and the magnitude of the force the sheet exerts on the proton is $F = eE$.

EXECUTE: $E = \dfrac{\sigma}{2\varepsilon_0} = \dfrac{2.34\times 10^{-9} \text{ C/m}^2}{2(8.854\times 10^{-12} \text{ C}^2/(\text{N}\cdot\text{m}^2))} = 132.1$ N/C. Newton's second law gives

$a_y = \dfrac{Eq}{m} = \dfrac{(132.1 \text{ N/C})(1.602\times 10^{-19} \text{ C})}{1.673\times 10^{-27} \text{ kg}} = 1.265\times 10^{10}$ m/s^2. Kinematics gives

$v_y = v_{0y} + a_y y = (1.265\times 10^{10}$ m/s$^2)(5.00\times 10^{-8}$ s$) = 632.7$ m/s. The speed of the proton is the magnitude of its velocity, so $v = \sqrt{v_x^2 + v_y^2} = \sqrt{(9.70\times 10^2 \text{ m/s})^2 + (632.7 \text{ m/s})^2} = 1.16\times 10^3$ m/s.

EVALUATE: We can use the constant-acceleration kinematics equations because the uniform electric field of the sheet exerts a constant force on the proton, giving it a constant acceleration. We could *not* use this approach if the sheet were replaced with a sphere, for example.

22.37. **IDENTIFY:** Find the net flux through the parallelepiped surface and then use that in Gauss's law to find the net charge within. Flux out of the surface is positive and flux into the surface is negative.

(a) SET UP: $\vec{E}_1$ gives flux out of the surface. See Figure 22.37a.

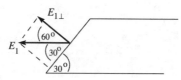

EXECUTE: $\Phi_1 = +E_{1\perp}A$.

$A = (0.0600 \text{ m})(0.0500 \text{ m}) = 3.00 \times 10^{-3} \text{ m}^2$.

$E_{1\perp} = E_1 \cos 60° = (2.50 \times 10^4 \text{ N/C}) \cos 60°$.

$E_{1\perp} = 1.25 \times 10^4 \text{ N/C}$.

Figure 22.37a

$\Phi_{E_1} = +E_{1\perp}A = +(1.25 \times 10^4 \text{ N/C})(3.00 \times 10^{-3} \text{ m}^2) = 37.5 \text{ N} \cdot \text{m}^2/\text{C}$.

SET UP: $\vec{E}_2$ gives flux into the surface. See Figure 22.37b.

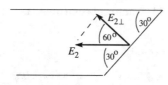

EXECUTE: $\Phi_2 = -E_{2\perp}A$.

$A = (0.0600 \text{ m})(0.0500 \text{ m}) = 3.00 \times 10^{-3} \text{ m}^2$.

$E_{2\perp} = E_2 \cos 60° = (7.00 \times 10^4 \text{ N/C}) \cos 60°$.

$E_{2\perp} = 3.50 \times 10^4 \text{ N/C}$.

Figure 22.37b

$\Phi_{E_2} = -E_{2\perp}A = -(3.50 \times 10^4 \text{ N/C})(3.00 \times 10^{-3} \text{ m}^2) = -105.0 \text{ N} \cdot \text{m}^2/\text{C}$.

The net flux is $\Phi_E = \Phi_{E_1} + \Phi_{E_2} = +37.5 \text{ N} \cdot \text{m}^2/\text{C} - 105.0 \text{ N} \cdot \text{m}^2/\text{C} = -67.5 \text{ N} \cdot \text{m}^2/\text{C}$.

The net flux is negative (inward), so the net charge enclosed is negative.

Apply Gauss's law: $\Phi_E = \dfrac{Q_{\text{encl}}}{\varepsilon_0}$

$Q_{\text{encl}} = \Phi_E \varepsilon_0 = (-67.5 \text{ N} \cdot \text{m}^2/\text{C})(8.854 \times 10^{-12} \text{ C}^2/\text{N} \cdot \text{m}^2) = -5.98 \times 10^{-10} \text{ C}$.

EVALUATE: **(b)** If there were no charge within the parallelepiped the net flux would be zero. This is not the case, so there is charge inside. The electric field lines that pass out through the surface of the parallelepiped must terminate on charges, so there also must be charges outside the parallelepiped.

22.41. **IDENTIFY:** Apply Gauss's law.

SET UP: Use a Gaussian surface that is a cylinder of radius r and length l, and that is coaxial with the cylindrical charge distributions. The volume of the Gaussian cylinder is $\pi r^2 l$ and the area of its curved surface is $2\pi r l$. The charge on a length l of the charge distribution is $q = \lambda l$, where $\lambda = \rho \pi R^2$.

EXECUTE: **(a)** For $r < R$, $Q_{\text{encl}} = \rho \pi r^2 l$ and Gauss's law gives $E(2\pi r l) = \dfrac{Q_{\text{encl}}}{\varepsilon_0} = \dfrac{\rho \pi r^2 l}{\varepsilon_0}$ and $E = \dfrac{\rho r}{2\varepsilon_0}$, radially outward.

(b) For $r > R$, $Q_{\text{encl}} = \lambda l = \rho \pi R^2 l$ and Gauss's law gives $E(2\pi r l) = \dfrac{Q_{\text{encl}}}{\varepsilon_0} = \dfrac{\rho \pi R^2 l}{\varepsilon_0}$ and

$E = \dfrac{\rho R^2}{2\varepsilon_0 r} = \dfrac{\lambda}{2\pi \varepsilon_0 r}$, radially outward.

(c) At $r = R$, the electric field for *both* regions is $E = \dfrac{\rho R}{2\varepsilon_0}$, so they are consistent.

(d) The graph of E versus r is sketched in Figure 22.41 (next page).

EVALUATE: For $r > R$ the field is the same as for a line of charge along the axis of the cylinder.

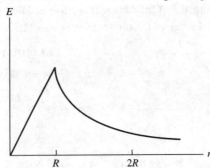

Figure 22.41

22.49. IDENTIFY: We apply Gauss's law in (a) and take a spherical Gaussian surface because of the spherical symmetry of the charge distribution. In (b), the net field is the vector sum of the field due to q and the field due to the sphere.

(a) SET UP: $\rho(r) = \dfrac{\alpha}{r}$, $dV = 4\pi r^2 dr$, and $Q = \int_a^r \rho(r')dV$.

EXECUTE: For a Gaussian sphere of radius r, $Q_{encl} = \int_a^r \rho(r')dV = 4\pi\alpha \int_a^r r'dr' = 4\pi\alpha \dfrac{1}{2}(r^2 - a^2)$. Gauss's law says that $E(4\pi r^2) = \dfrac{2\pi\alpha(r^2 - a^2)}{\varepsilon_0}$, which gives $E = \dfrac{\alpha}{2\varepsilon_0}\left(1 - \dfrac{a^2}{r^2}\right)$.

(b) SET UP and EXECUTE: The electric field of the point charge is $E_q = \dfrac{q}{4\pi\varepsilon_0 r^2}$. The total electric field is $E_{total} = \dfrac{\alpha}{2\varepsilon_0} - \dfrac{\alpha}{2\varepsilon_0}\dfrac{a^2}{r^2} + \dfrac{q}{4\pi\varepsilon_0 r^2}$. For E_{total} to be constant, $-\dfrac{\alpha a^2}{2\varepsilon_0} + \dfrac{q}{4\pi\varepsilon_0} = 0$ and $q = 2\pi\alpha a^2$. The constant electric field is $\dfrac{\alpha}{2\varepsilon_0}$.

EVALUATE: The net field is constant, but not zero.

22.51. IDENTIFY: There is a force on each electron due to the other electron and a force due to the sphere of charge. Use Coulomb's law for the force between the electrons. Example 22.9 gives E inside a uniform sphere and $F = \dfrac{1}{4\pi\varepsilon_0}\dfrac{|q_1 q_2|}{r^2}$ gives the force.

SET UP: The positions of the electrons are sketched in Figure 22.51a.

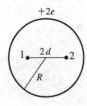

If the electrons are in equilibrium the net force on each one is zero.

Figure 22.51a

EXECUTE: Consider the forces on electron 2. There is a repulsive force F_1 due to the other electron, electron 1.

$$F_1 = \dfrac{1}{4\pi\varepsilon_0}\dfrac{e^2}{(2d)^2}$$

The electric field inside the uniform distribution of positive charge is $E = \dfrac{Qr}{4\pi\varepsilon_0 R^3}$ (Example 22.9), where $Q = +2e$. At the position of electron 2, $r = d$. The force F_{cd} exerted by the positive charge distribution is $F_{cd} = eE = \dfrac{e(2e)d}{4\pi\varepsilon_0 R^3}$ and is attractive.

The force diagram for electron 2 is given in Figure 22.51b.

Figure 22.51b

Net force equals zero implies $F_1 = F_{cd}$ and $\dfrac{1}{4\pi\varepsilon_0}\dfrac{e^2}{4d^2} = \dfrac{2e^2 d}{4\pi\varepsilon_0 R^3}$.

Thus $(1/4d^2) = 2d/R^3$, so $d^3 = R^3/8$ and $d = R/2$.

EVALUATE: The electric field of the sphere is radially outward; it is zero at the center of the sphere and increases with distance from the center. The force this field exerts on one of the electrons is radially inward and increases as the electron is farther from the center. The force from the other electron is radially outward, is infinite when $d = 0$ and decreases as d increases. It is reasonable therefore for there to be a value of d for which these forces balance.

22.55. **(a) IDENTIFY and SET UP:** Consider the direction of the field for x slightly greater than and slightly less than zero. The slab is sketched in Figure 22.55a.

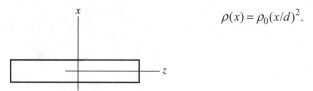

Figure 22.55a

EXECUTE: The charge distribution is symmetric about $x = 0$, so by symmetry $E(x) = E(-x)$. But for $x > 0$ the field is in the $+x$-direction and for $x < 0$ the field is in the $-x$-direction. At $x = 0$ the field can't be both in the $+x$- and $-x$-directions so must be zero. That is, $E_x(x) = -E_x(-x)$. At point $x = 0$ this gives $E_x(0) = -E_x(0)$ and this equation is satisfied only for $E_x(0) = 0$.

(b) IDENTIFY and SET UP: $|x| > d$ (outside the slab).

Apply Gauss's law to a cylindrical Gaussian surface whose axis is perpendicular to the slab and whose end caps have area A and are the same distance $|x| > d$ from $x = 0$, as shown in Figure 22.55b.

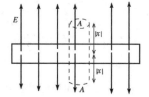

EXECUTE: $\Phi_E = 2EA$.

Figure 22.55b

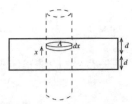

To find Q_{encl} consider a thin disk at coordinate x and with thickness dx, as shown in Figure 22.55c. The charge within this disk is

$$dq = \rho dV = \rho A dx = (\rho_0 A/d^2) x^2 dx.$$

Figure 22.55c

The total charge enclosed by the Gaussian cylinder is

$$Q_{encl} = 2\int_0^d dq = (2\rho_0 A/d^2)\int_0^d x^2 dx = (2\rho_0 A/d^2)(d^3/3) = \tfrac{2}{3}\rho_0 Ad.$$

Then $\Phi_E = \dfrac{Q_{encl}}{\varepsilon_0}$ gives $2EA = 2\rho_0 Ad/3\varepsilon_0$. This gives $E = \rho_0 d/3\varepsilon_0$.

$\vec{E}$ is directed away from $x = 0$, so $\vec{E} = (\rho_0 d/3\varepsilon_0)(x/|x|)\hat{\imath}$.

(c) IDENTIFY and **SET UP**: $|x| < d$ (inside the slab).

Apply Gauss's law to a cylindrical Gaussian surface whose axis is perpendicular to the slab and whose end caps have area A and are the same distance $|x| < d$ from $x = 0$, as shown in Figure 22.55d.

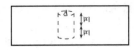

EXECUTE: $\Phi_E = 2EA$.

Figure 22.55d

Q_{encl} is found as above, but now the integral on dx is only from 0 to x instead of 0 do d.

$$Q_{encl} = 2\int_0^x dq = (2\rho_0 A/d^2)\int_0^x x^2 dx = (2\rho_0 A/d^2)(x^3/3).$$

Then $\Phi_E = \dfrac{Q_{encl}}{\varepsilon_0}$ gives $2EA = 2\rho_0 Ax^3/3\varepsilon_0 d^2$. This gives $E = \rho_0 x^3/3\varepsilon_0 d^2$.

$\vec{E}$ is directed away from $x = 0$, so $\vec{E} = (\rho_0 x^3/3\varepsilon_0 d^2)\hat{\imath}$.

EVALUATE: Note that $E = 0$ at $x = 0$ as stated in part (a). Note also that the expressions for $|x| > d$ and $|x| < d$ agree for $x = d$.

22.57. **(a) IDENTIFY:** Use $\vec{E}(\vec{r})$ from Example (22.9) (inside the sphere) and relate the position vector of a point inside the sphere measured from the origin to that measured from the center of the sphere.

SET UP: For an insulating sphere of uniform charge density ρ and centered at the origin, the electric field inside the sphere is given by $E = Qr'/4\pi\varepsilon_0 R^3$ (Example 22.9), where $\vec{r}'$ is the vector from the center of the sphere to the point where E is calculated.

But $\rho = 3Q/4\pi R^3$ so this may be written as $E = \rho r/3\varepsilon_0$. And $\vec{E}$ is radially outward, in the direction of $\vec{r}'$, so $\vec{E} = \rho\vec{r}'/3\varepsilon_0$.

For a sphere whose center is located by vector $\vec{b}$, a point inside the sphere and located by $\vec{r}$ is located by the vector $\vec{r}' = \vec{r} - \vec{b}$ relative to the center of the sphere, as shown in Figure 22.57.

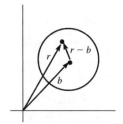

EXECUTE: Thus $\vec{E} = \dfrac{\rho(\vec{r}-\vec{b})}{3\varepsilon_0}$.

Figure 22.57

EVALUATE: When $b=0$ this reduces to the result of Example 22.9. When $\vec{r}=\vec{b}$, this gives $E=0$, which is correct since we know that $E=0$ at the center of the sphere.

(b) IDENTIFY: The charge distribution can be represented as a uniform sphere with charge density ρ and centered at the origin added to a uniform sphere with charge density $-\rho$ and centered at $\vec{r}=\vec{b}$.

SET UP: $\vec{E} = \vec{E}_{\text{uniform}} + \vec{E}_{\text{hole}}$, where $\vec{E}_{\text{uniform}}$ is the field of a uniformly charged sphere with charge density ρ and $\vec{E}_{\text{hole}}$ is the field of a sphere located at the hole and with charge density $-\rho$. (Within the spherical hole the net charge density is $+\rho-\rho=0$.)

EXECUTE: $\vec{E}_{\text{uniform}} = \dfrac{\rho\vec{r}}{3\varepsilon_0}$, where $\vec{r}$ is a vector from the center of the sphere.

$\vec{E}_{\text{hole}} = \dfrac{-\rho(\vec{r}-\vec{b})}{3\varepsilon_0}$, at points inside the hole. Then $\vec{E} = \dfrac{\rho\vec{r}}{3\varepsilon_0} + \left(\dfrac{-\rho(\vec{r}-\vec{b})}{3\varepsilon_0}\right) = \dfrac{\rho\vec{b}}{3\varepsilon_0}$.

EVALUATE: $\vec{E}$ is independent of $\vec{r}$ so is uniform inside the hole. The direction of $\vec{E}$ inside the hole is in the direction of the vector $\vec{b}$, the direction from the center of the insulating sphere to the center of the hole.

22.59. IDENTIFY and SET UP: For a uniformly charged sphere, $E = k\dfrac{|Q|}{r^2}$, so $Er^2 = k|Q| = $ constant. For a long uniform line of charge, $E = \dfrac{\lambda}{2\pi\varepsilon_0 r}$, so $Er = \dfrac{\lambda}{2\pi\varepsilon_0} = $ constant.

EXECUTE: **(a)** Figure 22.59a shows the graphs for data set A. We see that the graph of Er versus r is a horizontal line, which means that $Er = $ constant. Therefore data set A is for a uniform straight line of charge.

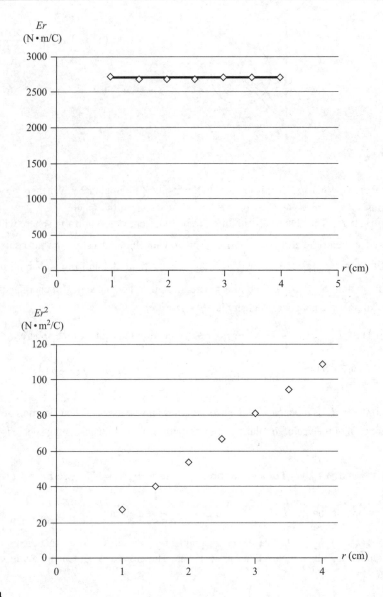

Figure 22.59a

Figure 22.59b shows the graphs for data set B. We see that the graph of Er^2 versus r is a horizontal line, so Er^2 = constant. Thus data set B is for a uniformly charged sphere.

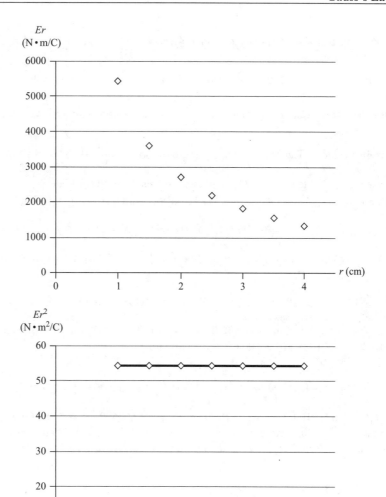

Figure 22.59b

(b) For A: $E = \dfrac{\lambda}{2\pi\varepsilon_0 r}$, so $\lambda = 2\pi\varepsilon_0 Er$. From our graph in Figure 22.59a, $Er = $ constant $= 2690$ N·m/C. Therefore

$\lambda = 2\pi\varepsilon_0 Er = 2\pi(8.854\times 10^{-12}$ C^2/N·m$^2)(2690$ N·m/C$) = 1.50\times 10^{-7}$ C/m $= 0.150$ μC/m.

For B: $E = k\dfrac{|Q|}{r^2}$, so $kQ = Er^2 = $ constant, which means that $Q = $ (constant)/k. From our graph in Figure 22.59b, $Er^2 = $ constant $= 54.1$ N·m^2/C. Therefore

$Q = (54.1$ N·m^2/C$)/(8.99\times 10^9$ N·m^2/C$^2) = 6.0175\times 10^{-9}$ C.

The charge density ρ is $\rho = \dfrac{Q}{V} = \dfrac{Q}{\frac{4}{3}\pi R^3} = (6.0175\times 10^{-9}$ C$)/[(4\pi/3)(0.00800$ m$)^3] = 2.81\times 10^{-3}$ C/m^3.

EVALUATE: A linear charge density of 0.150 C/m and a volume charge density of 2.81×10^{-3} C/m^3 are both physically reasonable and could be achieved in a normal laboratory.

22.61. **IDENTIFY** and **SET UP:** Apply Gauss's law, $\oint \vec{E} \cdot d\vec{A} = \dfrac{Q_{encl}}{\varepsilon_0}$. The enclosed charge is $Q_{encl} = \rho V$, where $V = \dfrac{4}{3}\pi r^3$ for a sphere of radius r. Read the charge densities from the graph in the problem.

EXECUTE: Apply Gauss's law $\oint \vec{E} \cdot d\vec{A} = \dfrac{Q_{encl}}{\varepsilon_0}$. As a Gaussian surface, use a sphere of radius r centered on the given sphere. This gives $E(4\pi r^2) = Q_{encl}/\varepsilon_0$, so $E = \dfrac{1}{4\pi\varepsilon_0} \dfrac{Q_{encl}}{r^2} = k\dfrac{Q_{encl}}{r^2}$. In each case, we must first use $Q_{encl} = \rho V$ to calculate Q_{encl} and then use that result to calculate E.

(i) First find Q_{encl}: $Q_{encl} = \rho V = (10.0 \times 10^{-6}\ \text{C/m}^3)(4\pi/3)(0.00100\ \text{m})^3 = 4.19 \times 10^{-14}\ \text{C}$.

Now calculate E: $E = k\dfrac{Q_{encl}}{r^2} = (8.99 \times 10^9\ \text{N} \cdot \text{m}^2/\text{C}^2)(4.19 \times 10^{-14}\ \text{C})/(0.00100\ \text{m})^2 = 377\ \text{N/C}$.

(ii) $Q_{encl} = (10.0 \times 10^{-6}\ \text{C/m}^3)(4\pi/3)(0.00200\ \text{m})^3 + (4.0 \times 10^{-6}\ \text{C/m}^3)(4\pi/3)[(0.00300\ \text{m})^3 - (0.00200\ \text{m})^3]$

$Q_{encl} = 6.534 \times 10^{-13}\ \text{C}$.

$E = k\dfrac{Q_{encl}}{r^2} = (8.99 \times 10^9\ \text{N} \cdot \text{m}^2/\text{C}^2)(6.534 \times 10^{-13}\ \text{C})/(0.00300\ \text{m})^2 = 653\ \text{N/C}$.

(iii) $Q_{encl} = (10.0 \times 10^{-6}\ \text{C/m}^3)(4\pi/3)(0.00200\ \text{m})^3 + (4.0 \times 10^{-6}\ \text{C/m}^3)(4\pi/3)[(0.00400\ \text{m})^3 - (0.00200\ \text{m})^3] + (-2.0 \times 10^{-6}\ \text{C/m}^3)(4\pi/3)[(0.00500\ \text{m})^3 - (0.00400\ \text{m})^3]$.

$Q_{encl} = 7.624 \times 10^{-13}\ \text{C}$.

$E = k\dfrac{Q_{encl}}{r^2} = (8.99 \times 10^9\ \text{N} \cdot \text{m}^2/\text{C}^2)(7.624 \times 10^{-13}\ \text{C})/(0.00500\ \text{m})^2 = 274\ \text{N/C}$.

(iv) $Q_{encl} = 7.624 \times 10^{-13}\ \text{C} + (-2.0 \times 10^{-6}\ \text{C/m}^3)(4\pi/3)[(0.00600\ \text{m})^3 - (0.00500\ \text{m})^3] = 0$, so $E = 0$.

EVALUATE: We found that $E = 0$ at $r = 7.00$ mm, but E is also zero at all points beyond $r = 6.00$ mm because the enclosed charge is zero for any Gaussian surface having a radius $r > 6.00$ mm.

ELECTRIC POTENTIAL

23.1. IDENTIFY: Apply $W_{a \to b} = U_a - U_b$ to calculate the work. The electric potential energy of a pair of point charges is given by $U = \dfrac{1}{4\pi\epsilon_0} \dfrac{q_1 q_2}{r}$.

SET UP: Let the initial position of q_2 be point a and the final position be point b, as shown in Figure 23.1.

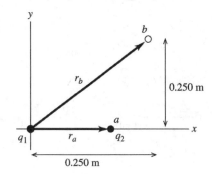

$r_a = 0.150$ m.
$r_b = \sqrt{(0.250 \text{ m})^2 + (0.250 \text{ m})^2}$.
$r_b = 0.3536$ m.

Figure 23.1

EXECUTE: $W_{a \to b} = U_a - U_b$.

$$U_a = \frac{1}{4\pi\epsilon_0} \frac{q_1 q_2}{r_a} = (8.988 \times 10^9 \text{ N} \cdot \text{m}^2/\text{C}^2) \frac{(+2.40 \times 10^{-6} \text{ C})(-4.30 \times 10^{-6} \text{ C})}{0.150 \text{ m}}.$$

$U_a = -0.6184$ J.

$$U_b = \frac{1}{4\pi\epsilon_0} \frac{q_1 q_2}{r_b} = (8.988 \times 10^9 \text{ N} \cdot \text{m}^2/\text{C}^2) \frac{(+2.40 \times 10^{-6} \text{ C})(-4.30 \times 10^{-6} \text{ C})}{0.3536 \text{ m}}.$$

$U_b = -0.2623$ J.

$$W_{a \to b} = U_a - U_b = -0.6184 \text{ J} - (-0.2623 \text{ J}) = -0.356 \text{ J}.$$

EVALUATE: The attractive force on q_2 is toward the origin, so it does negative work on q_2 when q_2 moves to larger r.

23.3. IDENTIFY: The work needed to assemble the nucleus is the sum of the electrical potential energies of the protons in the nucleus, relative to infinity.

SET UP: The total potential energy is the scalar sum of all the individual potential energies, where each potential energy is $U = (1/4\pi\epsilon_0)(qq_0/r)$. Each charge is e and the charges are equidistant from each other, so the total potential energy is $U = \dfrac{1}{4\pi\epsilon_0}\left(\dfrac{e^2}{r} + \dfrac{e^2}{r} + \dfrac{e^2}{r}\right) = \dfrac{3e^2}{4\pi\epsilon_0 r}$.

23-2 Chapter 23

EXECUTE: Adding the potential energies gives

$$U = \frac{3e^2}{4\pi\varepsilon_0 r} = \frac{3(1.60\times10^{-19}\text{ C})^2(9.00\times10^9\text{ N}\cdot\text{m}^2/\text{C}^2)}{2.00\times10^{-15}\text{ m}} = 3.46\times10^{-13}\text{ J} = 2.16\text{ MeV}.$$

EVALUATE: This is a small amount of energy on a macroscopic scale, but on the scale of atoms, 2 MeV is quite a lot of energy.

23.5. **(a) IDENTIFY:** Use conservation of energy: $K_a + U_a + W_{\text{other}} = K_b + U_b$. U for the pair of point charges is given by $U = \frac{1}{4\pi\varepsilon_0}\frac{q_1 q_2}{r}$.

SET UP:

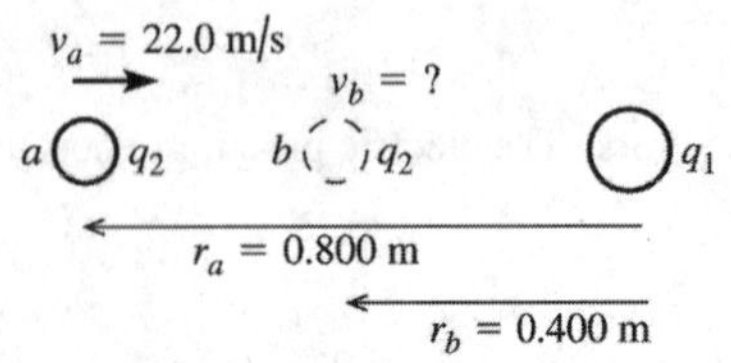

Let point a be where q_2 is 0.800 m from q_1 and point b be where q_2 is 0.400 m from q_1, as shown in Figure 23.5a.

Figure 23.5a

EXECUTE: Only the electric force does work, so $W_{\text{other}} = 0$ and $U = \frac{1}{4\pi\varepsilon_0}\frac{q_1 q_2}{r}$.

$$K_a = \tfrac{1}{2}mv_a^2 = \tfrac{1}{2}(1.50\times10^{-3}\text{ kg})(22.0\text{ m/s})^2 = 0.3630\text{ J}.$$

$$U_a = \frac{1}{4\pi\varepsilon_0}\frac{q_1 q_2}{r_a} = (8.988\times10^9\text{ N}\cdot\text{m}^2/\text{C}^2)\frac{(-2.80\times10^{-6}\text{ C})(-7.80\times10^{-6}\text{ C})}{0.800\text{ m}} = +0.2454\text{ J}.$$

$$K_b = \tfrac{1}{2}mv_b^2.$$

$$U_b = \frac{1}{4\pi\varepsilon_0}\frac{q_1 q_2}{r_b} = (8.988\times10^9\text{ N}\cdot\text{m}^2/\text{C}^2)\frac{(-2.80\times10^{-6}\text{ C})(-7.80\times10^{-6}\text{ C})}{0.400\text{ m}} = +0.4907\text{ J}.$$

The conservation of energy equation then gives $K_b = K_a + (U_a - U_b)$.

$$\tfrac{1}{2}mv_b^2 = +0.3630\text{ J} + (0.2454\text{ J} - 0.4907\text{ J}) = 0.1177\text{ J}.$$

$$v_b = \sqrt{\frac{2(0.1177\text{ J})}{1.50\times10^{-3}\text{ kg}}} = 12.5\text{ m/s}.$$

EVALUATE: The potential energy increases when the two positively charged spheres get closer together, so the kinetic energy and speed decrease.

(b) IDENTIFY: Let point c be where q_2 has its speed momentarily reduced to zero. Apply conservation of energy to points a and c: $K_a + U_a + W_{\text{other}} = K_c + U_c$.

SET UP: Points a and c are shown in Figure 23.5b.

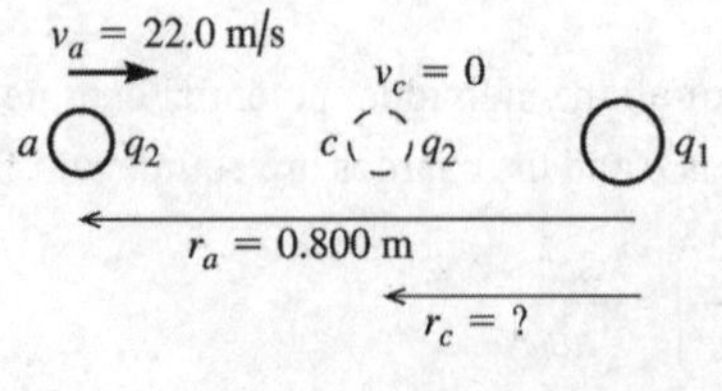

EXECUTE: $K_a = +0.3630\text{ J}$ (from part (a)).

$U_a = +0.2454\text{ J}$ (from part (a)).

Figure 23.5b

$K_c = 0$ (at distance of closest approach the speed is zero).

$$U_c = \frac{1}{4\pi\varepsilon_0}\frac{q_1 q_2}{r_c}.$$

Thus conservation of energy $K_a + U_a = U_c$ gives $\frac{1}{4\pi\varepsilon_0}\frac{q_1 q_2}{r_c} = +0.3630\text{ J} + 0.2454\text{ J} = 0.6084\text{ J}.$

$$r_c = \frac{1}{4\pi\varepsilon_0}\frac{q_1 q_2}{0.6084\text{ J}} = (8.988\times 10^9\text{ N}\cdot\text{m}^2/\text{C}^2)\frac{(-2.80\times 10^{-6}\text{ C})(-7.80\times 10^{-6}\text{ C})}{+0.6084\text{ J}} = 0.323\text{ m}.$$

EVALUATE: $U \to \infty$ as $r \to 0$ so q_2 will stop no matter what its initial speed is.

23.9. **IDENTIFY:** The protons repel each other and therefore accelerate away from one another. As they get farther and farther away, their kinetic energy gets greater and greater but their acceleration keeps decreasing. Conservation of energy and Newton's laws apply to these protons.
SET UP: Let a be the point when they are 0.750 nm apart and b be the point when they are very far apart. A proton has charge $+e$ and mass 1.67×10^{-27} kg. As they move apart the protons have equal kinetic energies and speeds. Their potential energy is $U = ke^2/r$ and $K = \tfrac{1}{2}mv^2$. $K_a + U_a = K_b + U_b$.
EXECUTE: **(a)** They have maximum speed when they are far apart and all their initial electrical potential energy has been converted to kinetic energy. $K_a + U_a = K_b + U_b$.
$K_a = 0$ and $U_b = 0$, so

$$K_b = U_a = k\frac{e^2}{r_a} = (8.99\times 10^9\text{ N}\cdot\text{m}^2/\text{C}^2)\frac{(1.60\times 10^{-19}\text{ C})^2}{0.750\times 10^{-9}\text{ m}} = 3.07\times 10^{-19}\text{ J}.$$

$K_b = \tfrac{1}{2}mv_b^2 + \tfrac{1}{2}mv_b^2$, so $K_b = mv_b^2$ and $v_b = \sqrt{\frac{K_b}{m}} = \sqrt{\frac{3.07\times 10^{-19}\text{ J}}{1.67\times 10^{-27}\text{ kg}}} = 1.36\times 10^4\text{ m/s}.$

(b) Their acceleration is largest when the force between them is largest and this occurs at $r = 0.750$ nm, when they are closest.

$$F = k\frac{e^2}{r^2} = (8.99\times 10^9\text{ N}\cdot\text{m}^2/\text{C}^2)\left(\frac{1.60\times 10^{-19}\text{ C}}{0.750\times 10^{-9}\text{ m}}\right)^2 = 4.09\times 10^{-10}\text{ N}.$$

$a = \frac{F}{m} = \frac{4.09\times 10^{-10}\text{ N}}{1.67\times 10^{-27}\text{ kg}} = 2.45\times 10^{17}\text{ m/s}^2.$

EVALUATE: The acceleration of the protons decreases as they move farther apart, but the force between them is repulsive so they continue to increase their speeds and hence their kinetic energies.

23.13. **IDENTIFY and SET UP:** Apply conservation of energy to points A and B.
EXECUTE: $K_A + U_A = K_B + U_B$.
$U = qV$, so $K_A + qV_A = K_B + qV_B$.
$K_B = K_A + q(V_A - V_B) = 0.00250\text{ J} + (-5.00\times 10^{-6}\text{ C})(200\text{ V} - 800\text{ V}) = 0.00550\text{ J}.$
$v_B = \sqrt{2K_B/m} = 7.42\text{ m/s}.$
EVALUATE: It is faster at B; a negative charge gains speed when it moves to higher potential.

23.17. **IDENTIFY:** The potential at any point is the scalar sum of the potentials due to individual charges.
SET UP: $V = kq/r$ and $W_{ab} = q(V_a - V_b)$.

EXECUTE: **(a)** $r_{a1} = r_{a2} = \tfrac{1}{2}\sqrt{(0.0300\text{ m})^2 + (0.0300\text{ m})^2} = 0.0212\text{ m}.$ $V_a = k\left(\frac{q_1}{r_{a1}} + \frac{q_2}{r_{a2}}\right) = 0.$

(b) $r_{b1} = 0.0424\text{ m}$, $r_{b2} = 0.0300\text{ m}.$

$V_b = k\left(\frac{q_1}{r_{b1}} + \frac{q_2}{r_{b2}}\right) = (8.99\times 10^9\text{ N}\cdot\text{m}^2/\text{C}^2)\left(\frac{+2.00\times 10^{-6}\text{ C}}{0.0424\text{ m}} + \frac{-2.00\times 10^{-6}\text{ C}}{0.0300\text{ m}}\right) = -1.75\times 10^5\text{ V}.$

(c) $W_{ab} = q_3(V_a - V_b) = (-5.00 \times 10^{-6} \text{ C})[0 - (-1.75 \times 10^5 \text{ V})] = -0.875 \text{ J}.$

EVALUATE: Since $V_b < V_a$, a positive charge would be pulled by the existing charges from a to b, so they would do positive work on this charge. But they would repel a negative charge and hence do negative work on it, as we found in part (c).

23.25. IDENTIFY: The potential at any point is the scalar sum of the potential due to each shell.

SET UP: $V = \dfrac{kq}{R}$ for $r \leq R$ and $V = \dfrac{kq}{r}$ for $r > R$.

EXECUTE: (a) (i) $r = 0$. This point is inside both shells so

$$V = k\left(\dfrac{q_1}{R_1} + \dfrac{q_2}{R_2}\right) = (8.99 \times 10^9 \text{ N} \cdot \text{m}^2/\text{C}^2)\left(\dfrac{6.00 \times 10^{-9} \text{ C}}{0.0300 \text{ m}} + \dfrac{-9.00 \times 10^{-9} \text{ C}}{0.0500 \text{ m}}\right).$$

$V = +1.798 \times 10^3 \text{ V} + (-1.618 \times 10^3 \text{ V}) = 180 \text{ V}.$

(ii) $r = 4.00$ cm. This point is outside shell 1 and inside shell 2.

$$V = k\left(\dfrac{q_1}{r} + \dfrac{q_2}{R_2}\right) = (8.99 \times 10^9 \text{ N} \cdot \text{m}^2/\text{C}^2)\left(\dfrac{6.00 \times 10^{-9} \text{ C}}{0.0400 \text{ m}} + \dfrac{-9.00 \times 10^{-9} \text{ C}}{0.0500 \text{ m}}\right).$$

$V = +1.348 \times 10^3 \text{ V} + (-1.618 \times 10^3 \text{ V}) = -270 \text{ V}.$

(iii) $r = 6.00$ cm. This point is outside both shells.

$$V = k\left(\dfrac{q_1}{r} + \dfrac{q_2}{r}\right) = \dfrac{k}{r}(q_1 + q_2) = \dfrac{8.99 \times 10^9 \text{ N} \cdot \text{m}^2/\text{C}^2}{0.0600 \text{ m}}\left[6.00 \times 10^{-9} \text{ C} + (-9.00 \times 10^{-9} \text{ C})\right]. \quad V = -450 \text{ V}.$$

(b) At the surface of the inner shell, $r = R_1 = 3.00$ cm. This point is inside the larger shell,

so $V_1 = k\left(\dfrac{q_1}{R_1} + \dfrac{q_2}{R_2}\right) = 180 \text{ V}.$ At the surface of the outer shell, $r = R_2 = 5.00$ cm. This point is outside the smaller shell, so

$$V = k\left(\dfrac{q_1}{r} + \dfrac{q_2}{R_2}\right) = (8.99 \times 10^9 \text{ N} \cdot \text{m}^2/\text{C}^2)\left(\dfrac{6.00 \times 10^{-9} \text{ C}}{0.0500 \text{ m}} + \dfrac{-9.00 \times 10^{-9} \text{ C}}{0.0500 \text{ m}}\right).$$

$V_2 = +1.079 \times 10^3 \text{ V} + (-1.618 \times 10^3 \text{ V}) = -539 \text{ V}.$ The potential difference is $V_1 - V_2 = 719 \text{ V}.$ The inner shell is at higher potential. The potential difference is due entirely to the charge on the inner shell.

EVALUATE: Inside a uniform spherical shell, the electric field is zero so the potential is constant (but *not* necessarily zero).

23.27. (a) IDENTIFY and SET UP: The electric field on the ring's axis is given by $E_x = \dfrac{1}{4\pi\varepsilon_0}\dfrac{Qx}{(x^2 + a^2)^{3/2}}.$ The magnitude of the force on the electron exerted by this field is given by $F = eE$.

EXECUTE: When the electron is on either side of the center of the ring, the ring exerts an attractive force directed toward the center of the ring. This restoring force produces oscillatory motion of the electron along the axis of the ring, with amplitude 30.0 cm. The force on the electron is *not* of the form $F = -kx$ so the oscillatory motion is not simple harmonic motion.

(b) IDENTIFY: Apply conservation of energy to the motion of the electron.

SET UP: $K_a + U_a = K_b + U_b$ with a at the initial position of the electron and b at the center of the ring.

From Example 23.11, $V = \dfrac{1}{4\pi\varepsilon_0}\dfrac{Q}{\sqrt{x^2 + a^2}}$, where a is the radius of the ring.

EXECUTE: $x_a = 30.0$ cm, $x_b = 0$.

$K_a = 0$ (released from rest), $K_b = \tfrac{1}{2}mv^2$.

Thus $\tfrac{1}{2}mv^2 = U_a - U_b$.

And $U = qV = -eV$ so $v = \sqrt{\dfrac{2e(V_b - V_a)}{m}}.$

$$V_a = \frac{1}{4\pi\varepsilon_0} \frac{Q}{\sqrt{x_a^2 + a^2}} = (8.988 \times 10^9 \text{ N} \cdot \text{m}^2/\text{C}^2) \frac{24.0 \times 10^{-9} \text{ C}}{\sqrt{(0.300 \text{ m})^2 + (0.150 \text{ m})^2}}.$$

$$V_a = 643 \text{ V}.$$

$$V_b = \frac{1}{4\pi\varepsilon_0} \frac{Q}{\sqrt{x_b^2 + a^2}} = (8.988 \times 10^9 \text{ N} \cdot \text{m}^2/\text{C}^2) \frac{24.0 \times 10^{-9} \text{ C}}{0.150 \text{ m}} = 1438 \text{ V}.$$

$$v = \sqrt{\frac{2e(V_b - V_a)}{m}} = \sqrt{\frac{2(1.602 \times 10^{-19} \text{ C})(1438 \text{ V} - 643 \text{ V})}{9.109 \times 10^{-31} \text{ kg}}} = 1.67 \times 10^7 \text{ m/s}.$$

EVALUATE: The positively charged ring attracts the negatively charged electron and accelerates it. The electron has its maximum speed at this point. When the electron moves past the center of the ring the force on it is opposite to its motion and it slows down.

23.35. IDENTIFY: The electric field of the line of charge does work on the sphere, increasing its kinetic energy.

SET UP: $K_1 + U_1 = K_2 + U_2$ and $K_1 = 0$. $U = qV$ so $qV_1 = K_2 + qV_2$. $V = \frac{\lambda}{2\pi\varepsilon_0}\ln\left(\frac{r_0}{r}\right)$.

EXECUTE: $V_1 = \frac{\lambda}{2\pi\varepsilon_0}\ln\left(\frac{r_0}{r_1}\right)$. $V_2 = \frac{\lambda}{2\pi\varepsilon_0}\ln\left(\frac{r_0}{r_2}\right)$.

$$K_2 = q(V_1 - V_2) = \frac{q\lambda}{2\pi\varepsilon_0}\left(\ln\left(\frac{r_0}{r_1}\right) - \ln\left(\frac{r_0}{r_2}\right)\right) = \frac{\lambda q}{2\pi\varepsilon_0}(\ln r_2 - \ln r_1) = \frac{\lambda q}{2\pi\varepsilon_0}\ln\left(\frac{r_2}{r_1}\right).$$

$$K_2 = \frac{(3.00 \times 10^{-6} \text{ C/m})(8.00 \times 10^{-6} \text{ C})}{2\pi(8.854 \times 10^{-12} \text{ C}^2/(\text{N} \cdot \text{m}^2))}\ln\left(\frac{4.50}{1.50}\right) = 0.474 \text{ J}.$$

EVALUATE: The potential due to the line of charge does *not* go to zero at infinity but is defined to be zero at an arbitrary distance r_0 from the line.

23.39. IDENTIFY: The potential of a solid conducting sphere is the same at every point inside the sphere because $E = 0$ inside, and this potential has the value $V = q/4\pi\varepsilon_0 R$ at the surface. Use the given value of E to find q.

SET UP: For negative charge the electric field is directed toward the charge.
For points outside this spherical charge distribution the field is the same as if all the charge were concentrated at the center.

EXECUTE: $E = \frac{|q|}{4\pi\varepsilon_0 r^2}$ and $|q| = 4\pi\varepsilon_0 E r^2 = \frac{(3800 \text{ N/C})(0.200 \text{ m})^2}{8.99 \times 10^9 \text{ N} \cdot \text{m}^2/\text{C}^2} = 1.69 \times 10^{-8}$ C.

Since the field is directed inward, the charge must be negative. The potential of a point charge, taking ∞ as zero, is $V = \frac{q}{4\pi\varepsilon_0 r} = \frac{(8.99 \times 10^9 \text{ N} \cdot \text{m}^2/\text{C}^2)(-1.69 \times 10^{-8} \text{ C})}{0.200 \text{ m}} = -760$ V at the surface of the sphere.

Since the charge all resides on the surface of a conductor, the field inside the sphere due to this symmetrical distribution is zero. No work is therefore done in moving a test charge from just inside the surface to the center, and the potential at the center must also be –760 V.

EVALUATE: Inside the sphere the electric field is zero and the potential is constant.

23.43. IDENTIFY and SET UP: Use $E_x = -\frac{\partial V}{\partial x}$, $E_y = -\frac{\partial V}{\partial y}$, and $E_z = -\frac{\partial V}{\partial z}$ to calculate the components of $\vec{E}$.

EXECUTE: $V = Axy - Bx^2 + Cy$.

(a) $E_x = -\frac{\partial V}{\partial x} = -Ay + 2Bx$.

$E_y = -\frac{\partial V}{\partial y} = -Ax - C$.

$E_z = \dfrac{\partial V}{\partial z} = 0.$

(b) $E = 0$ requires that $E_x = E_y = E_z = 0.$

$E_z = 0$ everywhere.

$E_y = 0$ at $x = -C/A.$

And E_x is also equal to zero for this x, any value of z and $y = 2Bx/A = (2B/A)(-C/A) = -2BC/A^2.$

EVALUATE: V doesn't depend on z so $E_z = 0$ everywhere.

23.49. IDENTIFY and SET UP: Treat the gold nucleus as a point charge so that $V = k\dfrac{q}{r}$. According to conservation of energy we have $K_1 + U_1 = K_2 + U_2$, where $U = qV$.

EXECUTE: Assume that the alpha particle is at rest before it is accelerated and that it momentarily stops when it arrives at its closest approach to the surface of the gold nucleus. Thus we have $K_1 = K_2 = 0$, which implies that $U_1 = U_2$. Since $U = qV$ we conclude that the accelerating voltage must be equal to the voltage at its point of closest approach to the surface of the gold nucleus. Therefore

$V_a = V_b = k\dfrac{q}{r} = (8.99\times 10^9 \text{ N}\cdot\text{m}^2/\text{C}^2)\dfrac{79(1.60\times 10^{-19}\text{ C})}{(7.3\times 10^{-15}\text{ m} + 2.0\times 10^{-14}\text{ m})} = 4.2\times 10^6 \text{ V}.$

EVALUATE: Although the alpha particle has kinetic energy as it approaches the gold nucleus this is irrelevant to our solution since energy is conserved for the whole process.

23.51. IDENTIFY: The remaining nucleus (radium minus the ejected alpha particle) repels the alpha particle, giving it 4.79 MeV of kinetic energy when it is far from the nucleus. The mechanical energy of the system is conserved.

SET UP: $U = k\dfrac{q_1 q_2}{r}.$ $U_a + K_a = U_b + K_b.$ The charge of the alpha particle is $+2e$ and the charge of the radon nucleus is $+86e.$

EXECUTE: (a) The final energy of the alpha particle, 4.79 MeV, equals the electrical potential energy of the alpha-radon combination just before the decay. $U = 4.79 \text{ MeV} = 7.66\times 10^{-13} \text{ J}.$

(b) $r = \dfrac{kq_1 q_2}{U} = \dfrac{(8.99\times 10^9 \text{ N}\cdot\text{m}^2/\text{C}^2)(2)(86)(1.60\times 10^{-19}\text{ C})^2}{7.66\times 10^{-13}\text{ J}} = 5.17\times 10^{-14} \text{ m}.$

EVALUATE: Although we have made some simplifying assumptions (such as treating the atomic nucleus as a spherically symmetric charge, even when very close to it), this result gives a fairly reasonable estimate for the size of a nucleus.

23.55. IDENTIFY and SET UP: Calculate the components of $\vec{E}$ using $E_x = -\dfrac{\partial V}{\partial x}$, $E_y = -\dfrac{\partial V}{\partial y}$, and $E_z = -\dfrac{\partial V}{\partial z}$, and use $\vec{F} = q\vec{E}.$

EXECUTE: (a) $V = Cx^{4/3}.$

$C = V/x^{4/3} = 240 \text{ V}/(13.0\times 10^{-3}\text{ m})^{4/3} = 7.85\times 10^4 \text{ V/m}^{4/3}.$

(b) $E_x(x) = -\dfrac{\partial V}{\partial x} = -\dfrac{4}{3}Cx^{1/3} = -(1.05\times 10^5 \text{ V/m}^{4/3})x^{1/3}.$

The minus sign means that E_x is in the $-x$-direction, which says that $\vec{E}$ points from the positive anode toward the negative cathode.

(c) $\vec{F} = q\vec{E}$ so $F_x = -eE_x = \dfrac{4}{3}eCx^{1/3}.$

Halfway between the electrodes means $x = 6.50\times 10^{-3} \text{ m}.$

$F_x = \dfrac{4}{3}(1.602\times 10^{-19}\text{ C})(7.85\times 10^4 \text{ V/m}^{4/3})(6.50\times 10^{-3}\text{ m})^{1/3} = 3.13\times 10^{-15} \text{ N}.$

F_x is positive, so the force is directed toward the positive anode.

EVALUATE: V depends only on x, so $E_y = E_z = 0$. $\vec{E}$ is directed from high potential (anode) to low potential (cathode). The electron has negative charge, so the force on it is directed opposite to the electric field.

23.57. IDENTIFY: $U = \dfrac{kq_1q_2}{r}$.

SET UP: Eight charges means there are $8(8-1)/2 = 28$ pairs. There are 12 pairs of q and $-q$ separated by d, 12 pairs of equal charges separated by $\sqrt{2}d$ and 4 pairs of q and $-q$ separated by $\sqrt{3}d$.

EXECUTE: (a) $U = kq^2\left(-\dfrac{12}{d} + \dfrac{12}{\sqrt{2}d} - \dfrac{4}{\sqrt{3}d}\right) = -\dfrac{12kq^2}{d}\left(1 - \dfrac{1}{\sqrt{2}} + \dfrac{1}{3\sqrt{3}}\right) = -1.46q^2/\pi\varepsilon_0 d$.

EVALUATE: (b) The fact that the electric potential energy is less than zero means that it is energetically favorable for the crystal ions to be together.

23.59. IDENTIFY: Apply $\Sigma F_x = 0$ and $\Sigma F_y = 0$ to the sphere. The electric force on the sphere is $F_e = qE$. The potential difference between the plates is $V = Ed$.

SET UP: The free-body diagram for the sphere is given in Figure 23.59.

EXECUTE: $T\cos\theta = mg$ and $T\sin\theta = F_e$ gives

$F_e = mg\tan\theta = (1.50\times 10^{-3} \text{ kg})(9.80 \text{ m/s}^2)\tan(30°) = 0.0085$ N.

$F_e = Eq = \dfrac{Vq}{d}$ and $V = \dfrac{Fd}{q} = \dfrac{(0.0085 \text{ N})(0.0500 \text{ m})}{8.90\times 10^{-6} \text{ C}} = 47.8$ V.

EVALUATE: $E = V/d = 956$ V/m. $E = \sigma/\varepsilon_0$ and $\sigma = E\varepsilon_0 = 8.46\times 10^{-9}$ C/m^2.

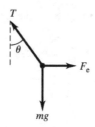

Figure 23.59

23.65. (a) IDENTIFY and SET UP: Problem 23.61 derived that $E = \dfrac{V_{ab}}{\ln(b/a)}\dfrac{1}{r}$, where a is the radius of the inner cylinder (wire) and b is the radius of the outer hollow cylinder. The potential difference between the two cylinders is V_{ab}. Use this expression to calculate E at the specified r.

EXECUTE: Midway between the wire and the cylinder wall is at a radius of

$r = (a+b)/2 = (90.0\times 10^{-6} \text{ m} + 0.140 \text{ m})/2 = 0.07004$ m.

$E = \dfrac{V_{ab}}{\ln(b/a)}\dfrac{1}{r} = \dfrac{50.0\times 10^3 \text{ V}}{\ln(0.140 \text{ m}/90.0\times 10^{-6} \text{ m})(0.07004 \text{ m})} = 9.71\times 10^4$ V/m.

(b) IDENTIFY and SET UP: The magnitude of the electric force is given by $F = |q|E$. Set this equal to ten times the weight of the particle and solve for $|q|$, the magnitude of the charge on the particle.

EXECUTE: $F_E = 10mg$.

$|q|E = 10mg$ and $|q| = \dfrac{10mg}{E} = \dfrac{10(30.0\times 10^{-9} \text{ kg})(9.80 \text{ m/s}^2)}{9.71\times 10^4 \text{ V/m}} = 3.03\times 10^{-11}$ C.

EVALUATE: It requires only this modest net charge for the electric force to be much larger than the weight.

23.67. **IDENTIFY:** We must integrate to find the total energy because the energy to bring in more charge depends on the charge already present.

SET UP: If ρ is the uniform volume charge density, the charge of a spherical shell of radius r and thickness dr is $dq = \rho 4\pi r^2\, dr$, and $\rho = Q/(4/3\,\pi R^3)$. The charge already present in a sphere of radius r is $q = \rho(4/3\,\pi r^3)$. The energy to bring the charge dq to the surface of the charge q is $V\,dq$, where V is the potential due to q, which is $q/4\pi\varepsilon_0 r$.

EXECUTE: The total energy to assemble the entire sphere of radius R and charge Q is sum (integral) of the tiny increments of energy.

$$U = \int V\,dq = \int \frac{q}{4\pi\varepsilon_0 r}\,dq = \int_0^R \frac{\rho\frac{4}{3}\pi r^3}{4\pi\varepsilon_0 r}(\rho 4\pi r^2\,dr) = \frac{3}{5}\left(\frac{1}{4\pi\varepsilon_0}\frac{Q^2}{R}\right)$$

where we have substituted $\rho = Q/(4/3\,\pi R^3)$ and simplified the result.

EVALUATE: For a point charge, $R \to 0$ so $U \to \infty$, which means that a point charge should have infinite self-energy. This suggests that either point charges are impossible, or that our present treatment of physics is not adequate at the extremely small scale, or both.

23.69. **IDENTIFY and SET UP:** The sphere no longer behaves as a point charge because we are inside of it. We know how the electric field varies with distance from the center of the sphere and want to use this to find the potential difference between the center and surface, which requires integration. $V_a - V_b = \int_a^b \vec{E} \cdot d\vec{l}$.

The electric field is radially outward, so $\vec{E} \cdot d\vec{l} = E\,dr$.

EXECUTE: For $r < R$: $E = \dfrac{kQr}{R^3}$. Integrating gives

$$V = -\int_\infty^R \vec{E}\cdot d\vec{r}' - \int_R^r \vec{E}\cdot d\vec{r}' = \frac{kQ}{R} - \frac{kQ}{R^3}\int_R^r r'\,dr' = \frac{kQ}{R} - \frac{kQ}{R^3}\frac{1}{2}r'^2\bigg|_R^r = \frac{kQ}{R} + \frac{kQ}{2R} - \frac{kQr^2}{2R^3} = \frac{kQ}{2R}\left[3 - \frac{r^2}{R^2}\right].$$

At the center of the sphere, $r = 0$ and $V_1 = \dfrac{3kQ}{2R}$. At the surface of the sphere, $r = R$ and $V_2 = \dfrac{kQ}{R}$. The potential difference is $V_1 - V_2 = \dfrac{kQ}{2R} = \dfrac{(8.99\times 10^9\ \text{N}\cdot\text{m}^2/\text{C}^2)(4.00\times 10^{-6}\ \text{C})}{2(0.0500\ \text{m})} = 3.60\times 10^5\ \text{V}$.

EVALUATE: To check our answer, we could actually do the integration. We can use the fact that $E = \dfrac{kQr}{R^3}$ so $V_1 - V_2 = \int_0^R E\,dr = \dfrac{kQ}{R^3}\int_0^R r\,dr = \dfrac{kQ}{R^3}\left(\dfrac{R^2}{2}\right) = \dfrac{kQ}{2R}$.

23.73. **IDENTIFY:** Slice the rod into thin slices and use $V = \dfrac{1}{4\pi\varepsilon_0}\dfrac{q}{r}$ to calculate the potential due to each slice.

Integrate over the length of the rod to find the total potential at each point.

(a) SET UP: An infinitesimal slice of the rod and its distance from point P are shown in Figure 23.73a.

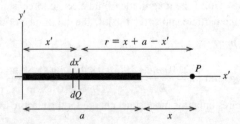

Figure 23.73a

Use coordinates with the origin at the left-hand end of the rod and one axis along the rod. Call the axes x' and y' so as not to confuse them with the distance x given in the problem.

EXECUTE: Slice the charged rod up into thin slices of width dx'. Each slice has charge $dQ = Q(dx'/a)$ and a distance $r = x + a - x'$ from point P. The potential at P due to the small slice dQ is

$$dV = \frac{1}{4\pi\varepsilon_0}\left(\frac{dQ}{r}\right) = \frac{1}{4\pi\varepsilon_0}\frac{Q}{a}\left(\frac{dx'}{x+a-x'}\right).$$

Compute the total V at P due to the entire rod by integrating dV over the length of the rod ($x' = 0$ to $x' = a$):

$$V = \int dV = \frac{Q}{4\pi\varepsilon_0 a}\int_0^a \frac{dx'}{(x+a-x')} = \frac{Q}{4\pi\varepsilon_0 a}[-\ln(x+a-x')]_0^a = \frac{Q}{4\pi\varepsilon_0 a}\ln\left(\frac{x+a}{x}\right).$$

EVALUATE: As $x \to \infty$, $V \to \frac{Q}{4\pi\varepsilon_0 a}\ln\left(\frac{x}{x}\right) = 0.$

(b) SET UP: An infinitesimal slice of the rod and its distance from point R are shown in Figure 23.73b.

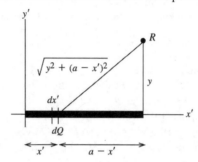

Figure 23.73b

$dQ = (Q/a)dx'$ as in part (a).

Each slice dQ is a distance $r = \sqrt{y^2 + (a-x')^2}$ from point R.

EXECUTE: The potential dV at R due to the small slice dQ is

$$dV = \frac{1}{4\pi\varepsilon_0}\left(\frac{dQ}{r}\right) = \frac{1}{4\pi\varepsilon_0}\frac{Q}{a}\frac{dx'}{\sqrt{y^2+(a-x')^2}}.$$

$$V = \int dV = \frac{Q}{4\pi\varepsilon_0 a}\int_0^a \frac{dx'}{\sqrt{y^2+(a-x')^2}}.$$

In the integral make the change of variable $u = a - x'$; $du = -dx'$

$$V = -\frac{Q}{4\pi\varepsilon_0 a}\int_a^0 \frac{du}{\sqrt{y^2+u^2}} = -\frac{Q}{4\pi\varepsilon_0 a}\left[\ln\left(u + \sqrt{y^2+u^2}\right)\right]_a^0.$$

$$V = -\frac{Q}{4\pi\varepsilon_0 a}\left[\ln y - \ln\left(a + \sqrt{y^2+a^2}\right)\right] = \frac{Q}{4\pi\varepsilon_0 a}\ln\left(\frac{a+\sqrt{a^2+y^2}}{y}\right).$$

(The expression for the integral was found in Appendix B.)

EVALUATE: As $y \to \infty$, $V \to \frac{Q}{4\pi\varepsilon_0 a}\ln\left(\frac{y}{y}\right) = 0.$

(c) SET UP: part (a): $V = \frac{Q}{4\pi\varepsilon_0 a}\ln\left(\frac{x+a}{x}\right) = \frac{Q}{4\pi\varepsilon_0 a}\ln\left(1+\frac{a}{x}\right).$

From Appendix B, $\ln(1+u) = u - u^2/2\ldots$, so $\ln(1+a/x) = a/x - a^2/2x^2$ and this becomes a/x when x is large.

EXECUTE: Thus $V \to \dfrac{Q}{4\pi\varepsilon_0 a}\left(\dfrac{a}{x}\right) = \dfrac{Q}{4\pi\varepsilon_0 x}$. For large x, V becomes the potential of a point charge.

part (b): $V = \dfrac{Q}{4\pi\varepsilon_0 a}\left[\ln\left(\dfrac{a+\sqrt{a^2+y^2}}{y}\right)\right] = \dfrac{Q}{4\pi\varepsilon_0 a}\ln\left(\dfrac{a}{y} + \sqrt{1+\dfrac{a^2}{y^2}}\right)$.

From Appendix B, $\sqrt{1+a^2/y^2} = (1+a^2/y^2)^{1/2} = 1 + a^2/2y^2 + \ldots$

Thus $a/y + \sqrt{1+a^2/y^2} \to 1 + a/y + a^2/2y^2 + \ldots \to 1 + a/y$. And then using $\ln(1+u) \approx u$ gives

$$V \to \dfrac{Q}{4\pi\varepsilon_0 a}\ln(1+a/y) \to \dfrac{Q}{4\pi\varepsilon_0 a}\left(\dfrac{a}{y}\right) = \dfrac{Q}{4\pi\varepsilon_0 y}.$$

EVALUATE: For large y, V becomes the potential of a point charge.

23.79. IDENTIFY and SET UP: We know that the potential is of the mathematical form $V(x,y,z) = Ax^l + By^m + Cz^n + D$. We also know that $E_x = -\dfrac{\partial V}{\partial x}$, $E_y = -\dfrac{\partial V}{\partial y}$, and $E_z = -\dfrac{\partial V}{\partial z}$. Various measurements are given in the table with the problem in the text.

EXECUTE: (a) First get A, B, C, and D using data from the table in the problem.
$V(0, 0, 0) = 10.0\text{ V} = 0 + 0 + 0 + D$, so $D = 10.0$ V.
$V(1.00, 0, 0) = A(1.00\text{ m})^l + 0 + 0 + 10.0\text{ V} = 4.00\text{ V}$, so $A = -6.0\text{ V}\cdot\text{m}^{-l}$.
$V(0, 1.00, 0) = B(1.00\text{ m})^m + 10.0\text{ V} = 6.0\text{ V}$, so $B = -4.0\text{ V}\cdot\text{m}^{-m}$.
$V(0, 0, 1.00\text{ m}) = C(1.00\text{ m})^n + 10.0\text{ V} = 8.0\text{ V}$, so $C = -2.0\text{ V}\cdot\text{m}^{-n}$.
Now get l, m, and n.

$E_x = -\dfrac{\partial V}{\partial x} = -lAx^{l-1}$, and from the table we know that $E_x(1.00, 0, 0) = 12.0$ V/m. Therefore

$-l(-6.0\text{ V}\cdot\text{m}^{-l})(1.00\text{ m})^{l-1} = 12.0$ V/m.
$l(6.0\text{ V}\cdot\text{m}^{-1}) = 12.0$ V/m.
$l = 2.0$.

$E_y = -\dfrac{\partial V}{\partial y} = -mBy^{m-1}$.

$E_y(0, 1.00, 0) = -m(-4.0\text{ V}\cdot\text{m}^{-m})(1.00\text{ m})^{m-1} = 12.0$ V/m.
$m = 3.0$.

$E_z = -\dfrac{\partial V}{\partial z} = -nCz^{n-1}$.

$E_z(0, 0, 1.00) = -n(-2.0\text{ V}\cdot\text{m}^{-n})(1.00\text{ m})^{n-1} = 12.0$ V/m.
$n = 6.0$.
Now that we have l, m, and n, we see the units of A, B, and C, so
$A = -6.0$ V/m^2.
$B = -4.0$ V/m^3.
$C = -2.0$ V/m^6.
Therefore the equation for $V(x,y,z)$ is
$V = (-6.0\text{ V/m}^2)x^2 + (-4.0\text{ V/m}^3)y^3 + (-2.0\text{ V/m}^6)z^6 + 10.0$ V.
(b) At $(0, 0, 0)$: $V = 0$ and $E = 0$ (from the table with the problem).
At $(0.50\text{ m}, 0.50\text{ m}, 0.50\text{ m})$:
$V = (-6.0\text{ V/m}^2)(0.50\text{ m})^2 + (-4.0\text{ V/m}^3)(0.50\text{ m})^3 + (-2.0\text{ V/m}^6)(0.50\text{ m})^6 + 10.0\text{ V} = 8.0$ V.

$E_x = -\dfrac{\partial V}{\partial x} = -(-12.0\text{ V/m}^2)x = (12.0\text{ V/m}^2)(0.50\text{ m}) = 6.0$ V/m.

$E_y = -\dfrac{\partial V}{\partial y} = -3(-4.0\text{ V/m}^3)y^2 = (12\text{ V/m}^3)(0.50\text{ m})^2 = 3.0$ V/m.

$E_z = -\dfrac{\partial V}{\partial z} = -(-12.0 \text{ V/m}^6)z^5 = (12.0 \text{ V/m}^6)(0.50 \text{ m})^5 = 0.375 \text{ V/m}.$

$E = \sqrt{E_x^2 + E_y^2 + E_z^2} = \sqrt{(6.0 \text{ V/m})^2 + (3.0 \text{ V/m})^2 + (0.375 \text{ V/m})^2} = 6.7 \text{ V/m}.$

At (1.00 m, 1.00 m, 1.00 m):
Follow the same procedure as above. The results are $V = -2.0$ V, $E = 21$ V/m.

EVALUATE: We know that l, m, and n must be greater that 1 because the components of the electric field are all zero at (0, 0, 0).

23.81. IDENTIFY: When the oil drop is at rest, the upward force $|q|E$ from the electric field equals the downward weight of the drop. When the drop is falling at its terminal speed, the upward viscous force equals the downward weight of the drop.

SET UP: The volume of the drop is related to its radius r by $V = \dfrac{4}{3}\pi r^3$.

EXECUTE: (a) $F_g = mg = \dfrac{4\pi r^3}{3}\rho g$. $F_e = |q|E = |q|V_{AB}/d$. $F_e = F_g$ gives $|q| = \dfrac{4\pi}{3}\dfrac{\rho r^3 g d}{V_{AB}}$.

(b) $\dfrac{4\pi r^3}{3}\rho g = 6\pi \eta r v_t$ gives $r = \sqrt{\dfrac{9\eta v_t}{2\rho g}}$. Using this result to replace r in the expression in part (a) gives

$|q| = \dfrac{4\pi}{3}\dfrac{\rho g d}{V_{AB}}\left[\sqrt{\dfrac{9\eta v_t}{2\rho g}}\right]^3 = 18\pi \dfrac{d}{V_{AB}}\sqrt{\dfrac{\eta^3 v_t^3}{2\rho g}}.$

(c) We use the values for V_{AB} and v_t given in the table in the problem and the formula $|q| = 18\pi\dfrac{d}{V_{AB}}\sqrt{\dfrac{\eta^3 v_t^3}{2\rho g}}$ from (c). For example, for drop 1 we get

$|q| = 18\pi\dfrac{1.00\times 10^{-3} \text{ m}}{9.16 \text{ V}}\sqrt{\dfrac{(1.81\times 10^{-5} \text{ N}\cdot\text{s/m}^2)^3(2.54\times 10^{-5} \text{ m/s})^3}{2(824 \text{ kg/m}^3)(9.80 \text{ m/s}^2)}} = 4.79\times 10^{-19}$ C. Similar calculations for

the remaining drops gives the following results:
Drop 1: 4.79×10^{-19} C
Drop 2: 1.59×10^{-19} C
Drop 3: 8.09×10^{-19} C
Drop 4: 3.23×10^{-19} C

(d) Use $n = q/e_2$ to find the number of excess electrons on each drop. Since all quantities have a power of 10^{-19} C, this factor will cancel, so all we need to do is divide the coefficients of 10^{-19} C. This gives
Drop 1: $n = q_1/q_2 = 4.79/1.59 = 3$ excess electrons
Drop 2: $n = q_2/q_2 = 1$ excess electron
Drop 3: $n = q_3/q_2 = 8.09/1.59 = 5$ excess electrons
Drop 4: $n = q_4/q_2 = 3.23/1.59 = 2$ excess electrons

(e) Using $q = -ne$ gives $e = -q/n$. All the charges are negative, so e will come out positive. Thus we get
Drop 1: $e_1 = q_1/n_1 = (4.79\times 10^{-19} \text{ C})/3 = 1.60\times 10^{-19}$ C
Drop 2: $e_2 = q_2/n_2 = (1.59\times 10^{-19} \text{ C})/1 = 1.59\times 10^{-19}$ C
Drop 3: $e_3 = q_3/n_3 = (8.09\times 10^{-19} \text{ C})/5 = 1.62\times 10^{-19}$ C
Drop 4: $e_4 = q_4/n_4 = (3.23\times 10^{-19} \text{ C})/2 = 1.61\times 10^{-19}$ C

The average is
$e_{av} = (e_1 + e_2 + e_3 + e_4)/4 = [(1.60 + 1.59 + 1.62 + 1.61)\times 10^{-19} \text{ C}]/4 = 1.61\times 10^{-19}$ C.

EVALUATE: The result $e = 1.61\times 10^{-19}$ C is very close to the well-established value of 1.60×10^{-19} C.

23.85. **IDENTIFY** and **SET UP:** Conservation of energy gives $K = U_{\text{electric}} = k\dfrac{q_1 q_2}{r}$.

EXECUTE: Solve for Q: $Q = rK/kq = (10 \times 10^{-15}\text{ m})(3.0\text{ MeV})/(2ek) = 1.67 \times 10^{-18}$ C. In terms of e, this is $Q = (1.67 \times 10^{-18}\text{ C})/(1.60 \times 10^{-19}\text{ C}) = 10.4e \approx 11e$, so choice (b) is best.

EVALUATE: If $Q = 11e$, the atom is sodium (Na), which has an atomic mass of 23, compared to 4 for He. So it is reasonable to assume that the nucleus does not move appreciably, since it is about 6 times more massive than the He.

CAPACITANCE AND DIELECTRICS

24.3. **IDENTIFY and SET UP:** It is a parallel-plate air capacitor, so we can apply the equations of Section 24.1.

EXECUTE: **(a)** $C = \dfrac{Q}{V_{ab}}$ so $V_{ab} = \dfrac{Q}{C} = \dfrac{0.148 \times 10^{-6} \text{ C}}{245 \times 10^{-12} \text{ F}} = 604 \text{ V}.$

(b) $C = \dfrac{\varepsilon_0 A}{d}$ so $A = \dfrac{Cd}{\varepsilon_0} = \dfrac{(245 \times 10^{-12} \text{ F})(0.328 \times 10^{-3} \text{ m})}{8.854 \times 10^{-12} \text{ C}^2/\text{N} \cdot \text{m}^2} = 9.08 \times 10^{-3} \text{ m}^2 = 90.8 \text{ cm}^2.$

(c) $V_{ab} = Ed$ so $E = \dfrac{V_{ab}}{d} = \dfrac{604 \text{ V}}{0.328 \times 10^{-3} \text{ m}} = 1.84 \times 10^6 \text{ V/m}.$

(d) $E = \dfrac{\sigma}{\varepsilon_0}$ so $\sigma = E\varepsilon_0 = (1.84 \times 10^6 \text{ V/m})(8.854 \times 10^{-12} \text{ C}^2/\text{N} \cdot \text{m}^2) = 1.63 \times 10^{-5} \text{ C/m}^2.$

EVALUATE: We could also calculate σ directly as Q/A. $\sigma = \dfrac{Q}{A} = \dfrac{0.148 \times 10^{-6} \text{ C}}{9.08 \times 10^{-3} \text{ m}^2} = 1.63 \times 10^{-5} \text{ C/m}^2,$ which checks.

24.9. **IDENTIFY:** Apply the results of Example 24.4. $C = Q/V.$

SET UP: $r_a = 0.50 \text{ mm}$, $r_b = 5.00 \text{ mm}.$

EXECUTE: **(a)** $C = \dfrac{L 2\pi\varepsilon_0}{\ln(r_b/r_a)} = \dfrac{(0.180 \text{ m})2\pi\varepsilon_0}{\ln(5.00/0.50)} = 4.35 \times 10^{-12} \text{ F}.$

(b) $V = Q/C = (10.0 \times 10^{-12} \text{ C})/(4.35 \times 10^{-12} \text{ F}) = 2.30 \text{ V}.$

EVALUATE: $\dfrac{C}{L} = 24.2$ pF. This value is similar to those in Example 24.4. The capacitance is determined entirely by the dimensions of the cylinders.

24.11. **IDENTIFY:** We can use the definition of capacitance to find the capacitance of the capacitor, and then relate the capacitance to geometry to find the inner radius.

(a) SET UP: By the definition of capacitance, $C = Q/V.$

EXECUTE: $C = \dfrac{Q}{V} = \dfrac{3.30 \times 10^{-9} \text{ C}}{2.20 \times 10^2 \text{ V}} = 1.50 \times 10^{-11} \text{ F} = 15.0 \text{ pF}.$

(b) SET UP: The capacitance of a spherical capacitor is $C = 4\pi\varepsilon_0 \dfrac{r_a r_b}{r_b - r_a}.$

EXECUTE: Solve for r_a and evaluate using $C = 15.0$ pF and $r_b = 4.00$ cm, giving $r_a = 3.09$ cm.

(c) SET UP: We can treat the inner sphere as a point charge located at its center and use Coulomb's law, $E = \dfrac{1}{4\pi\varepsilon_0} \dfrac{q}{r^2}.$

EXECUTE: $E = \dfrac{(8.99 \times 10^9 \text{ N} \cdot \text{m}^2/\text{C}^2)(3.30 \times 10^{-9} \text{ C})}{(0.0309 \text{ m})^2} = 3.12 \times 10^4 \text{ N/C}.$

EVALUATE: Outside the capacitor, the electric field is zero because the charges on the spheres are equal in magnitude but opposite in sign.

24.17. IDENTIFY: Replace series and parallel combinations of capacitors by their equivalents. In each equivalent network apply the rules for Q and V for capacitors in series and parallel; start with the simplest network and work back to the original circuit.

SET UP: Do parts (a) and (b) together. The capacitor network is drawn in Figure 24.17a.

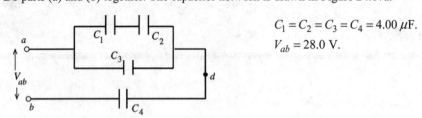

$C_1 = C_2 = C_3 = C_4 = 4.00\ \mu\text{F}.$
$V_{ab} = 28.0\text{ V}.$

Figure 24.17a

EXECUTE: Simplify the circuit by replacing the capacitor combinations by their equivalents: C_1 and C_2 are in series and are equivalent to C_{12} (Figure 24.17b).

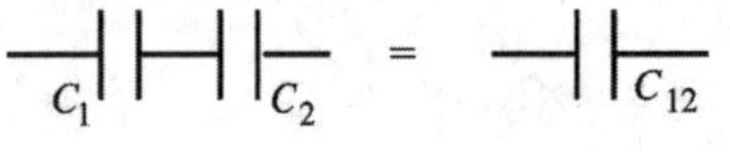

$$\frac{1}{C_{12}} = \frac{1}{C_1} + \frac{1}{C_2}.$$

Figure 24.17b

$$C_{12} = \frac{C_1 C_2}{C_1 + C_2} = \frac{(4.00\times10^{-6}\text{ F})(4.00\times10^{-6}\text{ F})}{4.00\times10^{-6}\text{ F} + 4.00\times10^{-6}\text{ F}} = 2.00\times10^{-6}\text{ F}.$$

C_{12} and C_3 are in parallel and are equivalent to C_{123} (Figure 24.17c).

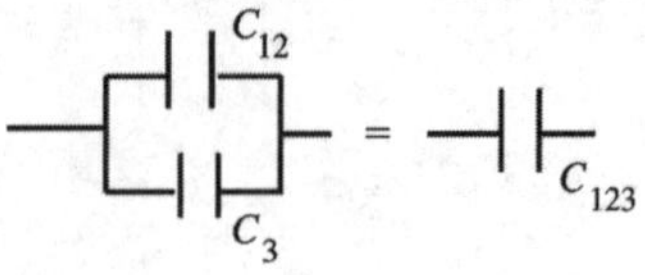

$C_{123} = C_{12} + C_3.$
$C_{123} = 2.00\times10^{-6}\text{ F} + 4.00\times10^{-6}\text{ F}.$
$C_{123} = 6.00\times10^{-6}\text{ F}.$

Figure 24.17c

C_{123} and C_4 are in series and are equivalent to C_{1234} (Figure 24.17d).

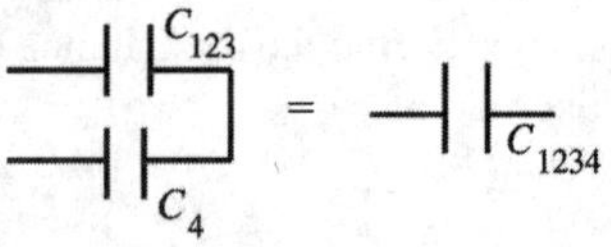

$$\frac{1}{C_{1234}} = \frac{1}{C_{123}} + \frac{1}{C_4}.$$

Figure 24.17d

$$C_{1234} = \frac{C_{123} C_4}{C_{123} + C_4} = \frac{(6.00\times10^{-6}\text{ F})(4.00\times10^{-6}\text{ F})}{6.00\times10^{-6}\text{ F} + 4.00\times10^{-6}\text{ F}} = 2.40\times10^{-6}\text{ F}.$$

The circuit is equivalent to the circuit shown in Figure 24.17e.

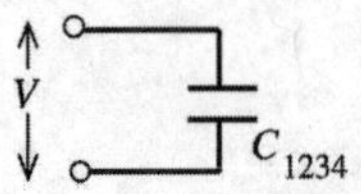

$V_{1234} = V = 28.0\text{ V}.$
$Q_{1234} = C_{1234} V = (2.40\times10^{-6}\text{ F})(28.0\text{ V}) = 67.2\ \mu\text{C}.$

Figure 24.17e

Now build back up the original circuit, step by step. C_{1234} represents C_{123} and C_4 in series (Figure 24.17f).

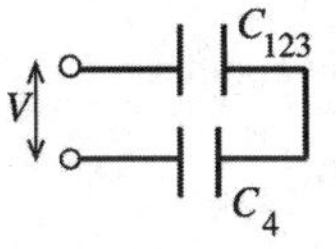

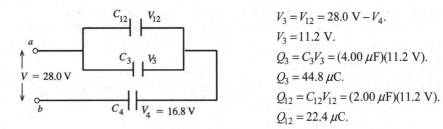

$Q_{123} = Q_4 = Q_{1234} = 67.2\ \mu C$
(charge same for capacitors in series).

Figure 24.17f

Then $V_{123} = \dfrac{Q_{123}}{C_{123}} = \dfrac{67.2\ \mu C}{6.00\ \mu F} = 11.2\ V$.

$V_4 = \dfrac{Q_4}{C_4} = \dfrac{67.2\ \mu C}{4.00\ \mu F} = 16.8\ V$.

Note that $V_4 + V_{123} = 16.8\ V + 11.2\ V = 28.0\ V$, as it should.
Next consider the circuit as written in Figure 24.17g.

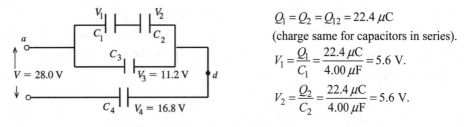

$V_3 = V_{12} = 28.0\ V - V_4$.
$V_3 = 11.2\ V$.
$Q_3 = C_3 V_3 = (4.00\ \mu F)(11.2\ V)$.
$Q_3 = 44.8\ \mu C$.
$Q_{12} = C_{12}V_{12} = (2.00\ \mu F)(11.2\ V)$.
$Q_{12} = 22.4\ \mu C$.

Figure 24.17g

Finally, consider the original circuit, as shown in Figure 24.17h.

$Q_1 = Q_2 = Q_{12} = 22.4\ \mu C$
(charge same for capacitors in series).

$V_1 = \dfrac{Q_1}{C_1} = \dfrac{22.4\ \mu C}{4.00\ \mu F} = 5.6\ V$.

$V_2 = \dfrac{Q_2}{C_2} = \dfrac{22.4\ \mu C}{4.00\ \mu F} = 5.6\ V$.

Figure 24.17h

Note that $V_1 + V_2 = 11.2\ V$, which equals V_3 as it should.
Summary: $Q_1 = 22.4\ \mu C$, $V_1 = 5.6\ V$.
$Q_2 = 22.4\ \mu C$, $V_2 = 5.6\ V$.
$Q_3 = 44.8\ \mu C$, $V_3 = 11.2\ V$.
$Q_4 = 67.2\ \mu C$, $V_4 = 16.8\ V$.
(c) $V_{ad} = V_3 = 11.2\ V$.
EVALUATE: $V_1 + V_2 + V_4 = V$, or $V_3 + V_4 = V$. $Q_1 = Q_2$, $Q_1 + Q_3 = Q_4$ and $Q_4 = Q_{1234}$.

24.23. **IDENTIFY and SET UP:** The energy density is given by $u = \tfrac{1}{2}\varepsilon_0 E^2$. Use $V = Ed$ to solve for E.

EXECUTE: Calculate E: $E = \dfrac{V}{d} = \dfrac{400 \text{ V}}{5.00 \times 10^{-3} \text{ m}} = 8.00 \times 10^4 \text{ V/m}$.

Then $u = \tfrac{1}{2}\varepsilon_0 E^2 = \tfrac{1}{2}(8.854 \times 10^{-12} \text{ C}^2/\text{N} \cdot \text{m}^2)(8.00 \times 10^4 \text{ V/m})^2 = 0.0283 \text{ J/m}^3$.

EVALUATE: E is smaller than the value in Example 24.8 by about a factor of 6 so u is smaller by about a factor of $6^2 = 36$.

24.31. IDENTIFY: $U = \tfrac{1}{2}QV$. Solve for Q. $C = Q/V$.

SET UP: Example 24.4 shows that for a cylindrical capacitor, $\dfrac{C}{L} = \dfrac{2\pi\varepsilon_0}{\ln(r_b/r_a)}$.

EXECUTE: (a) $U = \tfrac{1}{2}QV$ gives $Q = \dfrac{2U}{V} = \dfrac{2(3.20 \times 10^{-9} \text{ J})}{4.00 \text{ V}} = 1.60 \times 10^{-9} \text{ C}$.

(b) $\dfrac{C}{L} = \dfrac{2\pi\varepsilon_0}{\ln(r_b/r_a)}$. Solving for r_b/r_a gives

$\dfrac{r_b}{r_a} = \exp(2\pi\varepsilon_0 L/C) = \exp(2\pi\varepsilon_0 LV/Q) = \exp[2\pi\varepsilon_0(15.0 \text{ m})(4.00 \text{ V})/(1.60 \times 10^{-9} \text{ C})] = 8.05$.

The radius of the outer conductor is 8.05 times the radius of the inner conductor.

EVALUATE: When the ratio r_b/r_a increases, C/L decreases and less charge is stored for a given potential difference.

24.33. IDENTIFY: $C = KC_0$. $U = \tfrac{1}{2}CV^2$.

SET UP: $C_0 = 12.5 \text{ }\mu\text{F}$ is the value of the capacitance without the dielectric present.

EXECUTE: (a) With the dielectric, $C = (3.75)(12.5 \text{ }\mu\text{F}) = 46.9 \text{ }\mu\text{F}$.

Before: $U = \tfrac{1}{2}C_0 V^2 = \tfrac{1}{2}(12.5 \times 10^{-6} \text{ F})(24.0 \text{ V})^2 = 3.60 \text{ mJ}$.

After: $U = \tfrac{1}{2}CV^2 = \tfrac{1}{2}(46.9 \times 10^{-6} \text{ F})(24.0 \text{ V})^2 = 13.5 \text{ mJ}$.

(b) $\Delta U = 13.5 \text{ mJ} - 3.6 \text{ mJ} = 9.9 \text{ mJ}$. The energy increased.

EVALUATE: The power supply must put additional charge on the plates to maintain the same potential difference when the dielectric is inserted. $U = \tfrac{1}{2}QV$, so the stored energy increases.

24.37. IDENTIFY and SET UP: For a parallel-plate capacitor with a dielectric we can use the equation $C = K\varepsilon_0 A/d$. Minimum A means smallest possible d. d is limited by the requirement that E be less than $1.60 \times 10^7 \text{ V/m}$ when V is as large as 5500 V.

EXECUTE: $V = Ed$ so $d = \dfrac{V}{E} = \dfrac{5500 \text{ V}}{1.60 \times 10^7 \text{ V/m}} = 3.44 \times 10^{-4} \text{ m}$.

Then $A = \dfrac{Cd}{K\varepsilon_0} = \dfrac{(1.25 \times 10^{-9} \text{ F})(3.44 \times 10^{-4} \text{ m})}{(3.60)(8.854 \times 10^{-12} \text{ C}^2/\text{N} \cdot \text{m}^2)} = 0.0135 \text{ m}^2$.

EVALUATE: The relation $V = Ed$ applies with or without a dielectric present. A would have to be larger if there were no dielectric.

24.39. IDENTIFY: $C = Q/V$. $C = KC_0$. $V = Ed$.

SET UP: Table 24.1 gives $K = 3.1$ for mylar.

EXECUTE: (a) $\Delta Q = Q - Q_0 = (K-1)Q_0 = (K-1)C_0 V_0 = (2.1)(2.5 \times 10^{-7} \text{ F})(12 \text{ V}) = 6.3 \times 10^{-6} \text{ C}$.

(b) $\sigma_i = \sigma(1 - 1/K)$ so $Q_i = Q(1 - 1/K) = (9.3 \times 10^{-6} \text{ C})(1 - 1/3.1) = 6.3 \times 10^{-6} \text{ C}$.

(c) The addition of the mylar doesn't affect the electric field since the induced charge cancels the additional charge drawn to the plates.

EVALUATE: $E = V/d$ and V is constant so E doesn't change when the dielectric is inserted.

24.43. **IDENTIFY:** Apply $\oint K\vec{E} \cdot d\vec{A} = \dfrac{Q_{\text{encl-free}}}{\varepsilon_0}$ to calculate E. $V = Ed$ and $C = Q/V$ apply whether there is a dielectric between the plates or not.

(a) SET UP: Apply $\oint K\vec{E} \cdot d\vec{A} = \dfrac{Q_{\text{encl-free}}}{\varepsilon_0}$ to the dashed surface in Figure 24.43.

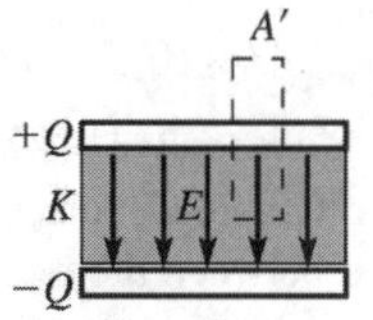

EXECUTE: $\oint K\vec{E} \cdot d\vec{A} = \dfrac{Q_{\text{encl-free}}}{\varepsilon_0}$.

$\oint K\vec{E} \cdot d\vec{A} = KEA'$.

since $E = 0$ outside the plates

$Q_{\text{encl-free}} = \sigma A' = (Q/A)A'$.

Figure 24.43

Thus $KEA' = \dfrac{(Q/A)A'}{\varepsilon_0}$ and $E = \dfrac{Q}{\varepsilon_0 AK}$.

SET UP and EXECUTE: (b) $V = Ed = \dfrac{Qd}{\varepsilon_0 AK}$.

(c) $C = \dfrac{Q}{V} = \dfrac{Q}{Qd/\varepsilon_0 AK} = K\dfrac{\varepsilon_0 A}{d} = KC_0$.

EVALUATE: Our result shows that $K = C/C_0$, which is Eq. (24.12).

24.47. **IDENTIFY:** $C = \dfrac{\varepsilon_0 A}{d}$.

SET UP: $A = 4.2 \times 10^{-5}$ m^2. The original separation between the plates is $d = 0.700 \times 10^{-3}$ m. d' is the separation between the plates at the new value of C.

EXECUTE: $C_0 = \dfrac{A\varepsilon_0}{d} = \dfrac{(4.20 \times 10^{-5} \text{ m}^2)\varepsilon_0}{7.00 \times 10^{-4} \text{ m}} = 5.31 \times 10^{-13}$ F. The new value of C is

$C = C_0 + 0.25$ pF $= 7.81 \times 10^{-13}$ F. But $C = \dfrac{A\varepsilon_0}{d'}$, so $d' = \dfrac{A\varepsilon_0}{C} = \dfrac{(4.20 \times 10^{-5} \text{ m}^2)\varepsilon_0}{7.81 \times 10^{-13} \text{ F}} = 4.76 \times 10^{-4}$ m.

Therefore the key must be depressed by a distance of 7.00×10^{-4} m $- 4.76 \times 10^{-4}$ m $= 0.224$ mm.

EVALUATE: When the key is depressed, d decreases and C increases.

24.49. **IDENTIFY:** Some of the charge from the original capacitor flows onto the uncharged capacitor until the potential differences across the two capacitors are the same.

SET UP: $C = \dfrac{Q}{V_{ab}}$. Let $C_1 = 20.0$ μF and $C_2 = 10.0$ μF. The energy stored in a capacitor is

$\tfrac{1}{2}QV_{ab} = \tfrac{1}{2}CV_{ab}^2 = \dfrac{Q^2}{2C}$.

EXECUTE: (a) The initial charge on the 20.0 μF capacitor is

$Q = C_1(800 \text{ V}) = (20.0 \times 10^{-6} \text{ F})(800 \text{ V}) = 0.0160$ C.

(b) In the final circuit, charge Q is distributed between the two capacitors and $Q_1 + Q_2 = Q$. The final circuit contains only the two capacitors, so the voltage across each is the same, $V_1 = V_2$. $V = \dfrac{Q}{C}$ so $V_1 = V_2$

gives $\dfrac{Q_1}{C_1} = \dfrac{Q_2}{C_2}$. $Q_1 = \dfrac{C_1}{C_2}Q_2 = 2Q_2$. Using this in $Q_1 + Q_2 = 0.0160$ C gives $3Q_2 = 0.0160$ C and

$Q_2 = 5.33 \times 10^{-3}$ C. $Q_1 = 2Q_2 = 1.066 \times 10^{-2}$ C. $V_1 = \dfrac{Q_1}{C_1} = \dfrac{1.066 \times 10^{-2} \text{ C}}{20.0 \times 10^{-6} \text{ F}} = 533$ V.

$V_2 = \dfrac{Q_2}{C_2} = \dfrac{5.33 \times 10^{23}\ \text{C}}{10.0 \times 10^{26}\ \text{F}} = 533$ V. The potential differences across the capacitors are the same, as they should be.

(c) Energy $= \tfrac{1}{2} C_1 V^2 + \tfrac{1}{2} C_2 V^2 = \tfrac{1}{2}(C_1 + C_2) V^2$ gives

Energy $= \tfrac{1}{2}(20.0 \times 10^{-6}\ \text{F} + 10.0 \times 10^{-6}\ \text{F})(533\ \text{V})^2 = 4.26$ J.

(d) The 20.0 μF capacitor initially has energy $= \tfrac{1}{2} C_1 V^2 = \tfrac{1}{2}(20.0 \times 10^{-6}\ \text{F})(800\ \text{V})^2 = 6.40$ J. The decrease in stored energy that occurs when the capacitors are connected is $6.40\ \text{J} - 4.26\ \text{J} = 2.14$ J.

EVALUATE: The decrease in stored energy is because of conversion of electrical energy to other forms during the motion of the charge when it becomes distributed between the two capacitors. Thermal energy is generated by the current in the wires and energy is emitted in electromagnetic waves.

24.51. IDENTIFY: Simplify the network by replacing series and parallel combinations by their equivalent. The stored energy in a capacitor is $U = \tfrac{1}{2} C V^2$.

SET UP: For capacitors in series the voltages add and the charges are the same; $\dfrac{1}{C_{eq}} = \dfrac{1}{C_1} + \dfrac{1}{C_2} + \ldots$ For capacitors in parallel the voltages are the same and the charges add; $C_{eq} = C_1 + C_2 + \ldots$ $C = \dfrac{Q}{V}$. $U = \tfrac{1}{2} C V^2$.

EXECUTE: **(a)** Find C_{eq} for the network by replacing each series or parallel combination by its equivalent. The successive simplified circuits are shown in Figure 24.51.

$U_{tot} = \tfrac{1}{2} C_{eq} V^2 = \tfrac{1}{2}(2.19 \times 10^{-6}\ \text{F})(12.0\ \text{V})^2 = 1.58 \times 10^{-4}\ \text{J} = 158\ \mu\text{J}$.

(b) From Figure 24.51c, $Q_{tot} = C_{eq} V = (2.19 \times 10^{-6}\ \text{F})(12.0\ \text{V}) = 2.63 \times 10^{-5}$ C. From Figure 24.51b,

$Q_{4.8} = 2.63 \times 10^{-5}$ C. $V_{4.8} = \dfrac{Q_{4.8}}{C_{4.8}} = \dfrac{2.63 \times 10^{-5}\ \text{C}}{4.80 \times 10^{-6}\ \text{F}} = 5.48$ V.

$U_{4.8} = \tfrac{1}{2} C V^2 = \tfrac{1}{2}(4.80 \times 10^{-6}\ \text{F})(5.48\ \text{V})^2 = 7.21 \times 10^{-5}\ \text{J} = 72.1\ \mu\text{J}$.

This one capacitor stores nearly half the total stored energy.

EVALUATE: $U = \dfrac{Q^2}{2C}$. For capacitors in series the capacitor with the smallest C stores the greatest amount of energy.

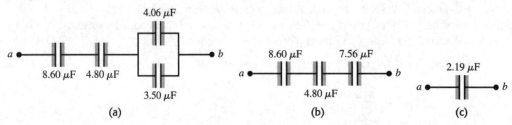

Figure 24.51

24.53. (a) IDENTIFY: Replace series and parallel combinations of capacitors by their equivalents.
SET UP: The network is sketched in Figure 24.53a.

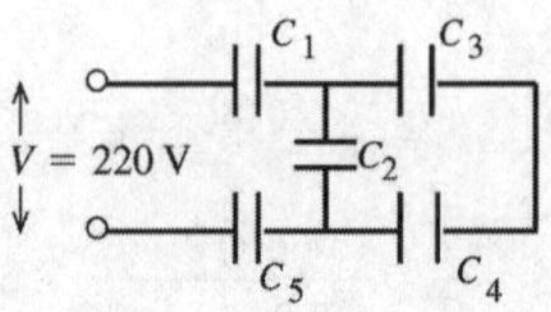

$C_1 = C_5 = 8.4\ \mu\text{F}$.
$C_2 = C_3 = C_4 = 4.2\ \mu\text{F}$.

Figure 24.53a

EXECUTE: Simplify the circuit by replacing the capacitor combinations by their equivalents: C_3 and C_4 are in series and can be replaced by C_{34} (Figure 24.53b):

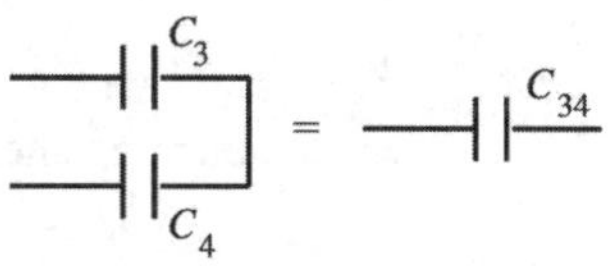

$$\frac{1}{C_{34}} = \frac{1}{C_3} + \frac{1}{C_4}.$$

$$\frac{1}{C_{34}} = \frac{C_3 + C_4}{C_3 C_4}.$$

Figure 24.53b

$$C_{34} = \frac{C_3 C_4}{C_3 + C_4} = \frac{(4.2\ \mu\text{F})(4.2\ \mu\text{F})}{4.2\ \mu\text{F} + 4.2\ \mu\text{F}} = 2.1\ \mu\text{F}.$$

C_2 and C_{34} are in parallel and can be replaced by their equivalent (Figure 24.53c):

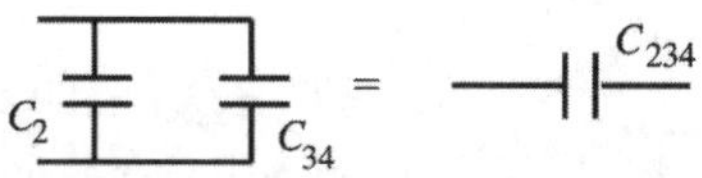

$$C_{234} = C_2 + C_{34}.$$
$$C_{234} = 4.2\ \mu\text{F} + 2.1\ \mu\text{F}.$$
$$C_{234} = 6.3\ \mu\text{F}.$$

Figure 24.53c

$C_1, C_5,$ and C_{234} are in series and can be replaced by C_{eq} (Figure 24.53d):

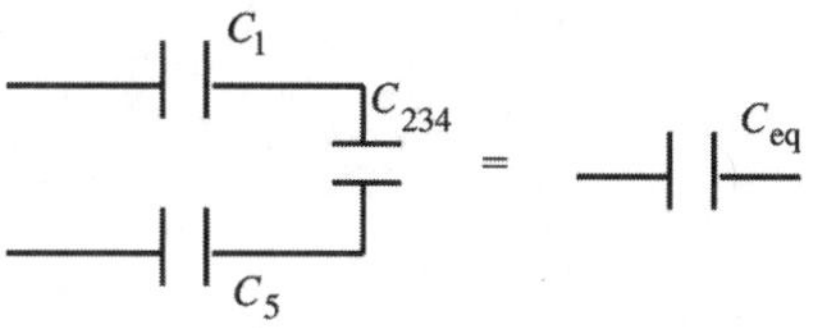

$$\frac{1}{C_{eq}} = \frac{1}{C_1} + \frac{1}{C_5} + \frac{1}{C_{234}}.$$

$$\frac{1}{C_{eq}} = \frac{2}{8.4\ \mu\text{F}} + \frac{1}{6.3\ \mu\text{F}}.$$

$$C_{eq} = 2.5\ \mu\text{F}.$$

Figure 24.53d

EVALUATE: For capacitors in series the equivalent capacitor is smaller than any of those in series. For capacitors in parallel the equivalent capacitance is larger than any of those in parallel.
(b) IDENTIFY and SET UP: In each equivalent network apply the rules for Q and V for capacitors in series and parallel; start with the simplest network and work back to the original circuit.
EXECUTE: The equivalent circuit is drawn in Figure 24.53e.

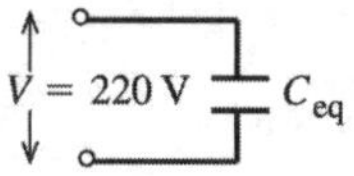

$$Q_{eq} = C_{eq} V.$$
$$Q_{eq} = (2.5\ \mu\text{F})(220\ \text{V}) = 550\ \mu\text{C}.$$

Figure 24.53e

$Q_1 = Q_5 = Q_{234} = 550\ \mu\text{C}$ (capacitors in series have same charge).

$$V_1 = \frac{Q_1}{C_1} = \frac{550\ \mu\text{C}}{8.4\ \mu\text{F}} = 65\ \text{V}.$$

$$V_5 = \frac{Q_5}{C_5} = \frac{550\ \mu\text{C}}{8.4\ \mu\text{F}} = 65\ \text{V}.$$

$V_{234} = \dfrac{Q_{234}}{C_{234}} = \dfrac{550 \ \mu C}{6.3 \ \mu F} = 87$ V.

Now draw the network as in Figure 24.53f.

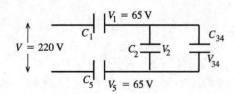

$V_2 = V_{34} = V_{234} = 87$ V

capacitors in parallel have the same potential.

Figure 24.53f

$Q_2 = C_2 V_2 = (4.2 \ \mu F)(87 \ V) = 370 \ \mu C$.

$Q_{34} = C_{34} V_{34} = (2.1 \ \mu F)(87 \ V) = 180 \ \mu C$.

Finally, consider the original circuit (Figure 24.53g).

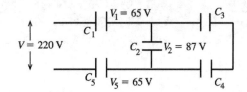

$Q_3 = Q_4 = Q_{34} = 180 \ \mu C$

capacitors in series have the same charge.

Figure 24.53g

$V_3 = \dfrac{Q_3}{C_3} = \dfrac{180 \ \mu C}{4.2 \ \mu F} = 43$ V.

$V_4 = \dfrac{Q_4}{C_4} = \dfrac{180 \ \mu C}{4.2 \ \mu F} = 43$ V.

Summary: $Q_1 = 550 \ \mu C$, $V_1 = 65$ V.

$Q_2 = 370 \ \mu C$, $V_2 = 87$ V.

$Q_3 = 180 \ \mu C$, $V_3 = 43$ V.

$Q_4 = 180 \ \mu C$, $V_4 = 43$ V.

$Q_5 = 550 \ \mu C$, $V_5 = 65$ V.

EVALUATE: $V_3 + V_4 = V_2$ and $V_1 + V_2 + V_5 = 220$ V (apart from some small rounding error)
$Q_1 = Q_2 + Q_3$ and $Q_5 = Q_2 + Q_4$.

24.55. IDENTIFY: Capacitors in series carry the same charge, while capacitors in parallel have the same potential difference across them.

SET UP: $V_{ab} = 150$ V, $Q_1 = 150 \ \mu C$, $Q_3 = 450 \ \mu C$, and $V = Q/C$.

EXECUTE: $C_1 = 3.00 \ \mu F$ so $V_1 = \dfrac{Q_1}{C_1} = \dfrac{150 \ \mu C}{3.00 \ \mu F} = 50.0$ V and $V_1 = V_2 = 50.0$ V. $V_1 + V_3 = V_{ab}$ so

$V_3 = 100$ V. $C_3 = \dfrac{Q_3}{V_3} = \dfrac{450 \ \mu C}{100 \ V} = 4.50 \ \mu F$. $Q_1 + Q_2 = Q_3$ so $Q_2 = Q_3 - Q_1 = 450 \ \mu C - 150 \ \mu C = 300 \ \mu C$

and $C_2 = \dfrac{Q_2}{V_2} = \dfrac{300 \ \mu C}{50.0 \ V} = 6.00 \ \mu F$.

EVALUATE: Capacitors in parallel only carry the same charge if they have the same capacitance.

24.59. IDENTIFY: Replace series and parallel combinations of capacitors by their equivalents. In each equivalent network apply the rules for Q and V for capacitors in series and parallel; start with the simplest network and work back to the original circuit.

(a) SET UP: The network is sketched in Figure 24.59a.

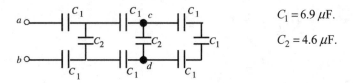

$C_1 = 6.9\ \mu F.$

$C_2 = 4.6\ \mu F.$

Figure 24.59a

EXECUTE: Simplify the network by replacing the capacitor combinations by their equivalents. Make the replacement shown in Figure 24.59b.

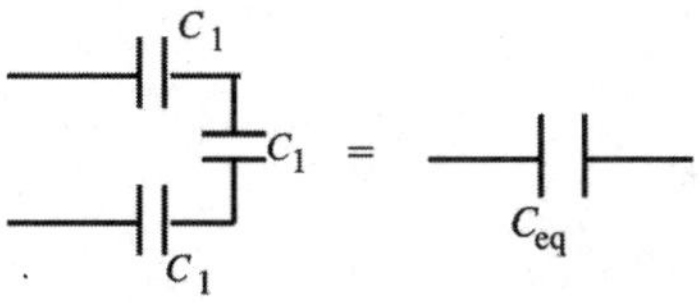

$$\frac{1}{C_{eq}} = \frac{3}{C_1}.$$

$$C_{eq} = \frac{C_1}{3} = \frac{6.9\ \mu F}{3} = 2.3\ \mu F.$$

Figure 24.59b

Next make the replacement shown in Figure 24.59c.

$C_{eq} = 2.3\ \mu F + C_2.$

$C_{eq} = 2.3\ \mu F + 4.6\ \mu F = 6.9\ \mu F.$

Figure 24.59c

Make the replacement shown in Figure 24.59d.

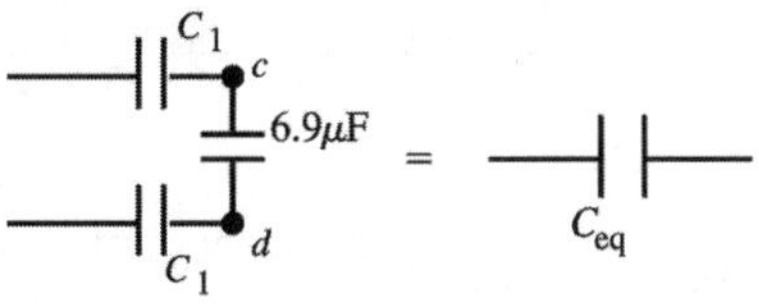

$$\frac{1}{C_{eq}} = \frac{2}{C_1} + \frac{1}{6.9\ \mu F} = \frac{3}{6.9\ \mu F}.$$

$C_{eq} = 2.3\ \mu F.$

Figure 24.59d

Make the replacement shown in Figure 24.59e.

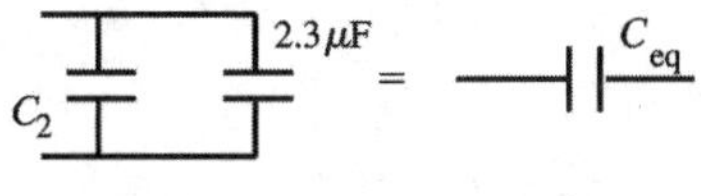

$C_{eq} = C_2 + 2.3\ \mu F = 4.6\ \mu F + 2.3\ \mu F.$

$C_{eq} = 6.9\ \mu F.$

Figure 24.59e

Make the replacement shown in Figure 24.59f.

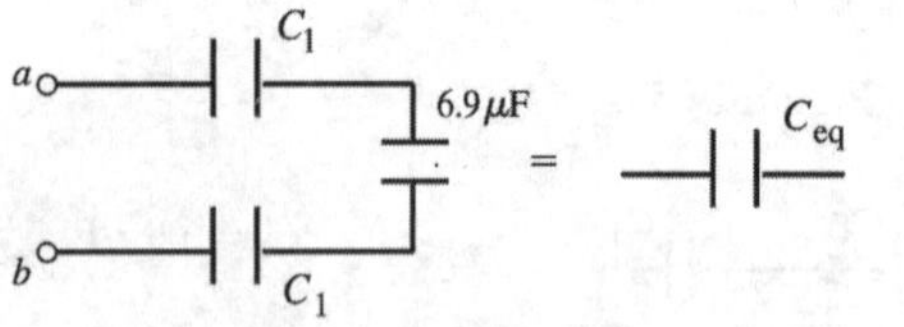

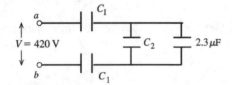

$$\frac{1}{C_{eq}} = \frac{2}{C_1} + \frac{1}{6.9\ \mu F} = \frac{3}{6.9\ \mu F}.$$

$$C_{eq} = 2.3\ \mu F.$$

Figure 24.59f

(b) SET UP and **EXECUTE:** Consider the network as drawn in Figure 24.59g.

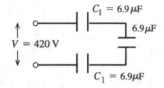

From part (a) $2.3\ \mu F$ is the equivalent capacitance of the rest of the network.

Figure 24.59g

The equivalent network is shown in Figure 24.59h.

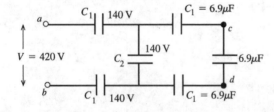

The capacitors are in series, so all three capacitors have the same Q.

Figure 24.59h

But here all three have the same C, so by $V = Q/C$ all three must have the same V. The three voltages must add to 420 V, so each capacitor has $V = 140$ V. The $6.9\ \mu F$ to the right is the equivalent of C_2 and the $2.3\ \mu F$ capacitor in parallel, so $V_2 = 140$ V. (Capacitors in parallel have the same potential difference.) Hence $Q_1 = C_1 V_1 = (6.9\ \mu F)(140\ V) = 9.7 \times 10^{-4}$ C and $Q_2 = C_2 V_2 = (4.6\ \mu F)(140\ V) = 6.4 \times 10^{-4}$ C.

(c) From the potentials deduced in part (b) we have the situation shown in Figure 24.59i.

From part (a) $6.9\ \mu F$ is the equivalent capacitance of the rest of the network.

Figure 24.59i

The three right-most capacitors are in series and therefore have the same charge. But their capacitances are also equal, so by $V = Q/C$ they each have the same potential difference. Their potentials must sum to 140 V, so the potential across each is 47 V and $V_{cd} = 47$ V.

EVALUATE: In each capacitor network the rules for combining V for capacitors in series and parallel are obeyed. Note that $V_{cd} < V$, in fact $V - 2(140\ V) - 2(47\ V) = V_{cd}$.

24.65. IDENTIFY: The two slabs of dielectric are in series with each other.

SET UP: The capacitor is equivalent to C_1 and C_2 in series, so $\frac{1}{C_1} + \frac{1}{C_2} = \frac{1}{C}$, which gives $C = \frac{C_1 C_2}{C_1 + C_2}$.

EXECUTE: With $d = 1.90$ mm, $C_1 = \frac{K_1 \varepsilon_0 A}{d}$ and $C_2 = \frac{K_2 \varepsilon_0 A}{d}$.

$C = \left(\frac{K_1 K_2}{K_1 + K_2}\right)\frac{\varepsilon_0 A}{d} = \left(\frac{(4.7)(2.6)}{4.7 + 2.6}\right)\frac{(8.854 \times 10^{-12} \text{ C}^2/\text{N} \cdot \text{m}^2)(0.0800 \text{ m})^2}{1.90 \times 10^{-3} \text{ m}} = 4.992 \times 10^{-11}$ F.

$U = \frac{1}{2}CV^2 = \frac{1}{2}(4.992 \times 10^{-11} \text{ F})(86.0 \text{ V})^2 = 1.85 \times 10^{-7}$ J.

EVALUATE: The dielectrics increase the capacitance, allowing the capacitor to store more energy than if it were air-filled.

24.67. IDENTIFY: The object is equivalent to two identical capacitors in parallel, where each has the same area A, plate separation d and dielectric with dielectric constant K.

SET UP: For each capacitor in the parallel combination, $C = \frac{\varepsilon_0 A}{d}$.

EXECUTE: **(a)** The charge distribution on the plates is shown in Figure 24.67.

(b) $C = 2\left(\frac{\varepsilon_0 A}{d}\right) = \frac{2(4.2)\varepsilon_0 (0.120 \text{ m})^2}{4.5 \times 10^{-4} \text{ m}} = 2.38 \times 10^{-9}$ F.

EVALUATE: If two of the plates are separated by both sheets of paper to form a capacitor,

$C = \frac{\varepsilon_0 A}{2d} = \frac{2.38 \times 10^{-9} \text{ F}}{4}$, smaller by a factor of 4 compared to the capacitor in the problem.

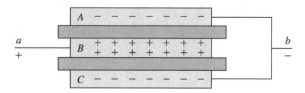

Figure 24.67

24.71. IDENTIFY and SET UP: For a parallel-plate capacitor, $C = \frac{\varepsilon_0 A}{d}$. The stored energy can be expressed as

$U = \frac{1}{2}CV^2$ or $U = \frac{Q^2}{2C}$.

EXECUTE: **(a)** If the battery remains connected, V remains constant, so it is useful to write the energy in terms of V and C:

$U = \frac{1}{2}CV^2 = \frac{1}{2}\left(\frac{\varepsilon_0 A}{d}\right)V^2 = \frac{\varepsilon_0 A V^2}{2} \cdot \frac{1}{d}$.

If the battery is disconnected, Q remains constant, so it is useful to write the energy in terms of Q and C:

$U = \frac{Q^2}{2C} = \frac{Q^2}{2\left(\frac{\varepsilon_0 A}{d}\right)} = \left(\frac{Q^2}{2\varepsilon_0 A}\right)d$.

The graph shows a linear relationship between U and $1/d$, so it must represent the case where the battery remains connected to the capacitor.

(b) In a graph of U versus $1/d$ for the equation $U = \frac{\varepsilon_0 A V^2}{2} \cdot \frac{1}{d}$, the slope should be equal to $\frac{\varepsilon_0 A V^2}{2}$.

Choosing points on the graph in the problem, the slope is $\frac{(73 - 18) \times 10^{-9} \text{ J}}{20.0 \text{ cm}^{-1} - 5.0 \text{ cm}^{-1}} = 3.67 \times 10^{-11}$ J·m.

Solving for A gives
$A = 2(\text{slope})/\varepsilon_0 V^2 = 2(3.67\times 10^{-11}\text{ J}\cdot\text{m})/(8.854\times 10^{-12}\text{ C}^2/\text{N}\cdot\text{m}^2)(24.0\text{ V})^2] = 0.014\text{ m}^2 = 144\text{ cm}^2$.

(c) <u>With the battery connected</u>: $U = \dfrac{\varepsilon_0 A V^2}{2}\cdot\dfrac{1}{d}$, so as we increase d from 0.0500 cm to 0.400 cm, the energy *decreases* since V remains constant.

<u>With the battery disconnected</u>: $U = \left(\dfrac{Q^2}{2\varepsilon_0 A}\right)d$, so as we increase d, the energy *increases* since Q does not change. Therefore there is more energy stored with the battery *disconnected* as d is increased.

EVALUATE: If this capacitor were square, its plates would be 12 cm×12 cm. This is a reasonable size for a piece of apparatus for use in a laboratory and could easily be manufactured.

24.73. **IDENTIFY** and **SET UP:** The potential difference is $V = 30\text{ mV} - (-70\text{ mV}) = 100\text{ mV}$, and $Q = CV$.
EXECUTE: $Q = CV$ gives $Q/\text{cm}^2 = (C/\text{cm}^2)V = (1\ \mu\text{F}/\text{cm}^2)(100\text{ mV})(1\text{ mol}/10^5\text{ C}) = 10^{-12}\text{ mol/cm}^2$, which is choice (c).
EVALUATE: This charge produces a potential difference of 100 mV = 0.1 V, which is certainly measurable using ordinary laboratory meters.

CURRENT, RESISTANCE, AND ELECTROMOTIVE FORCE

25.3. **IDENTIFY:** $I = Q/t$. $J = I/A$. $J = n|q|v_d$.

SET UP: $A = (\pi/4)D^2$, with $D = 2.05 \times 10^{-3}$ m. The charge of an electron has magnitude $+e = 1.60 \times 10^{-19}$ C.

EXECUTE: **(a)** $Q = It = (5.00 \text{ A})(1.00 \text{ s}) = 5.00$ C. The number of electrons is $\dfrac{Q}{e} = 3.12 \times 10^{19}$.

(b) $J = \dfrac{I}{(\pi/4)D^2} = \dfrac{5.00 \text{ A}}{(\pi/4)(2.05 \times 10^{-3} \text{ m})^2} = 1.51 \times 10^6 \text{ A/m}^2$.

(c) $v_d = \dfrac{J}{n|q|} = \dfrac{1.51 \times 10^6 \text{ A/m}^2}{(8.5 \times 10^{28} \text{ m}^{-3})(1.60 \times 10^{-19} \text{ C})} = 1.11 \times 10^{-4}$ m/s $= 0.111$ mm/s.

EVALUATE: **(d)** If I is the same, $J = I/A$ would decrease and v_d would decrease. The number of electrons passing through the light bulb in 1.00 s would not change.

25.7. **IDENTIFY and SET UP:** Apply $I = \dfrac{dQ}{dt}$ to find the charge dQ in time dt. Integrate to find the total charge in the whole time interval.

EXECUTE: **(a)** $dQ = I\,dt$.

$Q = \int_0^{8.0\text{s}} (55 \text{ A} - (0.65 \text{ A/s}^2)t^2)\,dt = \left[(55 \text{ A})t - (0.217 \text{ A/s}^2)t^3 \right]_0^{8.0 \text{ s}}$.

$Q = (55 \text{ A})(8.0 \text{ s}) - (0.217 \text{ A/s}^2)(8.0 \text{ s})^3 = 330$ C.

(b) $I = \dfrac{Q}{t} = \dfrac{330 \text{ C}}{8.0 \text{ s}} = 41$ A.

EVALUATE: The current decreases from 55 A to 13.4 A during the interval. The decrease is not linear and the average current is not equal to $(55\text{A} + 13.4 \text{ A})/2$.

25.9. **IDENTIFY and SET UP:** The number of ions that enter gives the charge that enters the axon in the specified time. $I = \dfrac{\Delta Q}{\Delta t}$.

EXECUTE: $\Delta Q = (5.6 \times 10^{11} \text{ ions})(1.60 \times 10^{-19} \text{ C/ion}) = 9.0 \times 10^{-8}$ C. $I = \dfrac{\Delta Q}{\Delta t} = \dfrac{9.0 \times 10^{-8} \text{ C}}{10 \times 10^{-3} \text{ s}} = 9.0 \, \mu\text{A}$.

EVALUATE: This current is much smaller than household currents but are comparable to many currents in electronic equipment.

25.11. **IDENTIFY:** First use Ohm's law to find the resistance at 20.0°C; then calculate the resistivity from the resistance. Finally use the dependence of resistance on temperature to calculate the temperature coefficient of resistance.

SET UP: Ohm's law is $R = V/I$, $R = \rho L/A$, $R = R_0[1 + \alpha(T - T_0)]$, and the radius is one-half the diameter.

EXECUTE: (a) At 20.0°C, $R = V/I = (15.0\text{ V})/(18.5\text{ A}) = 0.811\text{ }\Omega$. Using $R = \rho L/A$ and solving for ρ gives $\rho = RA/L = R\pi(D/2)^2/L = (0.811\text{ }\Omega)\pi[(0.00500\text{ m})/2]^2/(1.50\text{ m}) = 1.06\times 10^{-5}\text{ }\Omega\cdot\text{m}$.

(b) At 92.0°C, $R = V/I = (15.0\text{ V})/(17.2\text{ A}) = 0.872\text{ }\Omega$. Using $R = R_0[1+\alpha(T-T_0)]$ with T_0 taken as 20.0°C, we have $0.872\text{ }\Omega = (0.811\text{ }\Omega)[1+\alpha(92.0°\text{C}-20.0°\text{C})]$. This gives $\alpha = 0.00105\text{ (C°)}^{-1}$.

EVALUATE: The results are typical of ordinary metals.

25.15. (a) **IDENTIFY:** Start with the definition of resistivity and use its dependence on temperature to find the electric field.

SET UP: $E = \rho J = \rho_{20}[1+\alpha(T-T_0)]\dfrac{I}{\pi r^2}$.

EXECUTE: $E = (5.25\times 10^{-8}\text{ }\Omega\cdot\text{m})[1+(0.0045/\text{C}°)(120°\text{C}-20°\text{C})](12.5\text{ A})/[\pi(0.000500\text{ m})^2] = 1.21\text{ V/m}$.

(Note that the resistivity at 120°C turns out to be $7.61\times 10^{-8}\text{ }\Omega\cdot\text{m}$.)

EVALUATE: This result is fairly large because tungsten has a larger resistivity than copper.

(b) **IDENTIFY:** Relate resistance and resistivity.

SET UP: $R = \rho L/A = \rho L/\pi r^2$.

EXECUTE: $R = (7.61\times 10^{-8}\text{ }\Omega\cdot\text{m})(0.150\text{ m})/[\pi(0.000500\text{ m})^2] = 0.0145\text{ }\Omega$.

EVALUATE: Most metals have very low resistance.

(c) **IDENTIFY:** The potential difference is proportional to the length of wire.

SET UP: $V = EL$.

EXECUTE: $V = (1.21\text{ V/m})(0.150\text{ m}) = 0.182\text{ V}$.

EVALUATE: We could also calculate $V = IR = (12.5\text{ A})(0.0145\text{ }\Omega) = 0.181\text{ V}$, in agreement with part (c).

25.21. **IDENTIFY and SET UP:** The equation $\rho = E/J$ relates the electric field that is given to the current density. $V = EL$ gives the potential difference across a length L of wire and $V = IR$ allows us to calculate R.

EXECUTE: (a) $\rho = E/J$ so $J = E/\rho$.

From Table 25.1 the resistivity for gold is $2.44\times 10^{-8}\text{ }\Omega\cdot\text{m}$.

$$J = \dfrac{E}{\rho} = \dfrac{0.49\text{ V/m}}{2.44\times 10^{-8}\text{ }\Omega\cdot\text{m}} = 2.008\times 10^7\text{ A/m}^2.$$

$I = JA = J\pi r^2 = (2.008\times 10^7\text{ A/m}^2)\pi(0.42\times 10^{-3}\text{ m})^2 = 11\text{ A}$.

(b) $V = EL = (0.49\text{ V/m})(6.4\text{ m}) = 3.1\text{ V}$.

(c) We can use Ohm's law: $V = IR$.

$$R = \dfrac{V}{I} = \dfrac{3.1\text{ V}}{11\text{ A}} = 0.28\text{ }\Omega.$$

EVALUATE: We can also calculate R from the resistivity and the dimensions of the wire:

$$R = \dfrac{\rho L}{A} = \dfrac{\rho L}{\pi r^2} = \dfrac{(2.44\times 10^{-8}\text{ }\Omega\cdot\text{m})(6.4\text{ m})}{\pi(0.42\times 10^{-3}\text{ m})^2} = 0.28\text{ }\Omega,\text{ which checks.}$$

25.27. **IDENTIFY:** The terminal voltage of the battery is $V_{ab} = \varepsilon - Ir$. The voltmeter reads the potential difference between its terminals.

SET UP: An ideal voltmeter has infinite resistance.

EXECUTE: (a) Since an ideal voltmeter has infinite resistance, so there would be NO current through the 2.0 Ω resistor.

(b) $V_{ab} = \varepsilon = 5.0\text{ V}$; Since there is no current there is no voltage lost over the internal resistance.

(c) The voltmeter reading is therefore 5.0 V since with no current flowing there is no voltage drop across either resistor.

EVALUATE: This not the proper way to connect a voltmeter. If we wish to measure the terminal voltage of the battery in a circuit that does not include the voltmeter, then connect the voltmeter across the terminals of the battery.

25.29. **IDENTIFY:** The voltmeter reads the potential difference V_{ab} between the terminals of the battery.
SET UP: <u>open circuit:</u> $I = 0$. The circuit is sketched in Figure 25.29a.

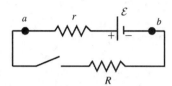

EXECUTE: $V_{ab} = \varepsilon = 3.08$ V.

Figure 25.29a

SET UP: <u>switch closed:</u> The circuit is sketched in Figure 25.29b.

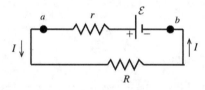

EXECUTE: $V_{ab} = \varepsilon - Ir = 2.97$ V.
$r = \dfrac{\varepsilon - 2.97 \text{ V}}{I}$.

$r = \dfrac{3.08 \text{ V} - 2.97 \text{ V}}{1.65 \text{ A}} = 0.067 \, \Omega$.

Figure 25.29b

And $V_{ab} = IR$ so $R = \dfrac{V_{ab}}{I} = \dfrac{2.97 \text{ V}}{1.65 \text{ A}} = 1.80 \, \Omega$.

EVALUATE: When current flows through the battery there is a voltage drop across its internal resistance and its terminal voltage V is less than its emf.

25.33. **IDENTIFY and SET UP:** There is a single current path so the current is the same at all points in the circuit. Assume the current is counterclockwise and apply Kirchhoff's loop rule.
EXECUTE: **(a)** Apply the loop rule, traveling around the circuit in the direction of the current.
$+16.0 \text{ V} - I(1.6 \, \Omega + 5.0 \, \Omega + 1.4 \, \Omega + 9.0 \, \Omega) - 8.0 \text{ V} = 0$. $I = \dfrac{16.0 \text{ V} - 8.0 \text{ V}}{17.0 \, \Omega} = 0.471$ A. Our calculated I is positive so I is counterclockwise, as we assumed.
(b) $V_b + 16.0 \text{ V} - I(1.6 \, \Omega) = V_a$. $V_{ab} = 16.0 \text{ V} - (0.471 \text{ A})(1.6 \, \Omega) = 15.2$ V.
EVALUATE: If we traveled around the circuit in the direction opposite to the current, the final answers would be the same.

25.37. **IDENTIFY:** A "100-W" European bulb dissipates 100 W when used across 220 V.
(a) SET UP: Take the ratio of the power in the U.S. to the power in Europe, as in the alternative method for Problem 25.36, using $P = V^2/R$.
EXECUTE: $\dfrac{P_{US}}{P_E} = \dfrac{V_{US}^2/R}{V_E^2/R} = \left(\dfrac{V_{US}}{V_E}\right)^2 = \left(\dfrac{120 \text{ V}}{220 \text{ V}}\right)^2$. This gives $P_{US} = (100 \text{ W})\left(\dfrac{120 \text{ V}}{220 \text{ V}}\right)^2 = 29.8$ W.
(b) SET UP: Use $P = IV$ to find the current.
EXECUTE: $I = P/V = (29.8 \text{ W})/(120 \text{ V}) = 0.248$ A.
EVALUATE: The bulb draws considerably less power in the U.S., so it would be much dimmer than in Europe.

25.39. **IDENTIFY:** Calculate the current in the circuit. The power output of a battery is its terminal voltage times the current through it. The power dissipated in a resistor is I^2R.
SET UP: The sum of the potential changes around the circuit is zero.
EXECUTE: **(a)** $I = \dfrac{8.0 \text{ V}}{17 \, \Omega} = 0.47$ A. Then $P_{5\Omega} = I^2R = (0.47 \text{ A})^2(5.0 \, \Omega) = 1.1$ W and
$P_{9\Omega} = I^2R = (0.47 \text{ A})^2(9.0 \, \Omega) = 2.0$ W, so the total is 3.1 W.

(b) $P_{16V} = \varepsilon I - I^2 r = (16 \text{ V})(0.47 \text{ A}) - (0.47 \text{ A})^2 (1.6 \text{ }\Omega) = 7.2 \text{ W}.$

(c) $P_{8V} = \varepsilon I + Ir^2 = (8.0 \text{ V})(0.47 \text{ A}) + (0.47 \text{ A})^2 (1.4 \text{ }\Omega) = 4.1 \text{ W}.$

EVALUATE: **(d)** (b) = (a) + (c). The rate at which the 16.0-V battery delivers electrical energy to the circuit equals the rate at which it is consumed in the 8.0-V battery and the 5.0-Ω and 9.0-Ω resistors.

25.43. **(a) IDENTIFY** and **SET UP:** $P = VI$ and energy = (power) × (time).

EXECUTE: $P = VI = (12 \text{ V})(60 \text{ A}) = 720 \text{ W}.$

The battery can provide this for 1.0 h, so the energy the battery has stored is
$U = Pt = (720 \text{ W})(3600 \text{ s}) = 2.6 \times 10^6 \text{ J}.$

(b) IDENTIFY and **SET UP:** For gasoline the heat of combustion is $L_c = 46 \times 10^6 \text{ J/kg}.$ Solve for the mass m required to supply the energy calculated in part (a) and use density $\rho = m/V$ to calculate V.

EXECUTE: The mass of gasoline that supplies $2.6 \times 10^6 \text{ J}$ is $m = \dfrac{2.6 \times 10^6 \text{ J}}{46 \times 10^6 \text{ J/kg}} = 0.0565 \text{ kg}.$

The volume of this mass of gasoline is
$$V = \frac{m}{\rho} = \frac{0.0565 \text{ kg}}{900 \text{ kg/m}^3} = 6.3 \times 10^{-5} \text{ m}^3 \left(\frac{1000 \text{ L}}{1 \text{ m}^3}\right) = 0.063 \text{ L}.$$

(c) IDENTIFY and **SET UP:** Energy = (power) × (time); the energy is that calculated in part (a).

EXECUTE: $U = Pt, t = \dfrac{U}{P} = \dfrac{2.6 \times 10^6 \text{ J}}{450 \text{ W}} = 5800 \text{ s} = 97 \text{ min} = 1.6 \text{ h}.$

EVALUATE: The battery discharges at a rate of 720 W (for 1.0 h) and is charged at a rate of 450 W (for 1.6 h), so it takes longer to charge than to discharge.

25.45. **IDENTIFY:** Some of the power generated by the internal emf of the battery is dissipated across the battery's internal resistance, so it is not available to the bulb.

SET UP: Use $P = I^2 R$ and take the ratio of the power dissipated in the internal resistance r to the total power.

EXECUTE: $\dfrac{P_r}{P_{\text{Total}}} = \dfrac{I^2 r}{I^2(r+R)} = \dfrac{r}{r+R} = \dfrac{3.5 \text{ }\Omega}{28.5 \text{ }\Omega} = 0.123 = 12.3\%.$

EVALUATE: About 88% of the power of the battery goes to the bulb. The rest appears as heat in the internal resistance.

25.49. **IDENTIFY:** The resistivity is $\rho = \dfrac{m}{ne^2 \tau}.$

SET UP: For silicon, $\rho = 2300 \text{ }\Omega \cdot \text{m}.$

EXECUTE: **(a)** $\tau = \dfrac{m}{ne^2 \rho} = \dfrac{9.11 \times 10^{-31} \text{ kg}}{(1.0 \times 10^{16} \text{ m}^{-3})(1.60 \times 10^{-19} \text{ C})^2 (2300 \text{ }\Omega \cdot \text{m})} = 1.55 \times 10^{-12} \text{ s}.$

EVALUATE: **(b)** The number of free electrons in copper $(8.5 \times 10^{28} \text{ m}^{-3})$ is much larger than in pure silicon $(1.0 \times 10^{16} \text{ m}^{-3}).$ A smaller density of current carriers means a higher resistivity.

25.53. **IDENTIFY** and **SET UP:** With the voltmeter connected across the terminals of the battery there is no current through the battery and the voltmeter reading is the battery emf; $\varepsilon = 12.6 \text{ V}.$
With a wire of resistance R connected to the battery current I flows and $\varepsilon - Ir - IR = 0$, where r is the internal resistance of the battery. Apply this equation to each piece of wire to get two equations in the two unknowns.

EXECUTE: Call the resistance of the 20.0-m piece R_1; then the resistance of the 40.0-m piece is $R_2 = 2R_1.$

$\varepsilon - I_1 r - I_1 R_1 = 0; \quad 12.6 \text{ V} - (7.00 \text{ A})r - (7.00 \text{ A})R_1 = 0.$

$\varepsilon - I_2 r - I_2 (2R_2) = 0; \quad 12.6 \text{ V} - (4.20 \text{ A})r - (4.20 \text{ A})(2R_1) = 0.$

Solving these two equations in two unknowns gives $R_1 = 1.20 \text{ }\Omega.$ This is the resistance of 20.0 m, so the resistance of one meter is $[1.20 \text{ }\Omega/(20.0 \text{ m})](1.00 \text{ m}) = 0.060 \text{ }\Omega.$

EVALUATE: We can also solve for r and we get $r = 0.600\ \Omega$. When measuring small resistances, the internal resistance of the battery has a large effect.

25.55. IDENTIFY: Conservation of charge requires that the current be the same in both sections of the wire. $E = \rho J = \dfrac{\rho I}{A}$. For each section, $V = IR = JAR = \left(\dfrac{EA}{\rho}\right)\left(\dfrac{\rho L}{A}\right) = EL$. The voltages across each section add.

SET UP: $A = (\pi/4)D^2$, where D is the diameter.

EXECUTE: (a) The current must be the same in both sections of the wire, so the current in the thin end is 2.5 mA.

(b) $E_{1.6\text{mm}} = \rho J = \dfrac{\rho I}{A} = \dfrac{(1.72\times 10^{-8}\ \Omega\cdot \text{m})(2.5\times 10^{-3}\ \text{A})}{(\pi/4)(1.6\times 10^{-3}\ \text{m})^2} = 2.14\times 10^{-5}$ V/m.

(c) $E_{0.8\text{mm}} = \rho J = \dfrac{\rho I}{A} = \dfrac{(1.72\times 10^{-8}\ \Omega\cdot \text{m})(2.5\times 10^{-3}\ \text{A})}{(\pi/4)(0.80\times 10^{-3}\ \text{m})^2} = 8.55\times 10^{-5}$ V/m. This is $4 E_{1.6\text{mm}}$.

(d) $V = E_{1.6\text{mm}}L_{1.6\text{ mm}} + E_{0.8\text{ mm}}L_{0.8\text{ mm}}$. $V = (2.14\times 10^{-5}\text{ V/m})(1.20\text{ m}) + (8.55\times 10^{-5}\text{ V/m})(1.80\text{ m}) = 1.80\times 10^{-4}$ V.

EVALUATE: The currents are the same but the current density is larger in the thinner section and the electric field is larger there.

25.59. (a) **IDENTIFY:** Apply $R = \dfrac{\rho L}{A}$ to calculate the resistance of each thin disk and then integrate over the truncated cone to find the total resistance.

SET UP:

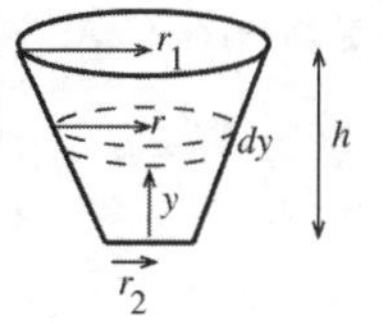

EXECUTE: The radius of a truncated cone a distance y above the bottom is given by $r = r_2 + (y/h)(r_1 - r_2) = r_2 + y\beta$ with $\beta = (r_1 - r_2)/h$.

Figure 25.59

Consider a thin slice a distance y above the bottom. The slice has thickness dy and radius r (see Figure 25.59). The resistance of the slice is $dR = \dfrac{\rho\, dy}{A} = \dfrac{\rho\, dy}{\pi r^2} = \dfrac{\rho\, dy}{\pi(r_2 + \beta y)^2}$.

The total resistance of the cone if obtained by integrating over these thin slices:

$R = \int dR = \dfrac{\rho}{\pi}\int_0^h \dfrac{dy}{(r_2 + \beta y)^2} = \dfrac{\rho}{\pi}\left[-\dfrac{1}{\beta}(r_2 + y\beta)^{-1}\right]_0^h = -\dfrac{\rho}{\pi\beta}\left[\dfrac{1}{r_2 + h\beta} - \dfrac{1}{r_2}\right]$.

But $r_2 + h\beta = r_1$.

$R = \dfrac{\rho}{\pi\beta}\left[\dfrac{1}{r_2} - \dfrac{1}{r_1}\right] = \dfrac{\rho}{\pi}\left(\dfrac{h}{r_1 - r_2}\right)\left(\dfrac{r_1 - r_2}{r_1 r_2}\right) = \dfrac{\rho h}{\pi r_1 r_2}$.

(b) **EVALUATE:** Let $r_1 = r_2 = r$. Then $R = \rho h/\pi r^2 = \rho L/A$ where $A = \pi r^2$ and $L = h$. This agrees with $R = \dfrac{\rho L}{A}$.

25.61. IDENTIFY: In each case write the terminal voltage in terms of $\varepsilon, I,$ and r. Since I is known, this gives two equations in the two unknowns ε and r.

SET UP: The battery with the 1.50-A current is sketched in Figure 25.61a.

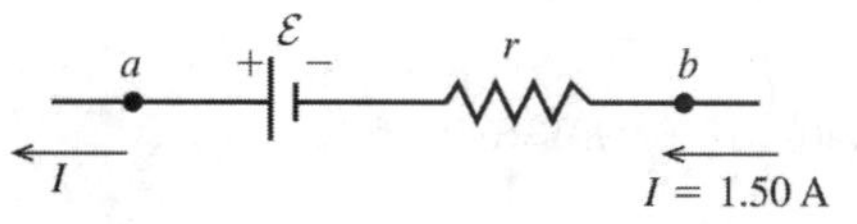

$V_{ab} = 8.40$ V.
$V_{ab} = \varepsilon - Ir$.
$\varepsilon - (1.50\text{ A})r = 8.40$ V.

Figure 25.61a

The battery with the 3.50-A current is sketched in Figure 25.61b.

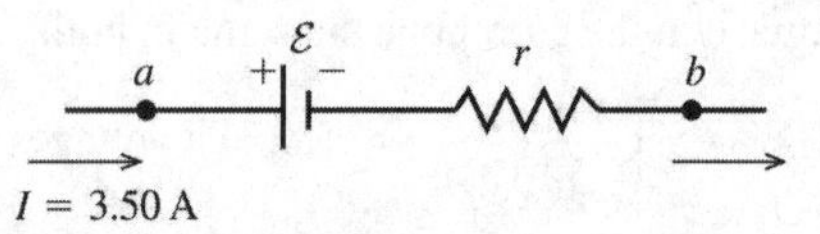

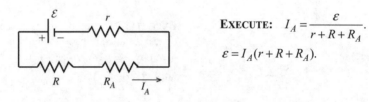

$V_{ab} = 10.2$ V.
$V_{ab} = \varepsilon + Ir$.
$\varepsilon + (3.50 \text{ A})r = 10.2$ V.

Figure 25.61b

EXECUTE: (a) Solve the first equation for ε and use that result in the second equation:
$\varepsilon = 8.40 \text{ V} + (1.50 \text{ A})r$.
$8.40 \text{ V} + (1.50 \text{ A})r + (3.50 \text{ A})r = 10.2 \text{ V}$.
$(5.00 \text{ A})r = 1.8 \text{ V}$ so $r = \dfrac{1.8 \text{ V}}{5.00 \text{ A}} = 0.36 \ \Omega$.

(b) Then $\varepsilon = 8.40 \text{ V} + (1.50 \text{ A})r = 8.40 \text{ V} + (1.50 \text{ A})(0.36 \ \Omega) = 8.94 \text{ V}$.

EVALUATE: When the current passes through the emf in the direction from − to +, the terminal voltage is less than the emf and when it passes through from + to −, the terminal voltage is greater than the emf.

25.67. IDENTIFY: The ammeter acts as a resistance in the circuit loop. Set the sum of the potential rises and drops around the circuit equal to zero.

(a) **SET UP:** The circuit with the ammeter is sketched in Figure 25.67a.

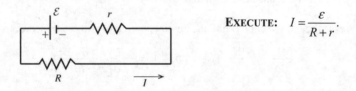

EXECUTE: $I_A = \dfrac{\varepsilon}{r + R + R_A}$.

$\varepsilon = I_A(r + R + R_A)$.

Figure 25.67a

SET UP: The circuit with the ammeter removed is sketched in Figure 25.67b.

EXECUTE: $I = \dfrac{\varepsilon}{R + r}$.

Figure 25.67b

Combining the two equations gives

$I = \left(\dfrac{1}{R+r}\right) I_A(r + R + R_A) = I_A \left(1 + \dfrac{R_A}{r + R}\right)$.

(b) Want $I_A = 0.990 I$. Use this in the result for part (a).

$I = 0.990 I \left(1 + \dfrac{R_A}{r + R}\right)$.

$0.010 = 0.990 \left(\dfrac{R_A}{r + R}\right)$.

$R_A = (r + R)(0.010/0.990) = (0.45 \ \Omega + 3.80 \ \Omega)(0.010/0.990) = 0.0429 \ \Omega$.

(c) $I - I_A = \dfrac{\varepsilon}{r+R} - \dfrac{\varepsilon}{r+R+R_A}$.

$I - I_A = \varepsilon\left(\dfrac{r+R+R_A - r - R}{(r+R)(r+R+R_A)}\right) = \dfrac{\varepsilon R_A}{(r+R)(r+R+R_A)}$.

EVALUATE: The difference between I and I_A increases as R_A increases. If R_A is larger than the value calculated in part (b) then I_A differs from I by more than 1.0%.

25.69. **(a) IDENTIFY:** Since the resistivity is a function of the position along the length of the cylinder, we must integrate to find the resistance.

SET UP: The resistance of a cross-section of thickness dx is $dR = \rho dx/A$.

EXECUTE: Using the given function for the resistivity and integrating gives

$$R = \int \dfrac{\rho dx}{A} = \int_0^L \dfrac{(a+bx^2)dx}{\pi r^2} = \dfrac{aL + bL^3/3}{\pi r^2}.$$

Now get the constants a and b: $\rho(0) = a = 2.25 \times 10^{-8}\ \Omega \cdot m$ and $\rho(L) = a + bL^2$ gives $8.50 \times 10^{-8}\ \Omega \cdot m = 2.25 \times 10^{-8}\ \Omega \cdot m + b(1.50\ m)^2$ which gives $b = 2.78 \times 10^{-8}\ \Omega/m$. Now use the above result to find R.

$$R = \dfrac{(2.25 \times 10^{-8}\ \Omega \cdot m)(1.50\ m) + (2.78 \times 10^{-8}\ \Omega/m)(1.50\ m)^3/3}{\pi(0.0110\ m)^2} = 1.71 \times 10^{-4}\ \Omega = 171\ \mu\Omega.$$

(b) IDENTIFY: Use the definition of resistivity to find the electric field at the midpoint of the cylinder, where $x = L/2$.

SET UP: $E = \rho J$. Evaluate the resistivity, using the given formula, for $x = L/2$.

EXECUTE: At the midpoint, $x = L/2$, giving $E = \dfrac{\rho I}{\pi r^2} = \dfrac{[a + b(L/2)^2]I}{\pi r^2}$.

$$E = \dfrac{[2.25 \times 10^{-8}\ \Omega \cdot m + (2.78 \times 10^{-8}\ \Omega/m)(0.750\ m)^2](1.75\ A)}{\pi(0.0110\ m)^2} = 1.76 \times 10^{-4}\ V/m = 176\ \mu V/m$$

(c) IDENTIFY: For the first segment, the result is the same as in part (a) except that the upper limit of the integral is $L/2$ instead of L.

SET UP: Integrating using the upper limit of $L/2$ gives $R_1 = \dfrac{a(L/2) + (b/3)(L^3/8)}{\pi r^2}$.

EXECUTE: Substituting the numbers gives

$$R_1 = \dfrac{(2.25 \times 10^{-8}\ \Omega \cdot m)(0.750\ m) + (2.78 \times 10^{-8}\ \Omega/m)/3((1.50\ m)^3/8)}{\pi(0.0110\ m)^2} = 5.47 \times 10^{-5}\ \Omega = 54.7\ \mu\Omega.$$

The resistance R_2 of the second half is equal to the total resistance minus the resistance of the first half.

$R_2 = R - R_1 = 1.71 \times 10^{-4}\ \Omega - 5.47 \times 10^{-5}\ \Omega = 1.16 \times 10^{-4}\ \Omega = 116\ \mu\Omega$.

EVALUATE: The second half has a greater resistance than the first half because the resistance increases with distance along the cylinder.

25.73. **IDENTIFY:** No current flows through the capacitor when it is fully charged.

SET UP: With the capacitor fully charged, $I = \dfrac{\varepsilon}{R_1 + R_2}$. $V_R = IR$ and $V_C = Q/C$.

EXECUTE: $V_C = \dfrac{Q}{C} = \dfrac{36.0\ \mu C}{9.00\ \mu F} = 4.00\ V$. $V_{R_1} = V_C = 4.00\ V$ and $I = \dfrac{V_{R_1}}{R_1} = \dfrac{4.00\ V}{6.00\ \Omega} = 0.667\ A$.

$V_{R_2} = IR_2 = (0.667\ A)(4.00\ \Omega) = 2.668\ V$. $\varepsilon = V_{R_1} + V_{R_2} = 4.00\ V + 2.668\ V = 6.67\ V$.

EVALUATE: When a capacitor is fully charged, it acts like an open circuit and prevents any current from flowing though it.

25.75. **IDENTIFY:** According to Ohm's law, $R = \dfrac{V_{ab}}{I}$ = constant, and a graph of V_{ab} versus I will be a straight line with positive slope passing through the origin.

SET UP and EXECUTE: (a) Figure 25.75a shows the graphs of V_{ab} versus I and R versus I for resistor A. Figure 25.75b shows these graphs for resistor B.

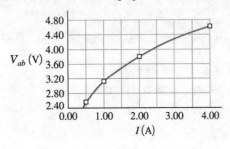
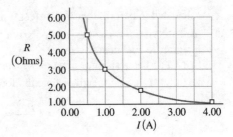

Figure 25.75a

(b) In Figure 25.75a, the graph of V_{ab} versus I is not a straight line so resistor A does not obey Ohm's law. In the graph of R versus I, R is not constant; it decreases as I increases.

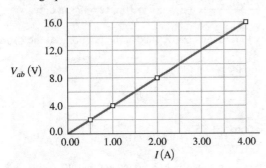
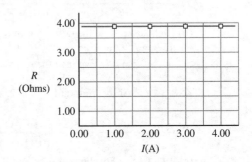

Figure 25.75b

(c) In Figure 25.75b, the graph of V_{ab} versus I is a straight line with positive slope passing through the origin, so resistor B obeys Ohm's law. The graph of R versus I is a horizontal line. This means that R is constant, which is consistent with Ohm's law.

(d) We use $P = IV$. From the graph of V_{ab} versus I in Figure 25.75a, we read that $I = 2.35$ A when $V = 4.00$ V. Therefore $P = IV = (2.35\ \text{A})(4.00\ \text{V}) = 9.40$ W.

(e) We use $P = V^2/R$. From the graph of R versus I in Figure 25.75b, we find that $R = 3.88\ \Omega$. Thus $P = V^2/R = (4.00\ \text{V})^2/(3.88\ \Omega) = 4.12$ W.

EVALUATE: Since resistor B obeys Ohm's law $V_{ab} = RI$, R is the slope of the graph of V_{ab} versus I in Figure 25.75b. The given data points lie on the line, so we use them to calculate the slope.

slope = $R = \dfrac{15.52\ \text{V} - 1.94\ \text{V}}{4.00\ \text{A} - 0.50\ \text{A}} = 3.88\ \Omega$. This value is the same as the one we got from the graph of R versus I in Figure 25.75b, so our results agree.

DIRECT-CURRENT CIRCUITS

26.1. **IDENTIFY:** The newly-formed wire is a combination of series and parallel resistors.
SET UP: Each of the three linear segments has resistance $R/3$. The circle is two $R/6$ resistors in parallel.
EXECUTE: The resistance of the circle is $R/12$ since it consists of two $R/6$ resistors in parallel. The equivalent resistance is two $R/3$ resistors in series with an $R/12$ resistor, giving
$R_{equiv} = R/3 + R/3 + R/12 = 3R/4$.
EVALUATE: The equivalent resistance of the original wire has been reduced because the circle's resistance is less than it was as a linear wire.

26.7. **IDENTIFY:** First do as much series-parallel reduction as possible.
SET UP: The 45.0-Ω and 15.0-Ω resistors are in parallel, so first reduce them to a single equivalent resistance. Then find the equivalent series resistance of the circuit.
EXECUTE: $1/R_p = 1/(45.0\ \Omega) + 1/(15.0\ \Omega)$ and $R_p = 11.25\ \Omega$. The total equivalent resistance is $18.0\ \Omega + 11.25\ \Omega + 3.26\ \Omega = 32.5\ \Omega$. Ohm's law gives $I = (25.0\ \text{V})/(32.5\ \Omega) = 0.769$ A.
EVALUATE: The circuit appears complicated until we realize that the 45.0-Ω and 15.0-Ω resistors are in parallel.

26.13. **IDENTIFY:** For resistors in parallel, the voltages are the same and the currents add. $\frac{1}{R_{eq}} = \frac{1}{R_1} + \frac{1}{R_2}$ so $R_{eq} = \frac{R_1 R_2}{R_1 + R_2}$. For resistors in series, the currents are the same and the voltages add. $R_{eq} = R_1 + R_2$.
SET UP: The rules for combining resistors in series and parallel lead to the sequences of equivalent circuits shown in Figure 26.13.
EXECUTE: $R_{eq} = 5.00\ \Omega$. In Figure 26.13c, $I = \frac{60.0\ \text{V}}{5.00\ \Omega} = 12.0$ A. This is the current through each of the resistors in Figure 26.13b. $V_{12} = IR_{12} = (12.0\ \text{A})(2.00\ \Omega) = 24.0$ V.
$V_{34} = IR_{34} = (12.0\ \text{A})(3.00\ \Omega) = 36.0$ V. Note that $V_{12} + V_{34} = 60.0$ V. V_{12} is the voltage across R_1 and across R_2, so $I_1 = \frac{V_{12}}{R_1} = \frac{24.0\ \text{V}}{3.00\ \Omega} = 8.00$ A and $I_2 = \frac{V_{12}}{R_2} = \frac{24.0\ \text{V}}{6.00\ \Omega} = 4.00$ A. V_{34} is the voltage across R_3 and across R_4, so $I_3 = \frac{V_{34}}{R_3} = \frac{36.0\ \text{V}}{12.0\ \Omega} = 3.00$ A and $I_4 = \frac{V_{34}}{R_4} = \frac{36.0\ \text{V}}{4.00\ \Omega} = 9.00$ A.
EVALUATE: Note that $I_1 + I_2 = I_3 + I_4$.

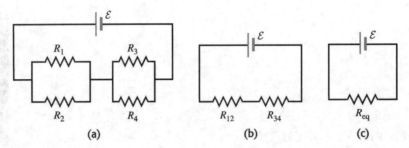

Figure 26.13

26.15. IDENTIFY: In both circuits, with and without R_4, replace series and parallel combinations of resistors by their equivalents. Calculate the currents and voltages in the equivalent circuit and infer from this the currents and voltages in the original circuit. Use $P = I^2 R$ to calculate the power dissipated in each bulb.
(a) SET UP: The circuit is sketched in Figure 26.15a.

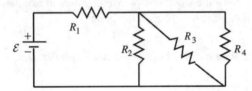

EXECUTE: R_2, R_3, and R_4 are in parallel, so their equivalent resistance R_{eq} is given by
$$\frac{1}{R_{eq}} = \frac{1}{R_2} + \frac{1}{R_3} + \frac{1}{R_4}.$$

Figure 26.15a

$\dfrac{1}{R_{eq}} = \dfrac{3}{4.50\,\Omega}$ and $R_{eq} = 1.50\,\Omega$.

The equivalent circuit is drawn in Figure 26.15b.

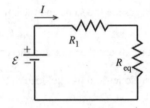

$\varepsilon - I(R_1 + R_{eq}) = 0.$

$I = \dfrac{\varepsilon}{R_1 + R_{eq}}.$

Figure 26.15b

$I = \dfrac{9.00\text{ V}}{4.50\,\Omega + 1.50\,\Omega} = 1.50\text{ A}$ and $I_1 = 1.50\text{ A}$.

Then $V_1 = I_1 R_1 = (1.50\text{ A})(4.50\,\Omega) = 6.75\text{ V}$.

$I_{eq} = 1.50\text{ A}$, $V_{eq} = I_{eq} R_{eq} = (1.50\text{ A})(1.50\,\Omega) = 2.25\text{ V}$.

For resistors in parallel the voltages are equal and are the same as the voltage across the equivalent resistor, so $V_2 = V_3 = V_4 = 2.25\text{ V}$.

$I_2 = \dfrac{V_2}{R_2} = \dfrac{2.25\text{ V}}{4.50\,\Omega} = 0.500\text{ A}$, $I_3 = \dfrac{V_3}{R_3} = 0.500\text{ A}$, $I_4 = \dfrac{V_4}{R_4} = 0.500\text{ A}$.

EVALUATE: Note that $I_2 + I_3 + I_4 = 1.50\text{ A}$, which is I_{eq}. For resistors in parallel the currents add and their sum is the current through the equivalent resistor.

(b) SET UP: $P = I^2 R$.

EXECUTE: $P_1 = (1.50\text{ A})^2 (4.50\,\Omega) = 10.1\text{ W}$.

$P_2 = P_3 = P_4 = (0.500\text{ A})^2 (4.50\,\Omega) = 1.125\text{ W}$, which rounds to 1.12 W. R_1 glows brightest.

Direct-Current Circuits 26-3

EVALUATE: Note that $P_2 + P_3 + P_4 = 3.37$ W. This equals $P_{eq} = I_{eq}^2 R_{eq} = (1.50 \text{ A})^2 (1.50 \, \Omega) = 3.37$ W, the power dissipated in the equivalent resistor.

(c) SET UP: With R_4 removed the circuit becomes the circuit in Figure 26.15c.

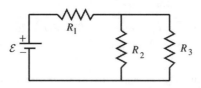

EXECUTE: R_2 and R_3 are in parallel and their equivalent resistance R_{eq} is given by

$$\frac{1}{R_{eq}} = \frac{1}{R_2} + \frac{1}{R_3} = \frac{2}{4.50 \, \Omega} \text{ and } R_{eq} = 2.25 \, \Omega.$$

Figure 26.15c

The equivalent circuit is shown in Figure 26.15d.

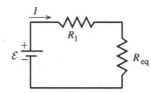

$\varepsilon - I(R_1 + R_{eq}) = 0.$

$I = \dfrac{\varepsilon}{R_1 + R_{eq}}.$

$I = \dfrac{9.00 \text{ V}}{4.50 \, \Omega + 2.25 \, \Omega} = 1.333 \text{ A}.$

Figure 26.15d

$I_1 = 1.33$ A, $V_1 = I_1 R_1 = (1.333 \text{ A})(4.50 \, \Omega) = 6.00$ V.

$I_{eq} = 1.33$ A, $V_{eq} = I_{eq} R_{eq} = (1.333 \text{ A})(2.25 \, \Omega) = 3.00$ V and $V_2 = V_3 = 3.00$ V.

$I_2 = \dfrac{V_2}{R_2} = \dfrac{3.00 \text{ V}}{4.50 \, \Omega} = 0.667$ A, $I_3 = \dfrac{V_3}{R_3} = 0.667$ A.

(d) SET UP: $P = I^2 R$.

EXECUTE: $P_1 = (1.333 \text{ A})^2 (4.50 \, \Omega) = 8.00$ W.

$P_2 = P_3 = (0.667 \text{ A})^2 (4.50 \, \Omega) = 2.00$ W.

EVALUATE: **(e)** When R_4 is removed, P_1 decreases and P_2 and P_3 increase. Bulb R_1 glows less brightly and bulbs R_2 and R_3 glow more brightly. When R_4 is removed the equivalent resistance of the circuit increases and the current through R_1 decreases. But in the parallel combination this current divides into two equal currents rather than three, so the currents through R_2 and R_3 increase. Can also see this by noting that with R_4 removed and less current through R_1 the voltage drop across R_1 is less so the voltage drop across R_2 and across R_3 must become larger.

26.19. IDENTIFY and SET UP: Replace series and parallel combinations of resistors by their equivalents until the circuit is reduced to a single loop. Use the loop equation to find the current through the 20.0-Ω resistor. Set $P = I^2 R$ for the 20.0-Ω resistor equal to the rate Q/t at which heat goes into the water and set $Q = mc\Delta T$.

EXECUTE: Replace the network by the equivalent resistor, as shown in Figure 26.19.

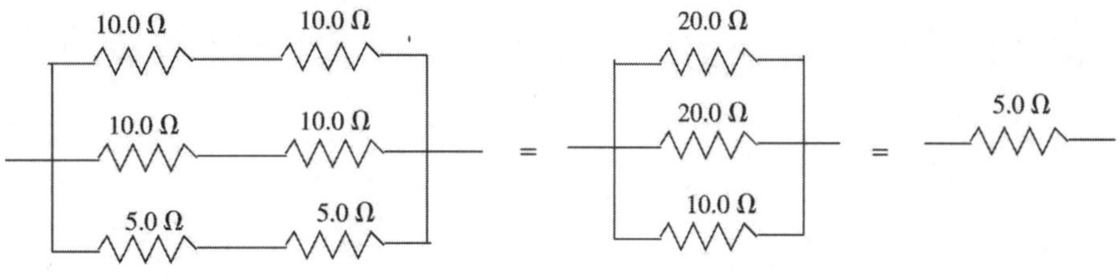

Figure 26.19

$30.0 \text{ V} - I(20.0 \text{ }\Omega + 5.0 \text{ }\Omega + 5.0 \text{ }\Omega) = 0; I = 1.00 \text{ A}.$

For the 20.0-Ω resistor thermal energy is generated at the rate $P = I^2 R = 20.0$ W. $Q = Pt$ and $Q = mc\Delta T$ gives $t = \dfrac{mc\Delta T}{P} = \dfrac{(0.100 \text{ kg})(4190 \text{ J/kg} \cdot \text{K})(48.0 \text{ C}°)}{20.0 \text{ W}} = 1.01 \times 10^3$ s.

EVALUATE: The battery is supplying heat at the rate $P = \varepsilon I = 30.0$ W. In the series circuit, more energy is dissipated in the larger resistor (20.0 Ω) than in the smaller ones (5.00 Ω).

26.23. IDENTIFY: Apply Kirchhoff's rules.

SET UP: Figure 26.23 shows the loops taken. When we go around loop (1) in the direction shown there is a potential rise across the 200.0 V battery, so there must be a drop across R and the current in R must be in the direction shown in the figure. Similar analysis of loops (2) and (3) tell us that currents I_2 and I_5 must be in the directions shown. The junction rule has been used to label the currents in all the other branches of the circuit.

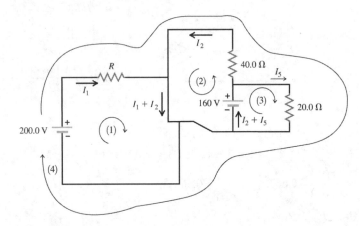

Figure 26.23

EXECUTE: (a) Apply the Kirchhoff loop rule to loop (1): $+200.0 \text{ V} - I_1 R = 0$. Solving for R gives

$R = \dfrac{+200.0 \text{ V}}{I_1} = \dfrac{+200.0 \text{ V}}{10.0 \text{ A}} = 20.0 \text{ }\Omega.$

(b) Loop (2): $+160.0 \text{ V} - I_2 (40.0 \text{ }\Omega) = 0$. $I_2 = \dfrac{160.0 \text{ V}}{40.0 \text{ }\Omega} = 4.00 \text{ A}.$

Loop (3): $+160.0 \text{ V} - I_5 (20.0 \text{ }\Omega) = 0$. $I_5 = \dfrac{160.0 \text{ V}}{20.0 \text{ }\Omega} = 8.00 \text{ A}.$

A_2 reads $I_2 = 4.00$ A. A_3 reads $I_2 + I_5 = 12.0$ A. A_4 reads $I_1 + I_2 = 14.0$ A. A_5 reads $I_5 = 8.00$ A.

EVALUATE: The sum of potential changes around the outer loop (4) is
$+200.0 \text{ V} - I_1 R + I_2 (40.0 \text{ }\Omega) - I_5 (20.0 \text{ }\Omega) = 200.0 \text{ V} - (10.0 \text{ A})(20.0 \text{ }\Omega) + (4.00 \text{ A})(40.0 \text{ }\Omega) -$
$(8.00 \text{ A})(20.0 \text{ }\Omega) = 200.0 \text{ V} - 200.0 \text{ V} - 160.0 \text{ V} = 0.$

The loop rule is satisfied for loop (4) and this is a good check of our calculations.

26.25. **IDENTIFY:** Apply Kirchhoff's junction rule at point a to find the current through R. Apply Kirchhoff's loop rule to loops (1) and (2) shown in Figure 26.25a to calculate R and ε. Travel around each loop in the direction shown.
SET UP:

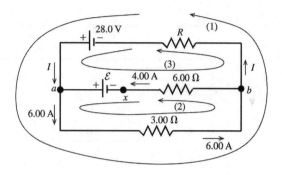

Figure 26.25a

EXECUTE: (a) Apply Kirchhoff's junction rule to point a: $\sum I = 0$ so $I + 4.00\text{ A} - 6.00\text{ A} = 0$
$I = 2.00\text{ A}$ (in the direction shown in the diagram).

(b) Apply Kirchhoff's loop rule to loop (1): $-(6.00\text{ A})(3.00\,\Omega) - (2.00\text{ A})R + 28.0\text{ V} = 0$
$-18.0\text{ V} - (2.00\,\Omega)R + 28.0\text{ V} = 0$.

$R = \dfrac{28.0\text{ V} - 18.0\text{ V}}{2.00\text{ A}} = 5.00\,\Omega.$

(c) Apply Kirchhoff's loop rule to loop (2): $-(6.00\text{ A})(3.00\,\Omega) - (4.00\text{ A})(6.00\,\Omega) + \varepsilon = 0.$
$\varepsilon = 18.0\text{ V} + 24.0\text{ V} = 42.0\text{ V}.$

EVALUATE: We can check that the loop rule is satisfied for loop (3), as a check of our work:
$28.0\text{ V} - \varepsilon + (4.00\text{ A})(6.00\,\Omega) - (2.00\text{ A})R = 0.$
$28.0\text{ V} - 42.0\text{ V} + 24.0\text{ V} - (2.00\text{ A})(5.00\,\Omega) = 0.$
$52.0\text{ V} = 42.0\text{ V} + 10.0\text{ V}.$
$52.0\text{ V} = 52.0\text{ V}$, so the loop rule is satisfied for this loop.

(d) **IDENTIFY:** If the circuit is broken at point x there can be no current in the 6.00-Ω resistor. There is now only a single current path and we can apply the loop rule to this path.
SET UP: The circuit is sketched in Figure 26.25b.

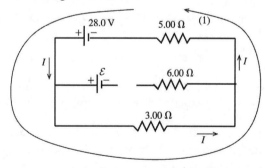

Figure 26.25b

EXECUTE: $+28.0\text{ V} - (3.00\,\Omega)I - (5.00\,\Omega)I = 0.$

$I = \dfrac{28.0\text{ V}}{8.00\,\Omega} = 3.50\text{ A}.$

EVALUATE: Breaking the circuit at x removes the 42.0-V emf from the circuit and the current through the 3.00-Ω resistor is reduced.

26.27. **IDENTIFY:** Apply Kirchhoff's junction rule at points *a*, *b*, *c*, and *d* to calculate the unknown currents. Then apply the loop rule to three loops to calculate $\varepsilon_1, \varepsilon_2$, and R.

SET UP: The circuit is sketched in Figure 26.27.

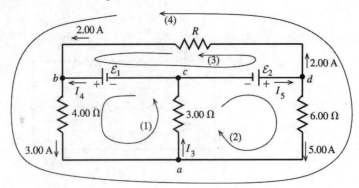

Figure 26.27

(a) EXECUTE: Apply the junction rule to point *a*: $3.00 \text{ A} + 5.00 \text{ A} - I_3 = 0$.

$I_3 = 8.00$ A.

Apply the junction rule to point *b*: $2.00 \text{ A} + I_4 - 3.00 \text{ A} = 0$.

$I_4 = 1.00$ A.

Apply the junction rule to point *c*: $I_3 - I_4 - I_5 = 0$.

$I_5 = I_3 - I_4 = 8.00 \text{ A} - 1.00 \text{ A} = 7.00$ A.

EVALUATE: As a check, apply the junction rule to point *d*: $I_5 - 2.00 \text{ A} - 5.00 \text{ A} = 0$.

$I_5 = 7.00$ A.

(b) EXECUTE: Apply the loop rule to loop (1): $\varepsilon_1 - (3.00 \text{ A})(4.00 \text{ }\Omega) - I_3(3.00 \text{ }\Omega) = 0$.

$\varepsilon_1 = 12.0 \text{ V} + (8.00 \text{ A})(3.00 \text{ }\Omega) = 36.0$ V.

Apply the loop rule to loop (2): $\varepsilon_2 - (5.00 \text{ A})(6.00 \text{ }\Omega) - I_3(3.00 \text{ }\Omega) = 0$.

$\varepsilon_2 = 30.0 \text{ V} + (8.00 \text{ A})(3.00 \text{ }\Omega) = 54.0$ V.

(c) EXECUTE: Apply the loop rule to loop (3): $-(2.00 \text{ A})R - \varepsilon_1 + \varepsilon_2 = 0$.

$R = \dfrac{\varepsilon_2 - \varepsilon_1}{2.00 \text{ A}} = \dfrac{54.0 \text{ V} - 36.0 \text{ V}}{2.00 \text{ A}} = 9.00 \text{ }\Omega$.

EVALUATE: Apply the loop rule to loop (4) as a check of our calculations:
$-(2.00 \text{ A})R - (3.00 \text{ A})(4.00 \text{ }\Omega) + (5.00 \text{ A})(6.00 \text{ }\Omega) = 0$.
$-(2.00 \text{ A})(9.00 \text{ }\Omega) - 12.0 \text{ V} + 30.0 \text{ V} = 0$.
$-18.0 \text{ V} + 18.0 \text{ V} = 0$.

26.31. (a) IDENTIFY: With the switch open, the circuit can be solved using series-parallel reduction.

SET UP: Find the current through the unknown battery using Ohm's law. Then use the equivalent resistance of the circuit to find the emf of the battery.

EXECUTE: The 30.0-Ω and 50.0-Ω resistors are in series, and hence have the same current. Using Ohm's law $I_{50} = (15.0 \text{ V})/(50.0 \text{ }\Omega) = 0.300 \text{ A} = I_{30}$. The potential drop across the 75.0-Ω resistor is the same as the potential drop across the 80.0-Ω series combination. We can use this fact to find the current through the 75.0-Ω resistor using Ohm's law: $V_{75} = V_{80} = (0.300 \text{ A})(80.0 \text{ }\Omega) = 24.0$ V and

$I_{75} = (24.0 \text{ V})/(75.0 \text{ }\Omega) = 0.320$ A.

The current through the unknown battery is the sum of the two currents we just found:

$$I_{\text{Total}} = 0.300 \text{ A} + 0.320 \text{ A} = 0.620 \text{ A}.$$

The equivalent resistance of the resistors in parallel is $1/R_p = 1/(75.0\ \Omega) + 1/(80.0\ \Omega)$. This gives $R_p = 38.7\ \Omega$. The equivalent resistance "seen" by the battery is $R_{equiv} = 20.0\ \Omega + 38.7\ \Omega = 58.7\ \Omega$. Applying Ohm's law to the battery gives $\varepsilon = R_{equiv} I_{Total} = (58.7\ \Omega)(0.620\ A) = 36.4\ V$.

(b) IDENTIFY: With the switch closed, the 25.0-V battery is connected across the 50.0-Ω resistor.
SET UP: Take a loop around the right part of the circuit.
EXECUTE: Ohm's law gives $I = (25.0\ V)/(50.0\ \Omega) = 0.500\ A$.

EVALUATE: The current through the 50.0-Ω resistor, and the rest of the circuit, depends on whether or not the switch is open.

26.35. IDENTIFY: To construct an ammeter, add a shunt resistor in parallel with the galvanometer coil. To construct a voltmeter, add a resistor in series with the galvanometer coil.
SET UP: The full-scale deflection current is 500 μA and the coil resistance is 25.0 Ω.
EXECUTE: (a) For a 20-mA ammeter, the two resistances are in parallel and the voltages across each are the same. $V_c = V_s$ gives $I_c R_c = I_s R_s$. $(500 \times 10^{-6}\ A)(25.0\ \Omega) = (20 \times 10^{-3}\ A - 500 \times 10^{-6}\ A) R_s$ and $R_s = 0.641\ \Omega$.

(b) For a 500-mV voltmeter, the resistances are in series and the current is the same through each:

$$V_{ab} = I(R_c + R_s) \text{ and } R_s = \frac{V_{ab}}{I} - R_c = \frac{500 \times 10^{-3}\ V}{500 \times 10^{-6}\ A} - 25.0\ \Omega = 975\ \Omega.$$

EVALUATE: The equivalent resistance of the voltmeter is $R_{eq} = R_s + R_c = 1000\ \Omega$. The equivalent resistance of the ammeter is given by $\frac{1}{R_{eq}} = \frac{1}{R_{sh}} + \frac{1}{R_c}$ and $R_{eq} = 0.625\ \Omega$. The voltmeter is a high-resistance device and the ammeter is a low-resistance device.

26.37. IDENTIFY: The meter introduces resistance into the circuit, which affects the current through the 5.00-kΩ resistor and hence the potential drop across it.
SET UP: Use Ohm's law to find the current through the 5.00-kΩ resistor and then the potential drop across it.
EXECUTE: (a) The parallel resistance with the voltmeter is 3.33 kΩ, so the total equivalent resistance across the battery is 9.33 kΩ, giving $I = (50.0\ V)/(9.33\ k\Omega) = 5.36\ mA$. Ohm's law gives the potential drop across the 5.00-kΩ resistor: $V_{5\ k\Omega} = (3.33\ k\Omega)(5.36\ mA) = 17.9\ V$.

(b) The current in the circuit is now $I = (50.0\ V)/(11.0\ k\Omega) = 4.55\ mA$. $V_{5\ k\Omega} = (5.00\ k\Omega)(4.55\ mA) = 22.7\ V$.

(c) % error = $(22.7\ V - 17.9\ V)/(22.7\ V) = 0.214 = 21.4$%. (We carried extra decimal places for accuracy since we had to subtract our answers.)

EVALUATE: The presence of the meter made a very large percent error in the reading of the "true" potential across the resistor.

26.41. IDENTIFY: An uncharged capacitor is placed into a circuit. Apply the loop rule at each time.
SET UP: The voltage across a capacitor is $V_C = q/C$.

EXECUTE: (a) At the instant the circuit is completed, there is no voltage across the capacitor, since it has no charge stored.

(b) Since $V_C = 0$, the full battery voltage appears across the resistor $V_R = \varepsilon = 245\ V$.

(c) There is no charge on the capacitor.

(d) The current through the resistor is $i = \dfrac{\varepsilon}{R_{total}} = \dfrac{245\ V}{7500\ \Omega} = 0.0327\ A = 32.7\ mA$.

(e) After a long time has passed the full battery voltage is across the capacitor and $i = 0$. The voltage across the capacitor balances the emf: $V_C = 245\ V$. The voltage across the resistor is zero. The capacitor's charge is $q = CV_C = (4.60 \times 10^{-6}\ F)(245\ V) = 1.13 \times 10^{-3}\ C$. The current in the circuit is zero.

EVALUATE: The current in the circuit starts at 0.0327 A and decays to zero. The charge on the capacitor starts at zero and rises to $q = 1.13 \times 10^{-3}\ C$.

26-8 Chapter 26

26.43. IDENTIFY: The capacitors, which are in parallel, will discharge exponentially through the resistors.
SET UP: Since V is proportional to Q, V must obey the same exponential equation as Q, $V = V_0 e^{-t/RC}$. The current is $I = (V_0/R) e^{-t/RC}$.
EXECUTE: (a) Solve for time when the potential across each capacitor is 10.0 V:
$$t = -RC \ln(V/V_0) = -(80.0\ \Omega)(35.0\ \mu F) \ln(10/45) = 4210\ \mu s = 4.21\ ms.$$
(b) $I = (V_0/R) e^{-t/RC}$. Using the above values, with $V_0 = 45.0$ V, gives $I = 0.125$ A.
EVALUATE: Since the current and the potential both obey the same exponential equation, they are both reduced by the same factor (0.222) in 4.21 ms.

26.45. IDENTIFY and SET UP: Apply Kirchhoff's loop rule. The voltage across the resistor depends on the current through it and the voltage across the capacitor depends on the charge on its plates.
EXECUTE: $\varepsilon - V_R - V_C = 0$.
$\varepsilon = 120$ V, $V_R = IR = (0.900\ \text{A})(80.0\ \Omega) = 72$ V, so $V_C = 48$ V.
$Q = CV = (4.00 \times 10^{-6}\ \text{F})(48\ \text{V}) = 192\ \mu C$.
EVALUATE: The initial charge is zero and the final charge is $C\varepsilon = 480\ \mu C$. Since current is flowing at the instant considered in the problem the capacitor is still being charged and its charge has not reached its final value.

26.47. IDENTIFY: The stored energy is proportional to the square of the charge on the capacitor, so it will obey an exponential equation, but not the same equation as the charge.
SET UP: The energy stored in the capacitor is $U = Q^2/2C$ and the charge on the plates is $Q_0 e^{-t/RC}$. The current is $I = I_0 e^{-t/RC}$.
EXECUTE: $U = Q^2/2C = (Q_0 e^{-t/RC})^2/2C = U_0 e^{-2t/RC}$. When the capacitor has lost 80% of its stored energy, the energy is 20% of the initial energy, which is $U_0/5$. $U_0/5 = U_0\ e^{-2t/RC}$ gives
$t = (RC/2) \ln 5 = (25.0\ \Omega)(4.62\ \text{pF})(\ln 5)/2 = 92.9$ ps.
At this time, the current is $I = I_0\ e^{-t/RC} = (Q_0/RC)\ e^{-t/RC}$, so
$$I = (3.5\ \text{nC})/[(25.0\ \Omega)(4.62\ \text{pF})]\ e^{-(92.9\ \text{ps})/[(25.0\ \Omega)(4.62\ \text{pF})]} = 13.6\ \text{A}.$$
EVALUATE: When the energy is reduced by 80%, neither the current nor the charge are reduced by that percent.

26.49. IDENTIFY: In both cases, simplify the complicated circuit by eliminating the appropriate circuit elements. The potential across an uncharged capacitor is initially zero, so it behaves like a short circuit. A fully charged capacitor allows no current to flow through it.
(a) SET UP: Just after closing the switch, the uncharged capacitors all behave like short circuits, so any resistors in parallel with them are eliminated from the circuit.
EXECUTE: The equivalent circuit consists of 50 Ω and 25 Ω in parallel, with this combination in series with 75 Ω, 15 Ω, and the 100-V battery. The equivalent resistance is $90\ \Omega + 16.7\ \Omega = 106.7\ \Omega$, which gives $I = (100\ \text{V})/(106.7\ \Omega) = 0.937$ A.
(b) SET UP: Long after closing the switch, the capacitors are essentially charged up and behave like open circuits since no charge can flow through them. They effectively eliminate any resistors in series with them since no current can flow through these resistors.
EXECUTE: The equivalent circuit consists of resistances of 75 Ω, 15 Ω, and three 25-Ω resistors, all in series with the 100-V battery, for a total resistance of 165 Ω. Therefore $I = (100\ \text{V})/(165\ \Omega) = 0.606$ A.
EVALUATE: The initial and final behavior of the circuit can be calculated quite easily using simple series-parallel circuit analysis. Intermediate times would require much more difficult calculations!

26.53. IDENTIFY and SET UP: The heater and hair dryer are in parallel so the voltage across each is 120 V and the current through the fuse is the sum of the currents through each appliance. As the power consumed by the dryer increases, the current through it increases. The maximum power setting is the highest one for which the current through the fuse is less than 20 A.

EXECUTE: Find the current through the heater. $P = VI$ so $I = P/V = (1500 \text{ W})/(120 \text{ V}) = 12.5$ A. The maximum total current allowed is 20 A, so the current through the dryer must be less than $20 \text{ A} - 12.5 \text{ A} = 7.5$ A. The power dissipated by the dryer if the current has this value is $P = VI = (120 \text{ V})(7.5 \text{ A}) = 900$ W. For P at this value or larger the circuit breaker trips.

EVALUATE: $P = V^2/R$ and for the dryer V is a constant 120 V. The higher power settings correspond to a smaller resistance R and larger current through the device.

26.57. (a) IDENTIFY: Break the circuit between points a and b means no current in the middle branch that contains the 3.00-Ω resistor and the 10.0-V battery. The circuit therefore has a single current path. Find the current, so that potential drops across the resistors can be calculated. Calculate V_{ab} by traveling from a to b, keeping track of the potential changes along the path taken.

SET UP: The circuit is sketched in Figure 26.57a.

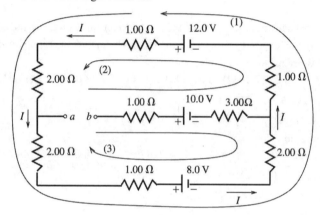

Figure 26.57a

EXECUTE: Apply Kirchhoff's loop rule to loop (1).
$+12.0 \text{ V} - I(1.00 \, \Omega + 2.00 \, \Omega + 2.00 \, \Omega + 1.00 \, \Omega) - 8.0 \text{ V} - I(2.00 \, \Omega + 1.00 \, \Omega) = 0.$

$$I = \frac{12.0 \text{ V} - 8.0 \text{ V}}{9.00 \, \Omega} = 0.4444 \text{ A}.$$

To find V_{ab} start at point b and travel to a, adding up the potential rises and drops. Travel on path (2) shown on the diagram. The 1.00-Ω and 3.00-Ω resistors in the middle branch have no current through them and hence no voltage across them. Therefore,
$V_b - 10.0 \text{ V} + 12.0 \text{ V} - I(1.00 \, \Omega + 1.00 \, \Omega + 2.00 \, \Omega) = V_a$; thus
$V_a - V_b = 2.0 \text{ V} - (0.4444 \text{ A})(4.00 \, \Omega) = +0.22$ V (point a is at higher potential).

EVALUATE: As a check on this calculation we also compute V_{ab} by traveling from b to a on path (3).
$V_b - 10.0 \text{ V} + 8.0 \text{ V} + I(2.00 \, \Omega + 1.00 \, \Omega + 2.00 \, \Omega) = V_a.$
$V_{ab} = -2.00 \text{ V} + (0.4444 \text{ A})(5.00 \, \Omega) = +0.22$ V, which checks.

(b) IDENTIFY and SET UP: With points a and b connected by a wire there are three current branches, as shown in Figure 26.57b.

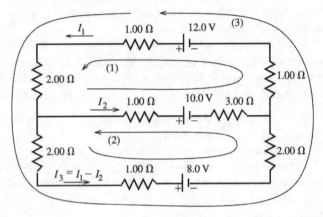

Figure 26.57b

The junction rule has been used to write the third current (in the 8.0-V battery) in terms of the other currents. Apply the loop rule to loops (1) and (2) to obtain two equations for the two unknowns I_1 and I_2.

EXECUTE: Apply the loop rule to loop (1).

$12.0 \text{ V} - I_1(1.00\,\Omega) - I_1(2.00\,\Omega) - I_2(1.00\,\Omega) - 10.0 \text{ V} - I_2(3.00\,\Omega) - I_1(1.00\,\Omega) = 0$

$2.0 \text{ V} - I_1(4.00\,\Omega) - I_2(4.00\,\Omega) = 0$

$(2.00\,\Omega)I_1 + (2.00\,\Omega)I_2 = 1.0 \text{ V}$ eq. (1)

Apply the loop rule to loop (2).

$-(I_1 - I_2)(2.00\,\Omega) - (I_1 - I_2)(1.00\,\Omega) - 8.0 \text{ V} - (I_1 - I_2)(2.00\,\Omega) + I_2(3.00\,\Omega) + 10.0 \text{ V} + I_2(1.00\,\Omega) = 0$

$2.0 \text{ V} - (5.00\,\Omega)I_1 + (9.00\,\Omega)I_2 = 0$ eq. (2)

Solve eq. (1) for I_2 and use this to replace I_2 in eq. (2).

$I_2 = 0.50 \text{ A} - I_1$

$2.0 \text{ V} - (5.00\,\Omega)I_1 + (9.00\,\Omega)(0.50 \text{ A} - I_1) = 0$

$(14.0\,\Omega)I_1 = 6.50 \text{ V}$ so $I_1 = (6.50 \text{ V})/(14.0\,\Omega) = 0.464 \text{ A}$

$I_2 = 0.500 \text{ A} - 0.464 \text{ A} = 0.036 \text{ A}$.

The current in the 12.0-V battery is $I_1 = 0.464 \text{ A}$

EVALUATE: We can apply the loop rule to loop (3) as a check.

$+12.0 \text{ V} - I_1(1.00\,\Omega + 2.00\,\Omega + 1.00\,\Omega) - (I_1 - I_2)(2.00\,\Omega + 1.00\,\Omega + 2.00\,\Omega) - 8.0 \text{ V} = 4.0 \text{ V} - 1.86 \text{ V} - 2.14 \text{ V} = 0$, as it should.

26.59. IDENTIFY: Apply Kirchhoff's junction rule to express the currents through the 5.00-Ω and 8.00-Ω resistors in terms of $I_1, I_2,$ and I_3. Apply the loop rule to three loops to get three equations in the three unknown currents.

SET UP: The circuit is sketched in Figure 26.59.

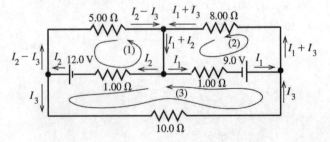

Figure 26.59

The current in each branch has been written in terms of $I_1, I_2,$ and I_3 such that the junction rule is satisfied at each junction point.

EXECUTE: Apply the loop rule to loop (1).
$-12.0 \text{ V} + I_2(1.00 \text{ }\Omega) + (I_2 - I_3)(5.00 \text{ }\Omega) = 0$
$I_2(6.00 \text{ }\Omega) - I_3(5.00 \text{ }\Omega) = 12.0 \text{ V}$ eq. (1)

Apply the loop rule to loop (2).
$-I_1(1.00 \text{ }\Omega) + 9.00 \text{ V} - (I_1 + I_3)(8.00 \text{ }\Omega) = 0$
$I_1(9.00 \text{ }\Omega) + I_3(8.00 \text{ }\Omega) = 9.00 \text{ V}$ eq. (2)

Apply the loop rule to loop (3).
$-I_3(10.0 \text{ }\Omega) - 9.00 \text{ V} + I_1(1.00 \text{ }\Omega) - I_2(1.00 \text{ }\Omega) + 12.0 \text{ V} = 0$
$-I_1(1.00 \text{ }\Omega) + I_2(1.00 \text{ }\Omega) + I_3(10.0 \text{ }\Omega) = 3.00 \text{ V}$ eq. (3)

Eq. (1) gives $I_2 = 2.00 \text{ A} + \frac{5}{6}I_3$; eq. (2) gives $I_1 = 1.00 \text{ A} - \frac{8}{9}I_3$.

Using these results in eq. (3) gives

$-(1.00 \text{ A} - \frac{8}{9}I_3)(1.00 \text{ }\Omega) + (2.00 \text{ A} + \frac{5}{6}I_3)(1.00 \text{ }\Omega) + I_3(10.0 \text{ }\Omega) = 3.00 \text{ V}.$

$\left(\frac{16+15+180}{18}\right)I_3 = 2.00 \text{ A}; \; I_3 = \frac{18}{211}(2.00 \text{ A}) = 0.171 \text{ A}.$

Then $I_2 = 2.00 \text{ A} + \frac{5}{6}I_3 = 2.00 \text{ A} + \frac{5}{6}(0.171 \text{ A}) = 2.14 \text{ A}$ and

$I_1 = 1.00 \text{ A} - \frac{8}{9}I_3 = 1.00 \text{ A} - \frac{8}{9}(0.171 \text{ A}) = 0.848 \text{ A}.$

EVALUATE: We could check that the loop rule is satisfied for a loop that goes through the 5.00-Ω, 8.00-Ω and 10.0-Ω resistors. Going around the loop clockwise:
$-(I_2 - I_3)(5.00 \text{ }\Omega) + (I_1 + I_3)(8.00 \text{ }\Omega) + I_3(10.0 \text{ }\Omega) = -9.85 \text{ V} + 8.15 \text{ V} + 1.71 \text{ V}$, which does equal zero, apart from rounding.

26.63. **IDENTIFY:** Simplify the resistor networks as much as possible using the rule for series and parallel combinations of resistors. Then apply Kirchhoff's laws.

SET UP: First do the series/parallel reduction. This gives the circuit in Figure 26.63. The rate at which the 10.0-Ω resistor generates thermal energy is $P = I^2R$.

EXECUTE: (a) Apply Kirchhoff's laws and solve for ε. $\Delta V_{adefa} = 0$: $-(20 \text{ }\Omega)(2 \text{ A}) - 5 \text{ V} - (20 \text{ }\Omega)I_2 = 0$.
This gives $I_2 = -2.25 \text{ A}$. Then $I_1 + I_2 = 2 \text{ A}$ gives $I_1 = 2 \text{ A} - (-2.25 \text{ A}) = 4.25 \text{ A}$.
$\Delta V_{abcdefa} = 0$: $(15 \text{ }\Omega)(4.25 \text{ A}) + \varepsilon - (20 \text{ }\Omega)(-2.25 \text{ A}) = 0$. This gives $\varepsilon = -109 \text{ V}$. Since ε is calculated to be negative, its polarity should be reversed.

(b) The parallel network that contains the 10.0-Ω resistor in one branch has an equivalent resistance of 10 Ω. The voltage across each branch of the parallel network is $V_{par} = RI = (10 \text{ }\Omega)(2 \text{ A}) = 20 \text{ V}$. The current in the upper branch is $I = \frac{V}{R} = \frac{20 \text{ V}}{30 \text{ }\Omega} = \frac{2}{3} \text{ A}$. $Pt = E$, so $I^2 Rt = E$, where $E = 60.0 \text{ J}$.

$\left(\frac{2}{3}\text{A}\right)^2 (10 \text{ }\Omega)t = 60 \text{ J}$, and $t = 13.5 \text{ s}$.

EVALUATE: For the 10.0-Ω resistor, $P = I^2R = 4.44 \text{ W}$. The total rate at which electrical energy is inputted to the circuit in the emf is $(5.0 \text{ V})(2.0 \text{ A}) + (109 \text{ V})(4.25 \text{ A}) = 473 \text{ J}$. Only a small fraction of the energy is dissipated in the 10.0-Ω resistor.

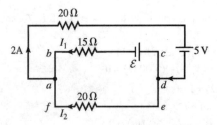

Figure 26.63

26.67. **(a) IDENTIFY** and **SET UP:** The circuit is sketched in Figure 26.67a.

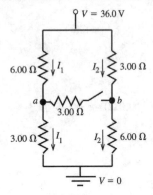

With the switch open there is no current through it and there are only the two currents I_1 and I_2 indicated in the sketch.

Figure 26.67a

The potential drop across each parallel branch is 36.0 V. Use this fact to calculate I_1 and I_2. Then travel from point a to point b and keep track of the potential rises and drops in order to calculate V_{ab}.

EXECUTE: $-I_1(6.00\ \Omega + 3.00\ \Omega) + 36.0\ V = 0.$

$$I_1 = \frac{36.0\ V}{6.00\ \Omega + 3.00\ \Omega} = 4.00\ A.$$

$-I_2(3.00\ \Omega + 6.00\ \Omega) + 36.0\ V = 0.$

$$I_2 = \frac{36.0\ V}{3.00\ \Omega + 6.00\ \Omega} = 4.00\ A.$$

To calculate $V_{ab} = V_a - V_b$ start at point b and travel to point a, adding up all the potential rises and drops along the way. We can do this by going from b up through the 3.00-Ω resistor:

$V_b + I_2(3.00\ \Omega) - I_1(6.00\ \Omega) = V_a.$

$V_a - V_b = (4.00\ A)(3.00\ \Omega) - (4.00\ A)(6.00\ \Omega) = 12.0\ V - 24.0\ V = -12.0\ V.$

$V_{ab} = -12.0\ V$ (point a is 12.0 V lower in potential than point b).

EVALUATE: Alternatively, we can go from point b down through the 6.00-Ω resistor.

$V_b - I_2(6.00\ \Omega) + I_1(3.00\ \Omega) = V_a.$

$V_a - V_b = -(4.00\ A)(6.00\ \Omega) + (4.00\ A)(3.00\ \Omega) = -24.0\ V + 12.0\ V = -12.0\ V,$ which checks.

(b) IDENTIFY: Now there are multiple current paths, as shown in Figure 26.67b. Use the junction rule to write the current in each branch in terms of three unknown currents $I_1, I_2,$ and I_3. Apply the loop rule to three loops to get three equations for the three unknowns. The target variable is I_3, the current through the switch. R_{eq} is calculated from $V = IR_{eq}$, where I is the total current that passes through the network.

SET UP:

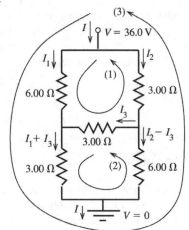

The three unknown currents $I_1, I_2,$ and I_3 are labeled on Figure 26.67b.

Figure 26.67b

EXECUTE: Apply the loop rule to loops (1), (2), and (3).
Loop (1): $-I_1(6.00\,\Omega) + I_3(3.00\,\Omega) + I_2(3.00\,\Omega) = 0$

$I_2 = 2I_1 - I_3$ eq. (1)

Loop (2): $-(I_1 + I_3)(3.00\,\Omega) + (I_2 - I_3)(6.00\,\Omega) - I_3(3.00\,\Omega) = 0$

$6I_2 - 12I_3 - 3I_1 = 0$ so $2I_2 - 4I_3 - I_1 = 0$

Use eq (1) to replace I_2:

$4I_1 - 2I_3 - 4I_3 - I_1 = 0$

$3I_1 = 6I_3$ and $I_1 = 2I_3$ eq. (2)

Loop (3): This loop is completed through the battery (not shown), in the direction from the $-$ to the $+$ terminal.

$-I_1(6.00\,\Omega) - (I_1 + I_3)(3.00\,\Omega) + 36.0\,\text{V} = 0$

$9I_1 + 3I_3 = 36.0\,\text{A}$ and $3I_1 + I_3 = 12.0\,\text{A}$ eq. (3)

Use eq. (2) in eq. (3) to replace I_1:

$3(2I_3) + I_3 = 12.0\,\text{A}$

$I_3 = 12.0\,\text{A}/7 = 1.71\,\text{A}$

$I_1 = 2I_3 = 3.42\,\text{A}$

$I_2 = 2I_1 - I_3 = 2(3.42\,\text{A}) - 1.71\,\text{A} = 5.13\,\text{A}$

The current through the switch is $I_3 = 1.71\,\text{A}$.

(c) SET UP and EXECUTE: From the results in part (a) the current through the battery is $I = I_1 + I_2 = 3.42\,\text{A} + 5.13\,\text{A} = 8.55\,\text{A}$. The equivalent circuit is a single resistor that produces the same current through the 36.0-V battery, as shown in Figure 26.67c.

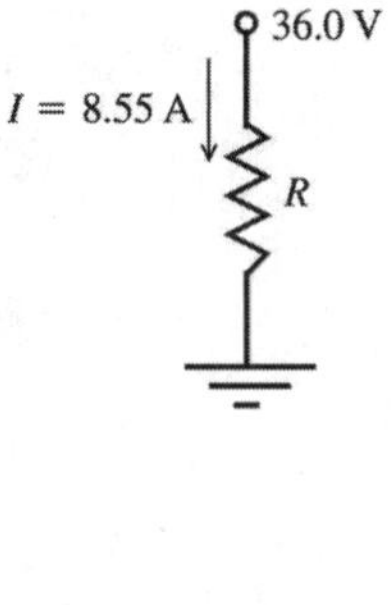

$-IR + 36.0\,\text{V} = 0.$

$R = \dfrac{36.0\,\text{V}}{I} = \dfrac{36.0\,\text{V}}{8.55\,\text{A}} = 4.21\,\Omega.$

Figure 26.67c

EVALUATE: With the switch open (part a), point b is at higher potential than point a, so when the switch is closed the current flows in the direction from b to a. With the switch closed the circuit cannot be simplified using series and parallel combinations but there is still an equivalent resistance that represents the network.

26.77. IDENTIFY and SET UP: Without the meter, the circuit consists of the two resistors in series. When the meter is connected, its resistance is added to the circuit in parallel with the resistor it is connected across.

(a) EXECUTE: $I = I_1 = I_2$.

$$I = \frac{90.0 \text{ V}}{R_1 + R_2} = \frac{90.0 \text{ V}}{224 \, \Omega + 589 \, \Omega} = 0.1107 \text{ A}.$$

$V_1 = I_1 R_1 = (0.1107 \text{ A})(224 \, \Omega) = 24.8 \text{ V}$; $V_2 = I_2 R_2 = (0.1107 \text{ A})(589 \, \Omega) = 65.2 \text{ V}$.

(b) SET UP: The resistor network is sketched in Figure 26.77a.

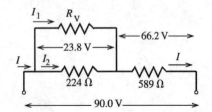

The voltmeter reads the potential difference across its terminals, which is 23.8 V. If we can find the current I_1 through the voltmeter then we can use Ohm's law to find its resistance.

Figure 26.77a

EXECUTE: The voltage drop across the 589-Ω resistor is $90.0 \text{ V} - 23.8 \text{ V} = 66.2 \text{ V}$, so

$$I = \frac{V}{R} = \frac{66.2 \text{ V}}{589 \, \Omega} = 0.1124 \text{ A}.$$ The voltage drop across the 224-Ω resistor is 23.8 V, so

$$I_2 = \frac{V}{R} = \frac{23.8 \text{ V}}{224 \, \Omega} = 0.1062 \text{ A}.$$ Then $I = I_1 + I_2$ gives $I_1 = I - I_2 = 0.1124 \text{ A} - 0.1062 \text{ A} = 0.0062 \text{ A}$.

$$R_V = \frac{V}{I_1} = \frac{23.8 \text{ V}}{0.0062 \text{ A}} = 3840 \, \Omega.$$

(c) SET UP: The circuit with the voltmeter connected is sketched in Figure 26.77b.

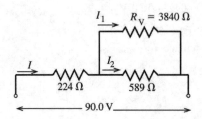

Figure 26.77b

EXECUTE: Replace the two resistors in parallel by their equivalent, as shown in Figure 26.77c.

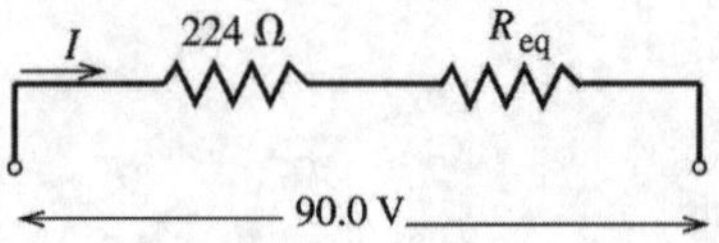

$$\frac{1}{R_{eq}} = \frac{1}{3840 \, \Omega} + \frac{1}{589 \, \Omega};$$

$$R_{eq} = \frac{(3840 \, \Omega)(589 \, \Omega)}{3840 \, \Omega + 589 \, \Omega} = 510.7 \, \Omega.$$

Figure 26.77c

$$I = \frac{90.0 \text{ V}}{224 \, \Omega + 510.7 \, \Omega} = 0.1225 \text{ A}.$$

The potential drop across the 224-Ω resistor then is $IR = (0.1225 \text{ A})(224 \text{ Ω}) = 27.4$ V, so the potential drop across the 589-Ω resistor and across the voltmeter (what the voltmeter reads) is
90.0 V − 27.4 V = 62.6 V.

EVALUATE: **(d)** No, any real voltmeter will draw some current and thereby reduce the current through the resistance whose voltage is being measured. Thus the presence of the voltmeter connected in parallel with the resistance lowers the voltage drop across that resistance. The resistance of the voltmeter in this problem is only about a factor of ten larger than the resistances in the circuit, so the voltmeter has a noticeable effect on the circuit.

26.81. **IDENTIFY and SET UP:** Kirchhoff's rules apply to the circuit. Taking a loop around the circuit gives $\varepsilon - Ri - q/C = 0$.

EXECUTE: **(a)** Solving the loop equation for q gives $q = q = \varepsilon C - RCi$. A graph of q as a function of i should be a straight line with slope equal to $-RC$ and y-intercept equal to εC. Figure 26.81 shows this graph.

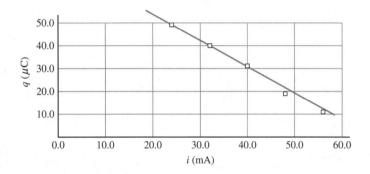

Figure 26.81

The best-fit slope of this graph is -1.233×10^{-3} C/A, and the y-intercept is 7.054×10^{-5} C.
(b) $RC = -\text{slope} = -(-1.233 \times 10^{-3}$ C/A), which gives
$R = (-1.233 \times 10^{-3}$ C/A)/(5.00 \times 10^{-6}$ F) = 246.6 Ω, which rounds to 247 Ω.
The y-intercept is εC, so
7.054×10^{-5} C = ε (5.00 \times 10^{-6}$ F).
ε = 15.9 V.
(c) $V_C = \varepsilon(1 - e^{-t/RC})$.
$V_C/\varepsilon = 1 - e^{-t/RC} = (10.0 \text{ V})(15.9 \text{ V})$.
Solving for t gives
$t = (247 \text{ Ω})(5.00 \text{ μF}) \ln(0.3714) = 1223$ μs, which rounds to 1.22 ms.
(d) $V_R = \varepsilon - V_C = 15.9$ V − 4.00 V = 11.9 V.

EVALUATE: As time increases, the potential difference across the capacitor increases as it gets charged, but the potential difference across the resistor decreases as the current decreases.

26.87. **IDENTIFY and SET UP:** The channels are all in parallel. For n identical resistors R in parallel,
$$\frac{1}{R_{eq}} = \frac{1}{R_1} + \frac{1}{R_2} + \ldots = \frac{1}{R} + \frac{1}{R} + \ldots = \frac{n}{R}, \text{ so } R_{eq} = R/n. \ I = jA.$$

EXECUTE: $I = jA = V/R_{eq} = V/(R/n) = nV/R$.

$jR/V = n/A = (5 \text{ mA/cm}^2)(10^{11} \text{ Ω})/(50 \text{ mV}) = 10^{10}/\text{cm}^2 = 100/\mu\text{m}^2$, which is choice (d).

EVALUATE: A density of 100 per μm^2 seems plausible, since these are microscopic structures.

MAGNETIC FIELD AND MAGNETIC FORCES

27.7. **IDENTIFY:** Apply $\vec{F} = q\vec{v} \times \vec{B}$.

SET UP: $\vec{v} = v_y \hat{j}$, with $v_y = -3.80 \times 10^3$ m/s. $F_x = +7.60 \times 10^{-3}$ N, $F_y = 0$, and $F_z = -5.20 \times 10^{-3}$ N.

EXECUTE: (a) $F_x = q(v_y B_z - v_z B_y) = qv_y B_z$.

$B_z = F_x/qv_y = (7.60 \times 10^{-3} \text{ N})/[(7.80 \times 10^{-6} \text{ C})(-3.80 \times 10^3 \text{ m/s})] = -0.256$ T.

$F_y = q(v_z B_x - v_x B_z) = 0$, which is consistent with $\vec{F}$ as given in the problem. There is no force component along the direction of the velocity.

$F_z = q(v_x B_y - v_y B_x) = -qv_y B_x$. $B_x = -F_z/qv_y = -0.175$ T.

(b) B_y is not determined. No force due to this component of $\vec{B}$ along $\vec{v}$; measurement of the force tells us nothing about B_y.

(c) $\vec{B} \cdot \vec{F} = B_x F_x + B_y F_y + B_z F_z = (-0.175 \text{ T})(+7.60 \times 10^{-3} \text{ N}) + (-0.256 \text{ T})(-5.20 \times 10^{-3} \text{ N})$

$\vec{B} \cdot \vec{F} = 0$. $\vec{B}$ and $\vec{F}$ are perpendicular (angle is 90°).

EVALUATE: The force is perpendicular to both $\vec{v}$ and $\vec{B}$, so $\vec{v} \cdot \vec{F}$ is also zero.

27.9. **IDENTIFY:** Apply $\vec{F} = q\vec{v} \times \vec{B}$ to the force on the proton and to the force on the electron. Solve for the components of $\vec{B}$ and use them to find its magnitude and direction.

SET UP: $\vec{F}$ is perpendicular to both $\vec{v}$ and $\vec{B}$. Since the force on the proton is in the $+y$-direction, $B_y = 0$ and $\vec{B} = B_x \hat{i} + B_z \hat{k}$. For the proton, $\vec{v}_p = (1.50 \text{ km/s})\hat{i} = v_p \hat{i}$ and $\vec{F}_p = (2.25 \times 10^{-16} \text{ N})\hat{j} = F_p \hat{j}$. For the electron, $\vec{v}_e = -(4.75 \text{ km/s})\hat{k} = -v_e \hat{k}$ and $\vec{F}_e = (8.50 \times 10^{-16} \text{ N})\hat{j} = F_e \hat{j}$. The magnetic force is $\vec{F} = q\vec{v} \times \vec{B}$.

EXECUTE: (a) For the proton, $\vec{F}_p = q\vec{v}_p \times \vec{B}$ gives $F_p \hat{j} = ev_p \hat{i} \times (B_x \hat{i} + B_z \hat{k}) = -ev_p B_z \hat{j}$. Solving for B_z

gives $B_z = -\dfrac{F_p}{ev_p} = -\dfrac{2.25 \times 10^{-16} \text{ N}}{(1.60 \times 10^{-19} \text{ C})(1500 \text{ m/s})} = -0.9375$ T. For the electron, $\vec{F}_e = -e\vec{v}_e \times \vec{B}$, which gives

$F_e \hat{j} = (-e)(-v_e \hat{k}) \times (B_x \hat{i} + B_z \hat{k}) = ev_e B_x \hat{j}$. Solving for B_x gives

$B_x = \dfrac{F_e}{ev_e} = \dfrac{8.50 \times 10^{-16} \text{ N}}{(1.60 \times 10^{-19} \text{ C})(4750 \text{ m/s})} = 1.118$ T. Therefore $\vec{B} = 1.118$ T$\hat{i} - 0.9375$ T$\hat{k}$. The magnitude of

the field is $B = \sqrt{B_x^2 + B_z^2} = \sqrt{(1.118 \text{ T})^2 + (-0.9375 \text{ T})^2} = 1.46$ T. Calling θ the angle that the magnetic

field makes with the $+x$-axis, we have $\tan\theta = \dfrac{B_z}{B_x} = \dfrac{-0.9375 \text{ T}}{1.118 \text{ T}} = -0.8386$, so $\theta = -40.0°$. Therefore the

magnetic field is in the xz-plane directed at 40.0° from the $+x$-axis toward the $-z$-axis, having a magnitude of 1.46 T.

(b) $\vec{B} = B_x\hat{i} + B_z\hat{k}$ and $\vec{v} = (3.2 \text{ km/s})(-\hat{j})$.

$\vec{F} = q\vec{v} \times \vec{B} = (-e)(3.2 \text{ km/s})(-\hat{j}) \times (B_x\hat{i} + B_z\hat{k}) = e(3.2 \times 10^3 \text{ m/s})[B_x(-\hat{k}) + B_z\hat{i}]$.

$\vec{F} = e(3.2 \times 10^3 \text{ m/s})(-1.118 \text{ T}\hat{k} - 0.9375 \text{ T}\hat{i}) = -4.80 \times 10^{-16} \text{ N}\hat{i} - 5.724 \times 10^{-16} \text{ N}\hat{k}$.

$F = \sqrt{F_x^2 + F_z^2} = 7.47 \times 10^{-16}$ N. Calling θ the angle that the force makes with the $-x$-axis, we have

$\tan\theta = \dfrac{F_z}{F_x} = \dfrac{-5.724 \times 10^{-16} \text{ N}}{-4.800 \times 10^{-16} \text{ N}}$, which gives $\theta = 50.0°$. The force is in the xz-plane and is directed at 50.0° from the $-x$-axis toward either the $-z$-axis.

EVALUATE: The force on the electrons in parts (a) and (b) are comparable in magnitude because the electron speeds are comparable in both cases.

27.13. IDENTIFY: The total flux through the bottle is zero because it is a closed surface.

SET UP: The total flux through the bottle is the flux through the plastic plus the flux through the open cap, so the sum of these must be zero. $\Phi_{\text{plastic}} + \Phi_{\text{cap}} = 0$.

$\Phi_{\text{plastic}} = -\Phi_{\text{cap}} = -BA\cos\phi = -B(\pi r^2)\cos\phi$.

EXECUTE: Substituting the numbers gives $\Phi_{\text{plastic}} = -(1.75 \text{ T})\pi(0.0125 \text{ m})^2 \cos 25° = -7.8 \times 10^{-4}$ Wb.

EVALUATE: It would be very difficult to calculate the flux through the plastic directly because of the complicated shape of the bottle, but with a little thought we can find this flux through a simple calculation.

27.15. (a) IDENTIFY: Apply $\vec{F} = q\vec{v} \times \vec{B}$ to relate the magnetic force $\vec{F}$ to the directions of $\vec{v}$ and $\vec{B}$. The electron has negative charge so $\vec{F}$ is opposite to the direction of $\vec{v} \times \vec{B}$. For motion in an arc of a circle the acceleration is toward the center of the arc so $\vec{F}$ must be in this direction. $a = v^2/R$.

SET UP:

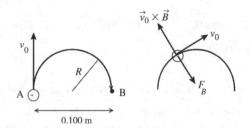

As the electron moves in the semicircle, its velocity is tangent to the circular path. The direction of $\vec{v}_0 \times \vec{B}$ at a point along the path is shown in Figure 27.15.

Figure 27.15

EXECUTE: For circular motion the acceleration of the electron $\vec{a}_{\text{rad}}$ is directed in toward the center of the circle. Thus the force $\vec{F}_B$ exerted by the magnetic field, since it is the only force on the electron, must be radially inward. Since q is negative, $\vec{F}_B$ is opposite to the direction given by the right-hand rule for $\vec{v}_0 \times \vec{B}$. Thus $\vec{B}$ is directed into the page. Apply Newton's second law to calculate the magnitude of $\vec{B}$:

$\Sigma\vec{F} = m\vec{a}$ gives $\Sigma F_{\text{rad}} = ma$ $F_B = m(v^2/R)$.

$F_B = |q|vB\sin\phi = |q|vB$, so $|q|vB = m(v^2/R)$.

$B = \dfrac{mv}{|q|R} = \dfrac{(9.109 \times 10^{-31} \text{ kg})(1.41 \times 10^6 \text{ m/s})}{(1.602 \times 10^{-19} \text{ C})(0.050 \text{ m})} = 1.60 \times 10^{-4}$ T.

(b) IDENTIFY and SET UP: The speed of the electron as it moves along the path is constant. ($\vec{F}_B$ changes the direction of $\vec{v}$ but not its magnitude.) The time is given by the distance divided by v_0.

EXECUTE: The distance along the semicircular path is πR, so $t = \dfrac{\pi R}{v_0} = \dfrac{\pi(0.050 \text{ m})}{1.41 \times 10^6 \text{ m/s}} = 1.11 \times 10^{-7}$ s.

EVALUATE: The magnetic field required increases when v increases or R decreases and also depends on the mass to charge ratio of the particle.

27.23. **IDENTIFY:** When a particle of charge $-e$ is accelerated through a potential difference of magnitude V, it gains kinetic energy eV. When it moves in a circular path of radius R, its acceleration is $\dfrac{v^2}{R}$.

SET UP: An electron has charge $q = -e = -1.60 \times 10^{-19}$ C and mass 9.11×10^{-31} kg.

EXECUTE: $\frac{1}{2}mv^2 = eV$ and $v = \sqrt{\dfrac{2eV}{m}} = \sqrt{\dfrac{2(1.60 \times 10^{-19}\ \text{C})(2.00 \times 10^3\ \text{V})}{9.11 \times 10^{-31}\ \text{kg}}} = 2.65 \times 10^7$ m/s. $\vec{F} = m\vec{a}$

gives $|q|vB\sin\phi = m\dfrac{v^2}{R}$. $\phi = 90°$ and $B = \dfrac{mv}{|q|R} = \dfrac{(9.11 \times 10^{-31}\ \text{kg})(2.65 \times 10^7\ \text{m/s})}{(1.60 \times 10^{-19}\ \text{C})(0.180\ \text{m})} = 8.38 \times 10^{-4}$ T.

EVALUATE: The smaller the radius of the circular path, the larger the magnitude of the magnetic field that is required.

27.25. **IDENTIFY and SET UP:** $\vec{F} = q(\vec{E} + \vec{v} \times \vec{B})$ gives the total force on the proton. At $t = 0$,

$\vec{F} = q\vec{v} \times \vec{B} = q(v_x\hat{i} + v_z\hat{k}) \times B_x\hat{i} = qv_zB_x\hat{j}$. **c)**

EXECUTE: **(a)** $\vec{F} = (1.60 \times 10^{-19}\ \text{C})(2.00 \times 10^5\ \text{m/s})(0.500\ \text{T})\hat{j} = (1.60 \times 10^{-14}\ \text{N})\hat{j}$.

(b) Yes. The electric field exerts a force in the direction of the electric field, since the charge of the proton is positive, and there is a component of acceleration in this direction.

(c) In the plane perpendicular to $\vec{B}$ (the yz-plane) the motion is circular. But there is a velocity component in the direction of $\vec{B}$, so the motion is a helix. The electric field in the $+\hat{i}$-direction exerts a force in the $+\hat{i}$-direction. This force produces an acceleration in the $+\hat{i}$-direction and this causes the pitch of the helix to vary. The force does not affect the circular motion in the yz-plane, so the electric field does not affect the radius of the helix.

(d) IDENTIFY and SET UP: Use $\omega = |q|B/m$ and $T = 2\pi/\omega$ to calculate the period of the motion. Calculate a_x produced by the electric force and use a constant acceleration equation to calculate the displacement in the x-direction in time $T/2$.

EXECUTE: Calculate the period T: $\omega = |q|B/m$.

$T = \dfrac{2\pi}{\omega} = \dfrac{2\pi m}{|q|B} = \dfrac{2\pi(1.67 \times 10^{-27}\ \text{kg})}{(1.60 \times 10^{-19}\ \text{C})(0.500\ \text{T})} = 1.312 \times 10^{-7}$ s. Then $t = T/2 = 6.56 \times 10^{-8}$ s.

$v_{0x} = 1.50 \times 10^5$ m/s.

$a_x = \dfrac{F_x}{m} = \dfrac{(1.60 \times 10^{-19}\ \text{C})(2.00 \times 10^4\ \text{V/m})}{1.67 \times 10^{-27}\ \text{kg}} = +1.916 \times 10^{12}$ m/s^2.

$x - x_0 = v_{0x}t + \frac{1}{2}a_xt^2$.

$x - x_0 = (1.50 \times 10^5\ \text{m/s})(6.56 \times 10^{-8}\ \text{s}) + \frac{1}{2}(1.916 \times 10^{12}\ \text{m/s}^2)(6.56 \times 10^{-8}\ \text{s})^2 = 1.40$ cm.

EVALUATE: The electric and magnetic fields are in the same direction but produce forces that are in perpendicular directions to each other.

27.29. **IDENTIFY:** For the alpha particles to emerge from the plates undeflected, the magnetic force on them must exactly cancel the electric force. The battery produces an electric field between the plates, which acts on the alpha particles.

SET UP: First use energy conservation to find the speed of the alpha particles as they enter the region between the plates: $qV = 1/2\ mv^2$. The electric field between the plates due to the battery is $E = V_b/d$. For the alpha particles not to be deflected, the magnetic force must cancel the electric force, so $qvB = qE$, giving $B = E/v$.

EXECUTE: Solve for the speed of the alpha particles just as they enter the region between the plates. Their charge is $2e$.

$v_\alpha = \sqrt{\dfrac{2(2e)V}{m}} = \sqrt{\dfrac{4(1.60 \times 10^{-19}\ \text{C})(1750\ \text{V})}{6.64 \times 10^{-27}\ \text{kg}}} = 4.11 \times 10^5$ m/s.

The electric field between the plates, produced by the battery, is
$$E = V_b/d = (150 \text{ V})/(0.00820 \text{ m}) = 18{,}300 \text{ V/m}.$$
The magnetic force must cancel the electric force:
$$B = E/v_\alpha = (18{,}300 \text{ V/m})/(4.11\times 10^5 \text{ m/s}) = 0.0445 \text{ T}.$$
The magnetic field is perpendicular to the electric field. If the charges are moving to the right and the electric field points upward, the magnetic field is out of the page.

EVALUATE: The sign of the charge of the alpha particle does not enter the problem, so negative charges of the same magnitude would also not be deflected.

27.33. IDENTIFY: A mass spectrometer separates ions by mass. Since ^{14}N and ^{15}N have different masses they will be separated and the relative amounts of these isotopes can be determined.

SET UP: $R = \dfrac{mv}{|q|B}$. For $m = 1.99\times 10^{-26}$ kg (^{12}C), $R_{12} = 12.5$ cm. The separation of the isotopes at the detector is $2(R_{15} - R_{14})$.

EXECUTE: Since $R = \dfrac{mv}{|q|B}$, $\dfrac{R}{m} = \dfrac{v}{|q|B} = $ constant. Therefore $\dfrac{R_{14}}{m_{14}} = \dfrac{R_{12}}{m_{12}}$ which gives

$$R_{14} = R_{12}\left(\dfrac{m_{14}}{m_{12}}\right) = (12.5 \text{ cm})\left(\dfrac{2.32\times 10^{-26} \text{ kg}}{1.99\times 10^{-26} \text{ kg}}\right) = 14.6 \text{ cm and}$$

$$R_{15} = R_{12}\left(\dfrac{m_{15}}{m_{12}}\right) = (12.5 \text{ cm})\left(\dfrac{2.49\times 10^{-26} \text{ kg}}{1.99\times 10^{-26} \text{ kg}}\right) = 15.6 \text{ cm}.$$ The separation of the isotopes at the detector is $2(R_{15} - R_{14}) = 2(15.6 \text{ cm} - 14.6 \text{ cm}) = 2.0 \text{ cm}$.

EVALUATE: The separation is large enough to be easily detectable. Since the diameter of the ion path is large, about 30 cm, the uniform magnetic field within the instrument must extend over a large area.

27.35. IDENTIFY: Apply $F = IlB\sin\phi$.

SET UP: Label the three segments in the field as a, b, and c. Let x be the length of segment a. Segment b has length 0.300 m and segment c has length $0.600 \text{ m} - x$. Figure 27.35a shows the direction of the force on each segment. For each segment, $\phi = 90°$. The total force on the wire is the vector sum of the forces on each segment.

EXECUTE: $F_a = IlB = (4.50 \text{ A})x(0.240 \text{ T})$. $F_c = (4.50 \text{ A})(0.600 \text{ m} - x)(0.240 \text{ T})$. Since $\vec{F}_a$ and $\vec{F}_c$ are in the same direction their vector sum has magnitude
$F_{ac} = F_a + F_c = (4.50 \text{ A})(0.600 \text{ m})(0.240 \text{ T}) = 0.648 \text{ N}$ and is directed toward the bottom of the page in Figure 27.35a. $F_b = (4.50 \text{ A})(0.300 \text{ m})(0.240 \text{ T}) = 0.324 \text{ N}$ and is directed to the right. The vector addition diagram for $\vec{F}_{ac}$ and $\vec{F}_b$ is given in Figure 27.35b.

$F = \sqrt{F_{ac}^2 + F_b^2} = \sqrt{(0.648 \text{ N})^2 + (0.324 \text{ N})^2} = 0.724 \text{ N}$. $\tan\theta = \dfrac{F_{ac}}{F_b} = \dfrac{0.648 \text{ N}}{0.324 \text{ N}}$ and $\theta = 63.4°$. The net force has magnitude 0.724 N and its direction is specified by $\theta = 63.4°$ in Figure 27.35b.

EVALUATE: All three current segments are perpendicular to the magnetic field, so $\phi = 90°$ for each in the force equation. The direction of the force on a segment depends on the direction of the current for that segment.

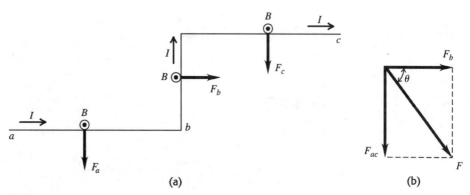

Figure 27.35

27.39. **IDENTIFY:** The magnetic force $\vec{F}_B$ must be upward and equal to mg. The direction of $\vec{F}_B$ is determined by the direction of I in the circuit.

SET UP: $F_B = IlB\sin\phi$, with $\phi = 90°$. $I = \dfrac{V}{R}$, where V is the battery voltage.

EXECUTE: **(a)** The forces are shown in Figure 27.39. The current I in the bar must be to the right to produce $\vec{F}_B$ upward. To produce current in this direction, point a must be the positive terminal of the battery.

(b) $F_B = mg$. $IlB = mg$. $m = \dfrac{IlB}{g} = \dfrac{VlB}{Rg} = \dfrac{(175\text{ V})(0.600\text{ m})(1.50\text{ T})}{(5.00\text{ }\Omega)(9.80\text{ m/s}^2)} = 3.21$ kg.

EVALUATE: If the battery had opposite polarity, with point a as the negative terminal, then the current would be clockwise and the magnetic force would be downward.

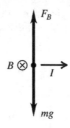

Figure 27.39

27.41. **IDENTIFY:** The wire segments carry a current in an external magnetic field. Only segments ab and cd will experience a magnetic force since the other two segments carry a current parallel (and antiparallel) to the magnetic field. Only the force on segment cd will produce a torque about the hinge.
SET UP: $F = IlB\sin\phi$. The direction of the magnetic force is given by the right-hand rule applied to the directions of I and $\vec{B}$. The torque due to a force equals the force times the moment arm, the perpendicular distance between the axis and the line of action of the force.
EXECUTE: **(a)** The direction of the magnetic force on each segment of the circuit is shown in Figure 27.41. For segments bc and da the current is parallel or antiparallel to the field and the force on these segments is zero.

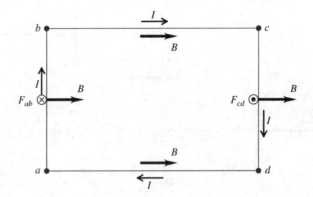

Figure 27.41

(b) $\vec{F}_{ab}$ acts at the hinge and therefore produces no torque. $\vec{F}_{cd}$ tends to rotate the loop about the hinge so it does produce a torque about this axis. $F_{cd} = IlB\sin\phi = (5.00\text{ A})(0.200\text{ m})(1.20\text{ T})\sin 90° = 1.20\text{ N}$

(c) $\tau = Fl = (1.20\text{ N})(0.350\text{ m}) = 0.420\text{ N}\cdot\text{m}$.

EVALUATE: The torque is directed so as to rotate side cd out of the plane of the page in Figure 27.41.

27.43. IDENTIFY: The magnetic field exerts a torque on the current-carrying coil, which causes it to turn. We can use the rotational form of Newton's second law to find the angular acceleration of the coil.

SET UP: The magnetic torque is given by $\vec{\tau} = \vec{\mu} \times \vec{B}$, and the rotational form of Newton's second law is $\Sigma\tau = I\alpha$. The magnetic field is parallel to the plane of the loop.

EXECUTE: **(a)** The coil rotates about axis A_2 because the only torque is along top and bottom sides of the coil.

(b) To find the moment of inertia of the coil, treat the two 1.00-m segments as point-masses (since all the points in them are 0.250 m from the rotation axis) and the two 0.500-m segments as thin uniform bars rotated about their centers. Since the coil is uniform, the mass of each segment is proportional to its fraction of the total perimeter of the coil. Each 1.00-m segment is 1/3 of the total perimeter, so its mass is $(1/3)(210\text{ g}) = 70\text{ g} = 0.070\text{ kg}$. The mass of each 0.500-m segment is half this amount, or 0.035 kg.
The result is

$$I = 2(0.070\text{ kg})(0.250\text{ m})^2 + 2\tfrac{1}{12}(0.035\text{ kg})(0.500\text{ m})^2 = 0.0102\text{ kg}\cdot\text{m}^2.$$

The torque is

$$|\vec{\tau}| = |\vec{\mu}\times\vec{B}| = IAB\sin 90° = (2.00\text{A})(0.500\text{m})(1.00\text{m})(3.00\text{T}) = 3.00\text{ N}\cdot\text{m}.$$

Using the above values, the rotational form of Newton's second law gives

$$\alpha = \frac{\tau}{I} = 290\text{ rad/s}^2.$$

EVALUATE: This angular acceleration will not continue because the torque changes as the coil turns.

27.47. IDENTIFY: The circuit consists of two parallel branches with the potential difference of 120 V applied across each. One branch is the rotor, represented by a resistance R_r and an induced emf that opposes the applied potential. Apply the loop rule to each parallel branch and use the junction rule to relate the currents through the field coil and through the rotor to the 4.82 A supplied to the motor.

SET UP: The circuit is sketched in Figure 27.47.

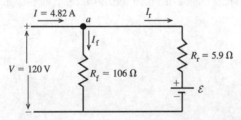

ε is the induced emf developed by the motor. It is directed so as to oppose the current through the rotor.

Figure 27.47

EXECUTE: (a) The field coils and the rotor are in parallel with the applied potential difference V, so $V = I_f R_f$. $I_f = \dfrac{V}{R_f} = \dfrac{120 \text{ V}}{106 \text{ }\Omega} = 1.13$ A.

(b) Applying the junction rule to point a in the circuit diagram gives $I - I_f - I_r = 0$.
$I_r = I - I_f = 4.82 \text{ A} - 1.13 \text{ A} = 3.69$ A.

(c) The potential drop across the rotor, $I_r R_r + \varepsilon$, must equal the applied potential difference V: $V = I_r R_r + \varepsilon$
$\varepsilon = V - I_r R_r = 120 \text{ V} - (3.69 \text{ A})(5.9 \text{ }\Omega) = 98.2$ V

(d) The mechanical power output is the electrical power input minus the rate of dissipation of electrical energy in the resistance of the motor:
electrical power input to the motor
$P_{\text{in}} = IV = (4.82 \text{ A})(120 \text{ V}) = 578$ W.
electrical power loss in the two resistances
$P_{\text{loss}} = I_f^2 R_f + I_r^2 R_r = (1.13 \text{ A})^2 (106 \text{ }\Omega) + (3.69 \text{ A})^2 (5.9 \text{ }\Omega) = 216$ W.
mechanical power output
$P_{\text{out}} = P_{\text{in}} - P_{\text{loss}} = 578 \text{ W} - 216 \text{ W} = 362$ W.
The mechanical power output is the power associated with the induced emf ε.
$P_{\text{out}} = P_\varepsilon = \varepsilon I_r = (98.2 \text{ V})(3.69 \text{ A}) = 362$ W, which agrees with the above calculation.

EVALUATE: The induced emf reduces the amount of current that flows through the rotor. This motor differs from the one described in Example 27.11. In that example the rotor and field coils are connected in series and in this problem they are in parallel.

27.49. **IDENTIFY:** The drift velocity is related to the current density by $J_x = n|q|v_d$. The electric field is determined by the requirement that the electric and magnetic forces on the current-carrying charges are equal in magnitude and opposite in direction.

SET UP and EXECUTE: (a) The section of the silver ribbon is sketched in Figure 27.49a.

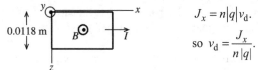

$J_x = n|q|v_d$.

so $v_d = \dfrac{J_x}{n|q|}$.

Figure 27.49a

EXECUTE: $J_x = \dfrac{I}{A} = \dfrac{I}{y_1 z_1} = \dfrac{120 \text{ A}}{(0.23 \times 10^{-3} \text{ m})(0.0118 \text{ m})} = 4.42 \times 10^7$ A/m^2.

$v_d = \dfrac{J_x}{n|q|} = \dfrac{4.42 \times 10^7 \text{ A/m}^2}{(5.85 \times 10^{28}/\text{m}^3)(1.602 \times 10^{-19} \text{ C})} = 4.7 \times 10^{-3}$ m/s = 4.7 mm/s.

(b) magnitude of $\vec{E}$:

$|q|E_z = |q|v_d B_y$.

$E_z = v_d B_y = (4.7 \times 10^{-3} \text{ m/s})(0.95 \text{ T}) = 4.5 \times 10^{-3}$ V/m.

direction of $\vec{E}$:

The drift velocity of the electrons is in the opposite direction to the current, as shown in Figure 27.49b.

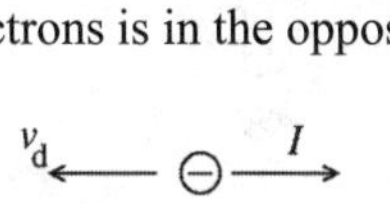

$\vec{v} \times \vec{B} \uparrow$.
$\vec{F}_B = q\vec{v} \times \vec{B} = -e\vec{v} \times \vec{B} \downarrow$.

Figure 27.49b

The directions of the electric and magnetic forces on an electron in the ribbon are shown in Figure 27.49c.

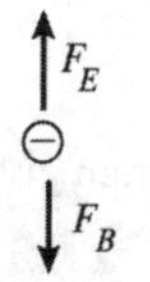

$\vec{F}_E$ must oppose $\vec{F}_B$ so $\vec{F}_E$ is in the $-z$-direction.

Figure 27.49c

$\vec{F}_E = q\vec{E} = -e\vec{E}$ so $\vec{E}$ is opposite to the direction of $\vec{F}_E$ and thus $\vec{E}$ is in the $+z$-direction.

(c) The Hall emf is the potential difference between the two edges of the strip (at $z = 0$ and $z = z_1$) that results from the electric field calculated in part (b). $\varepsilon_{\text{Hall}} = Ez_1 = (4.5\times10^{-3}\ \text{V/m})(0.0118\ \text{m}) = 53\ \mu\text{V}$.

EVALUATE: Even though the current is quite large the Hall emf is very small. Our calculated Hall emf is more than an order of magnitude larger than in Example 27.12. In this problem the magnetic field and current density are larger than in the example, and this leads to a larger Hall emf.

27.53. **IDENTIFY:** In part (a), apply conservation of energy to the motion of the two nuclei. In part (b) apply $|q|vB = mv^2/R$.

SET UP: In part (a), let point 1 be when the two nuclei are far apart and let point 2 be when they are at their closest separation.

EXECUTE: (a) $K_1 + U_1 = K_2 + U_2$. $U_1 = K_2 = 0$, so $K_1 = U_2$. There are two nuclei having equal kinetic energy, so $\tfrac{1}{2}mv^2 + \tfrac{1}{2}mv^2 = ke^2/r$. Solving for v gives

$$v = e\sqrt{\frac{k}{mr}} = (1.602\times10^{-19}\ \text{C})\sqrt{\frac{8.99\times10^9\ \text{N}\cdot\text{m}^2/\text{C}^2}{(3.34\times10^{-27}\ \text{kg})(1.0\times10^{-15}\ \text{m})}} = 8.3\times10^6\ \text{m/s}.$$

(b) $\Sigma \vec{F} = m\vec{a}$ gives $qvB = mv^2/r$. $B = \dfrac{mv}{qr} = \dfrac{(3.34\times10^{-27}\ \text{kg})(8.3\times10^6\ \text{m/s})}{(1.602\times10^{-19}\ \text{C})(1.25\ \text{m})} = 0.14\ \text{T}$.

EVALUATE: The speed calculated in part (a) is large, nearly 3% of the speed of light.

27.55. **IDENTIFY:** The sum of the magnetic, electrical and gravitational forces must be zero to aim at and hit the target.

SET UP: The magnetic field must point to the left when viewed in the direction of the target for no net force. The net force is zero, so $\Sigma F = F_B - F_E - mg = 0$ and $qvB - qE - mg = 0$.

EXECUTE: Solving for B gives

$$B = \frac{qE + mg}{qv} = \frac{(2500\times10^{-6}\ \text{C})(27.5\ \text{N/C}) + (0.00425\ \text{kg})(9.80\ \text{m/s}^2)}{(2500\times10^{-6}\ \text{C})(12.8\ \text{m/s})} = 3.45\ \text{T}.$$

The direction should be perpendicular to the initial velocity of the coin.

EVALUATE: This is a very strong magnetic field, but achievable in some labs.

27.61. **IDENTIFY:** The force exerted by the magnetic field is given by $F = IlB\sin\phi$. The net force on the wire must be zero.

SET UP: For the wire to remain at rest the force exerted on it by the magnetic field must have a component directed up the incline. To produce a force in this direction, the current in the wire must be directed from right to left in the figure with the problem in the textbook. Or, viewing the wire from its left-hand end the directions are shown in Figure 27.61a.

Figure 27.61a

The free-body diagram for the wire is given in Figure 27.61b.

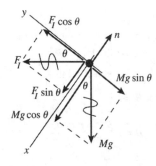

Figure 27.61b

EXECUTE: $\sum F_y = 0$.

$F_I \cos\theta - Mg\sin\theta = 0$.

$F_I = ILB\sin\phi$.

$\phi = 90°$ since $\vec{B}$ is perpendicular to the current direction.

Thus $(ILB)\cos\theta - Mg\sin\theta = 0$ and $I = \dfrac{Mg\tan\theta}{LB}$.

EVALUATE: The magnetic and gravitational forces are in perpendicular directions so their components parallel to the incline involve different trig functions. As the tilt angle θ increases there is a larger component of Mg down the incline and the component of F_I up the incline is smaller; I must increase with θ to compensate. As $\theta \to 0$, $I \to 0$ and as $\theta \to 90°$, $I \to \infty$.

27.65. **IDENTIFY:** The force exerted by the magnetic field is $F = ILB\sin\phi$. $a = F/m$ and is constant. Apply a constant acceleration equation to relate v and d.
SET UP: $\phi = 90°$. The direction of $\vec{F}$ is given by the right-hand rule.
EXECUTE: **(a)** $F = ILB$, to the right.

(b) $v_x^2 = v_{0x}^2 + 2a_x(x - x_0)$ gives $v^2 = 2ad$ and $d = \dfrac{v^2}{2a} = \dfrac{v^2 m}{2ILB}$.

(c) $d = \dfrac{(1.12\times 10^4 \text{ m/s})^2 (25 \text{ kg})}{2(2000 \text{ A})(0.50 \text{ m})(0.80 \text{ T})} = 1.96\times 10^6 \text{ m} = 1960 \text{ km}$.

EVALUATE: $a = \dfrac{ILB}{m} = \dfrac{(2.0\times 10^3 \text{ A})(0.50 \text{ m})(0.80 \text{ T})}{25 \text{ kg}} = 32 \text{ m/s}^2$. The acceleration due to gravity is not negligible. Since the bar would have to travel nearly 2000 km, this would not be a very effective launch mechanism using the numbers given.

27.71. **IDENTIFY:** Apply $\vec{F} = I\vec{l}\times\vec{B}$ to calculate the force on each side of the loop.
SET UP: The net force is the vector sum of the forces on each side of the loop.
EXECUTE: **(a)** $F_{PQ} = (5.00 \text{ A})(0.600 \text{ m})(3.00 \text{ T})\sin(0°) = 0 \text{ N}$.

$F_{RP} = (5.00 \text{ A})(0.800 \text{ m})(3.00 \text{ T})\sin(90°) = 12.0 \text{ N}$, into the page.

$F_{QR} = (5.00 \text{ A})(1.00 \text{ m})(3.00 \text{ T})(0.800/1.00) = 12.0 \text{ N}$, out of the page.

(b) The net force on the triangular loop of wire is zero.
(c) For calculating torque on a straight wire we can assume that the force on a wire is applied at the wire's center. Also, note that we are finding the torque with respect to the PR-axis (not about a point), and consequently the lever arm will be the distance from the wire's center to the x-axis. $\tau = rF\sin\phi$ gives $\tau_{PQ} = r(0 \text{ N}) = 0$, $\tau_{RP} = (0 \text{ m})F\sin\phi = 0$ and $\tau_{QR} = (0.300 \text{ m})(12.0 \text{ N})\sin(90°) = 3.60 \text{ N}\cdot\text{m}$. The net torque is $3.60 \text{ N}\cdot\text{m}$.

(d) Using $\tau = NIAB\sin\phi$ gives

$\tau = NIAB\sin\phi = (1)(5.00 \text{ A})\left(\tfrac{1}{2}\right)(0.600 \text{ m})(0.800 \text{ m})(3.00 \text{ T})\sin(90°) = 3.60 \text{ N}\cdot\text{m}$, which agrees with our result in part (c).

(e) Since F_{QR} is out of the page and since this is the force that produces the net torque, the point Q will be rotated out of the plane of the figure.
EVALUATE: In the expression $\tau = NIAB\sin\phi$, ϕ is the angle between the plane of the loop and the direction of $\vec{B}$. In this problem, $\phi = 90°$.

27.75. **IDENTIFY:** Apply $d\vec{F} = Id\vec{l} \times \vec{B}$ to each side of the loop.

SET UP: For each side of the loop, $d\vec{l}$ is parallel to that side of the loop and is in the direction of I. Since the loop is in the xy-plane, $z = 0$ at the loop and $B_y = 0$ at the loop.

EXECUTE: (a) The magnetic field lines in the yz-plane are sketched in Figure 27.75.

(b) Side 1, that runs from (0,0) to (0,L): $\vec{F} = \int_0^L Id\vec{l} \times \vec{B} = I\int_0^L \frac{B_0 y \, dy}{L}\hat{i} = \frac{1}{2}B_0 L I \hat{i}$.

Side 2, that runs from (0,L) to (L,L): $\vec{F} = \int_{0,y=L}^L Id\vec{l} \times \vec{B} = I\int_{0,y=L}^L \frac{B_0 y \, dx}{L}\hat{j} = -IB_0 L \hat{j}$.

Side 3, that runs from (L,L) to (L,0): $\vec{F} = \int_{L,x=L}^0 Id\vec{l} \times \vec{B} = I\int_{L,x=L}^0 \frac{B_0 y \, dy}{L}(-\hat{i}) = -\frac{1}{2}IB_0 L\hat{i}$.

Side 4, that runs from (L,0) to (0,0): $\vec{F} = \int_{L,y=0}^0 Id\vec{l} \times \vec{B} = I\int_{L,y=0}^0 \frac{B_0 y \, dx}{L}\hat{j} = 0$.

(c) The sum of all forces is $\vec{F}_{total} = -IB_0 L \hat{j}$.

EVALUATE: The net force on sides 1 and 3 is zero. The force on side 4 is zero, since $y = 0$ and $z = 0$ at that side and therefore $B = 0$ there. The net force on the loop equals the force on side 2.

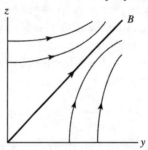

Figure 27.75

27.79. **IDENTIFY and SET UP:** The analysis in the text of the Thomson e/m experiment gives $\frac{e}{m} = \frac{E^2}{2VB^2}$. For a particle of charge e and mass m accelered through a potential V, $eV = \frac{1}{2}mv^2$.

EXECUTE: (a) Solving the equation $\frac{e}{m} = \frac{E^2}{2VB^2}$ for E^2 gives $E^2 = 2\left(\frac{e}{m}\right)B^2V$. Therefore a graph of E^2 versus V should be a straight line with slope equal to $2(e/m)B^2$.

(b) We can find the slope using two easily-read points on the graph. Using (100, 200) and (300, 600), we get $\frac{600\times 10^8 \text{ V}^2/\text{m}^2 - 200\times 10^8 \text{ V}^2/\text{m}^2}{300 \text{ V} - 100 \text{ V}} = 2.00\times 10^8 \text{ V}/\text{m}^2$ for the slope. This gives

e/m = (slope)/$2B^2$ = $(2.00\times 10^8 \text{ V}/\text{m}^2)/[2(0.340 \text{ T})^2]$ = 8.65×10^8 C/kg, which gives $m = 1.85\times 10^{-28}$ kg.

(c) $V = Ed = (2.00\times 10^5 \text{ V/m})(0.00600 \text{ m}) = 1.20$ kV.

(d) Using $eV = \frac{1}{2}mv^2$ to find the muon speed gives

$v = \sqrt{\frac{2eV}{m}} = \sqrt{2(8.65\times 10^8 \text{ C/kg})(400 \text{ V})} = 8.32\times 10^5$ m/s.

EVALUATE: Results may vary due to inaccuracies in determining the slope of the graph.

27.85. **IDENTIFY and SET UP:** Model the nerve as a current-carrying bar in a magnetic field. The resistance of the nerve is $R = \frac{\rho L}{A}$, the current through it is $I = V/R$ (by Ohm's law), and the maximum magnetic force on it is $F = ILB$.

EXECUTE: The resistance is $R = \dfrac{\rho L}{A} = (0.6\ \Omega \cdot \text{m})(0.001\ \text{m})/[\pi(0.0015/2\ \text{m})^2] = 340\ \Omega$.

The current is $I = V/R = (0.1\ \text{V})/(340\ \Omega) = 2.9 \times 10^{-4}\ \text{A}$.

The maximum force is $F = ILB = (2.9 \times 10^{-4}\ \text{A})(0.001\ \text{m})(2\ \text{T}) = 5.9 \times 10^{-7}\ \text{N} \approx 6 \times 10^{-7}\ \text{N}$, which is choice (a).

EVALUATE: This is the force on a 1-mm segment of nerve. The force on the entire nerve would be somewhat larger, depending on the length of the nerve.

28

SOURCES OF MAGNETIC FIELD

28.1. **IDENTIFY** and **SET UP:** Use $\vec{B} = \dfrac{\mu_0}{4\pi}\dfrac{q\vec{v}\times\hat{r}}{r^2}$ to calculate $\vec{B}$ at each point.

$\vec{B} = \dfrac{\mu_0}{4\pi}\dfrac{q\vec{v}\times\hat{r}}{r^2} = \dfrac{\mu_0}{4\pi}\dfrac{q\vec{v}\times\vec{r}}{r^3}$, since $\hat{r} = \dfrac{\vec{r}}{r}$.

$\vec{v} = (8.00\times 10^6 \text{ m/s})\hat{j}$ and $\vec{r}$ is the vector from the charge to the point where the field is calculated.

EXECUTE: (a) $\vec{r} = (0.500 \text{ m})\hat{i}, r = 0.500 \text{ m}$.

$\vec{v}\times\vec{r} = vr\hat{j}\times\hat{i} = -vr\hat{k}$.

$\vec{B} = -\dfrac{\mu_0}{4\pi}\dfrac{qv}{r^2}\hat{k} = -(1\times 10^{-7} \text{ T}\cdot\text{m/A})\dfrac{(6.00\times 10^{-6} \text{ C})(8.00\times 10^6 \text{ m/s})}{(0.500 \text{ m})^2}\hat{k}$.

$\vec{B} = -(1.92\times 10^{-5} \text{ T})\hat{k}$.

(b) $\vec{r} = -(0.500 \text{ m})\hat{j}, r = 0.500 \text{ m}$.

$\vec{v}\times\vec{r} = -vr\hat{j}\times\hat{j} = 0$ and $\vec{B} = 0$.

(c) $\vec{r} = (0.500 \text{ m})\hat{k}, r = 0.500 \text{ m}$.

$\vec{v}\times\vec{r} = vr\hat{j}\times\hat{k} = vr\hat{i}$.

$\vec{B} = (1\times 10^{-7} \text{ T}\cdot\text{m/A})\dfrac{(6.00\times 10^{-6} \text{ C})(8.00\times 10^6 \text{ m/s})}{(0.500 \text{ m})^2}\hat{i} = +(1.92\times 10^{-5} \text{ T})\hat{i}$.

(d) $\vec{r} = -(0.500 \text{ m})\hat{j} + (0.500 \text{ m})\hat{k}, r = \sqrt{(0.500 \text{ m})^2 + (0.500 \text{ m})^2} = 0.7071 \text{ m}$.

$\vec{v}\times\vec{r} = v(0.500 \text{ m})(-\hat{j}\times\hat{j} + \hat{j}\times\hat{k}) = (4.00\times 10^6 \text{ m}^2/\text{s})\hat{i}$.

$\vec{B} = (1\times 10^{-7} \text{ T}\cdot\text{m/A})\dfrac{(6.00\times 10^{-6} \text{ C})(4.00\times 10^6 \text{ m}^2/\text{s})}{(0.7071 \text{ m})^3}\hat{i} = +(6.79\times 10^{-6} \text{ T})\hat{i}$.

EVALUATE: At each point $\vec{B}$ is perpendicular to both $\vec{v}$ and $\vec{r}$. $B = 0$ along the direction of $\vec{v}$.

28.7. **IDENTIFY:** A moving charge creates a magnetic field.

SET UP: Apply $\vec{B} = \dfrac{\mu_0}{4\pi}\dfrac{q\vec{v}\times\vec{r}}{r^3}$. $\vec{r} = (0.200 \text{ m})\hat{i} + (-0.300 \text{ m})\hat{j}$, and $r = 0.3606 \text{ m}$.

EXECUTE: $\vec{v}\times\vec{r} = [(7.50\times 10^4 \text{ m/s})\hat{i} + (-4.90\times 10^4 \text{ m/s})\hat{j}]\times[(0.200 \text{ m})\hat{i} + (-0.300 \text{ m})\hat{j}]$, which simplifies to

$\vec{v}\times\vec{r} = (-2.25\times 10^4 \text{ m}^2/\text{s})\hat{k} + (9.80\times 10^3 \text{ m}^2/\text{s})\hat{k} = (-1.27\times 10^4 \text{ m}^2/\text{s})\hat{k}$.

$\vec{B} = (1.00\times 10^{-7} \text{ T}\cdot\text{m/A})\dfrac{(-3.00\times 10^{-6} \text{ C})(-1.27\times 10^4 \text{ m}^2/\text{s})}{(0.3606 \text{ m})^3}\hat{k} = (9.75\times 10^{-8} \text{ T})\hat{k}$.

EVALUATE: We can check the direction of the magnetic field using the right-hand rule, which shows that the field points in the +z-direction.

28.11. **IDENTIFY** and **SET UP:** The magnetic field produced by an infinitesimal current element is given by $d\vec{B} = \dfrac{\mu_0}{4\pi}\dfrac{I\vec{l}\times\hat{r}}{r^2}$.

As in Example 28.2, use $d\vec{B} = \dfrac{\mu_0}{4\pi}\dfrac{I\vec{l}\times\hat{r}}{r^2}$ for the finite 0.500-mm segment of wire since the $\Delta l = 0.500$-mm length is much smaller than the distances to the field points.

$\vec{B} = \dfrac{\mu_0}{4\pi}\dfrac{I\Delta\vec{l}\times\hat{r}}{r^2} = \dfrac{\mu_0}{4\pi}\dfrac{I\Delta\vec{l}\times\vec{r}}{r^3}$

I is in the $+z$-direction, so $\Delta\vec{l} = (0.500\times 10^{-3}\text{ m})\hat{k}$.

EXECUTE: (a) The field point is at $x = 2.00$ m, $y = 0$, $z = 0$ so the vector $\vec{r}$ from the source point (at the origin) to the field point is $\vec{r} = (2.00\text{ m})\hat{i}$.

$\Delta\vec{l}\times\vec{r} = (0.500\times 10^{-3}\text{ m})(2.00\text{ m})\hat{k}\times\hat{i} = +(1.00\times 10^{-3}\text{ m}^2)\hat{j}$.

$\vec{B} = \dfrac{(1\times 10^{-7}\text{ T}\cdot\text{m/A})(4.00\text{ A})(1.00\times 10^{-3}\text{ m}^2)}{(2.00\text{ m})^3}\hat{j} = (5.00\times 10^{-11}\text{ T})\hat{j}$.

(b) $\vec{r} = (2.00\text{ m})\hat{j}, r = 2.00$ m.

$\Delta\vec{l}\times\vec{r} = (0.500\times 10^{-3}\text{ m})(2.00\text{ m})\hat{k}\times\hat{j} = -(1.00\times 10^{-3}\text{ m}^2)\hat{i}$.

$\vec{B} = \dfrac{(1\times 10^{-7}\text{ T}\cdot\text{m/A})(4.00\text{ A})(-1.00\times 10^{-3}\text{ m}^2)}{(2.00\text{ m})^3}\hat{i} = -(5.00\times 10^{-11}\text{ T})\hat{i}$.

(c) $\vec{r} = (2.00\text{ m})(\hat{i}+\hat{j}), r = \sqrt{2}(2.00\text{ m})$.

$\Delta\vec{l}\times\vec{r} = (0.500\times 10^{-3}\text{ m})(2.00\text{ m})\hat{k}\times(\hat{i}+\hat{j}) = (1.00\times 10^{-3}\text{ m}^2)(\hat{j}-\hat{i})$.

$\vec{B} = \dfrac{(1\times 10^{-7}\text{ T}\cdot\text{m/A})(4.00\text{ A})(1.00\times 10^{-3}\text{ m}^2)}{[\sqrt{2}(2.00\text{ m})]^3}(\hat{j}-\hat{i}) = (-1.77\times 10^{-11}\text{ T})(\hat{i}-\hat{j})$.

(d) $\vec{r} = (2.00\text{ m})\hat{k}, r = 2.00$ m.

$\Delta\vec{l}\times\vec{r} = (0.500\times 10^{-3}\text{ m})(2.00\text{ m})\hat{k}\times\hat{k} = 0; \vec{B} = 0$.

EVALUATE: At each point $\vec{B}$ is perpendicular to both $\vec{r}$ and $\Delta\vec{l}$. $B = 0$ along the length of the wire.

28.13. **IDENTIFY:** A current segment creates a magnetic field.

SET UP: The law of Biot and Savart gives $dB = \dfrac{\mu_0}{4\pi}\dfrac{I\,dl\sin\phi}{r^2}$. Both fields are into the page, so their magnitudes add.

EXECUTE: Applying the Biot and Savart law, where $r = \tfrac{1}{2}\sqrt{(3.00\text{ cm})^2 + (3.00\text{ cm})^2} = 2.121$ cm, we have

$dB = 2\dfrac{4\pi\times 10^{-7}\text{ T}\cdot\text{m/A}}{4\pi}\dfrac{(28.0\text{ A})(0.00200\text{ m})\sin 45.0°}{(0.02121\text{ m})^2} = 1.76\times 10^{-5}$ T, into the paper.

EVALUATE: Even though the two wire segments are at right angles, the magnetic fields they create are in the same direction.

28.21. **IDENTIFY:** The total magnetic field is the vector sum of the constant magnetic field and the wire's magnetic field.

SET UP: For the wire, $B_{\text{wire}} = \dfrac{\mu_0 I}{2\pi r}$ and the direction of B_{wire} is given by the right-hand rule that is illustrated in Figure 28.6 in the textbook. $\vec{B}_0 = (1.50\times 10^{-6}\text{ T})\hat{i}$.

EXECUTE: (a) At $(0, 0, 1\text{ m})$, $\vec{B} = \vec{B}_0 - \dfrac{\mu_0 I}{2\pi r}\hat{i} = (1.50\times 10^{-6}\text{ T})\hat{i} - \dfrac{\mu_0(8.00\text{ A})}{2\pi(1.00\text{ m})}\hat{i} = -(1.0\times 10^{-7}\text{ T})\hat{i}$.

(b) At $(1\text{ m}, 0, 0)$, $\vec{B} = \vec{B}_0 + \dfrac{\mu_0 I}{2\pi r}\hat{k} = (1.50\times 10^{-6}\text{ T})\hat{i} + \dfrac{\mu_0(8.00\text{ A})}{2\pi(1.00\text{ m})}\hat{k}$.

$\vec{B} = (1.50 \times 10^{-6} \text{ T})\hat{i} + (1.6 \times 10^{-6} \text{ T})\hat{k} = 2.19 \times 10^{-6}$ T, at $\theta = 46.8°$ from x to z.

(c) At $(0, 0, -0.25 \text{ m})$, $\vec{B} = \vec{B}_0 + \dfrac{\mu_0 I}{2\pi r}\hat{i} = (1.50 \times 10^{-6} \text{ T})\hat{i} + \dfrac{\mu_0(8.00 \text{ A})}{2\pi(0.25 \text{ m})}\hat{i} = (7.9 \times 10^{-6} \text{ T})\hat{i}$.

EVALUATE: At point c the two fields are in the same direction and their magnitudes add. At point a they are in opposite directions and their magnitudes subtract. At point b the two fields are perpendicular.

28.27. IDENTIFY: The net magnetic field at any point is the vector sum of the magnetic fields of the two wires.

SET UP: For each wire $B = \dfrac{\mu_0 I}{2\pi r}$ and the direction of $\vec{B}$ is determined by the right-hand rule described in the text. Let the wire with 12.0 A be wire 1 and the wire with 10.0 A be wire 2.

EXECUTE: (a) Point Q: $B_1 = \dfrac{\mu_0 I_1}{2\pi r_1} = \dfrac{(4\pi \times 10^{-7} \text{ T} \cdot \text{m/A})(12.0 \text{ A})}{2\pi(0.15 \text{ m})} = 1.6 \times 10^{-5}$ T.

The direction of $\vec{B}_1$ is out of the page. $B_2 = \dfrac{\mu_0 I_2}{2\pi r_2} = \dfrac{(4\pi \times 10^{-7} \text{ T} \cdot \text{m/A})(10.0 \text{ A})}{2\pi(0.080 \text{ m})} = 2.5 \times 10^{-5}$ T.

The direction of $\vec{B}_2$ is out of the page. Since $\vec{B}_1$ and $\vec{B}_2$ are in the same direction,

$B = B_1 + B_2 = 4.1 \times 10^{-5}$ T and $\vec{B}$ is directed out of the page.

Point P: $B_1 = 1.6 \times 10^{-5}$ T, directed into the page. $B_2 = 2.5 \times 10^{-5}$ T, directed into the page.

$B = B_1 + B_2 = 4.1 \times 10^{-5}$ T and $\vec{B}$ is directed into the page.

(b) $\vec{B}_1$ is the same as in part (a), out of the page at Q and into the page at P. The direction of $\vec{B}_2$ is reversed from what it was in (a) so is into the page at Q and out of the page at P.

Point Q: $\vec{B}_1$ and $\vec{B}_2$ are in opposite directions so $B = B_2 - B_1 = 2.5 \times 10^{-5}$ T $- 1.6 \times 10^{-5}$ T $= 9.0 \times 10^{-6}$ T and $\vec{B}$ is directed into the page.

Point P: $\vec{B}_1$ and $\vec{B}_2$ are in opposite directions so $B = B_2 - B_1 = 9.0 \times 10^{-6}$ T and $\vec{B}$ is directed out of the page.

EVALUATE: Points P and Q are the same distances from the two wires. The only difference is that the fields point in either the same direction or in opposite directions.

28.29. IDENTIFY: Apply $\dfrac{F}{L} = \dfrac{\mu_0 I' I}{2\pi r}$.

SET UP: Two parallel conductors carrying current in the same direction attract each other. Parallel conductors carrying currents in opposite directions repel each other.

EXECUTE: (a) $F = \dfrac{\mu_0 I_1 I_2 L}{2\pi r} = \dfrac{\mu_0 (5.00 \text{ A})(2.00 \text{ A})(1.20 \text{ m})}{2\pi(0.400 \text{ m})} = 6.00 \times 10^{-6}$ N, and the force is repulsive since the currents are in opposite directions.

(b) Doubling the currents makes the force increase by a factor of four to $F = 2.40 \times 10^{-5}$ N.

EVALUATE: Doubling the current in a wire doubles the magnetic field of that wire. For fixed magnetic field, doubling the current in a wire doubles the force that the magnetic field exerts on the wire.

28.35. IDENTIFY: Calculate the magnetic field vector produced by each wire and add these fields to get the total field.

SET UP: First consider the field at P produced by the current I_1 in the upper semicircle of wire. See Figure 28.35a.

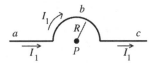

Consider the three parts of this wire:
a: long straight section
b: semicircle
c: long, straight section

Figure 28.35a

Apply the Biot-Savart law $d\vec{B} = \dfrac{\mu_0}{4\pi} \dfrac{I d\vec{l} \times \hat{r}}{r^2} = \dfrac{\mu_0}{4\pi} \dfrac{I d\vec{l} \times \vec{r}}{r^3}$ to each piece.

EXECUTE: part *a*: See Figure 28.35b.

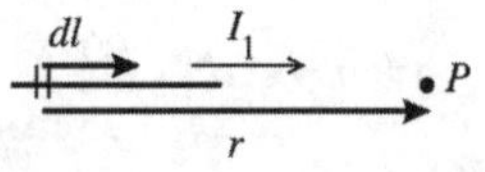

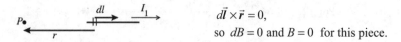

$d\vec{l} \times \vec{r} = 0$,
so $dB = 0$.

Figure 28.35b

The same is true for all the infinitesimal segments that make up this piece of the wire, so $B = 0$ for this piece.

part *c*: See Figure 28.35c.

$d\vec{l} \times \vec{r} = 0$,
so $dB = 0$ and $B = 0$ for this piece.

Figure 28.35c

part *b*: See Figure 28.35d.

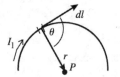

$d\vec{l} \times \vec{r}$ is directed into the paper for all infinitesimal segments that make up this semicircular piece, so $\vec{B}$ is directed into the paper and $B = \int dB$ (the vector sum of the $d\vec{B}$ is obtained by adding their magnitudes since they are in the same direction).

Figure 28.35d

$|d\vec{l} \times \vec{r}| = r\,dl\sin\theta$. The angle θ between $d\vec{l}$ and $\vec{r}$ is 90° and $r = R$, the radius of the semicircle. Thus $|d\vec{l} \times \vec{r}| = R\,dl$.

$$dB = \frac{\mu_0}{4\pi}\frac{I|d\vec{l} \times \vec{r}|}{r^3} = \frac{\mu_0 I_1}{4\pi}\frac{R}{R^3}dl = \left(\frac{\mu_0 I_1}{4\pi R^2}\right)dl.$$

$$B = \int dB = \left(\frac{\mu_0 I_1}{4\pi R^2}\right)\int dl = \left(\frac{\mu_0 I_1}{4\pi R^2}\right)(\pi R) = \frac{\mu_0 I_1}{4R}.$$

(We used that $\int dl$ is equal to πR, the length of wire in the semicircle.) We have shown that the two straight sections make zero contribution to $\vec{B}$, so $B_1 = \mu_0 I_1/4R$ and is directed into the page.

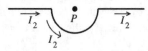

For current in the direction shown in Figure 28.35e, a similar analysis gives $B_2 = \mu_0 I_2/4R$, out of the paper.

Figure 28.35e

$\vec{B}_1$ and $\vec{B}_2$ are in opposite directions, so the magnitude of the net field at P is $B = |B_1 - B_2| = \dfrac{\mu_0|I_1 - I_2|}{4R}$.

EVALUATE: When $I_1 = I_2$, $B = 0$.

28.37. IDENTIFY: We use the equation for the magnetic field at the center of a single circular loop and then use the equation for the magnetic field inside a solenoid.

SET UP: The magnetic field at the center of a circular loop is $B_{\text{loop}} = \dfrac{\mu_0 I}{2R}$. The magnetic field at the center of a solenoid is $B_{\text{solenoid}} = \mu_0 n I$, where $n = \dfrac{N}{L}$ is the number of turns per meter.

EXECUTE: (a) $B_{\text{loop}} = \dfrac{\mu_0 I}{2R} = \dfrac{(4\pi \times 10^{-7}\ \text{T}\cdot\text{m/A})(2.00\ \text{A})}{2(0.050\ \text{m})} = 2.51 \times 10^{-5}\ \text{T}$.

(b) $n = \dfrac{N}{L} = \dfrac{1000}{5.00\ \text{m}} = 200\ \text{m}^{-1}$.

$B_{\text{solenoid}} = \mu_0 n I = (4\pi \times 10^{-7}\ \text{T}\cdot\text{m/A})(200\ \text{m}^{-1})(2.00\ \text{A}) = 5.03 \times 10^{-4}\ \text{T}$. $B_{\text{solenoid}} = 20 B_{\text{loop}}$. The field at the center of a circular loop depends on the radius of the loop. The field at the center of a solenoid depends on the length of the solenoid, not on its radius.

EVALUATE: The equation $B = \mu_0 n I$ for the field at the center of a solenoid is only correct for a very long solenoid, one whose length L is much greater than its radius R. We cannot consider the limit that L gets small and expect the expression for the solenoid to go over to the expression for N circular loops.

28.43. IDENTIFY: Apply Ampere's law.

SET UP: To calculate the magnetic field at a distance r from the center of the cable, apply Ampere's law to a circular path of radius r. By symmetry, $\oint \vec{B} \cdot d\vec{l} = B(2\pi r)$ for such a path.

EXECUTE: (a) For $a < r < b$, $I_{\text{encl}} = I \Rightarrow \oint \vec{B} \cdot d\vec{l} = \mu_0 I \Rightarrow B 2\pi r = \mu_0 I \Rightarrow B = \dfrac{\mu_0 I}{2\pi r}$.

(b) For $r > c$, the enclosed current is zero, so the magnetic field is also zero.

EVALUATE: A useful property of coaxial cables for many applications is that the current carried by the cable doesn't produce a magnetic field outside the cable.

28.45. IDENTIFY: We treat the solenoid as being ideal.

SET UP: At the center of an ideal solenoid, $B_{\text{solenoid}} = \mu_0 n I = \mu_0 \dfrac{N}{L} I$. A distance r from a long straight wire, $B_{\text{wire}} = \dfrac{\mu_0 I}{2\pi r}$.

EXECUTE: (a) $B_{\text{solenoid}} = (4\pi \times 10^{-7}\ \text{T}\cdot\text{m/A})\left(\dfrac{450}{0.35\ \text{m}}\right)(1.75\ \text{A}) = 2.83 \times 10^{-3}\ \text{T}$.

(b) $B_{\text{wire}} = \dfrac{(4\pi \times 10^{-7}\ \text{T}\cdot\text{m/A})(1.75\ \text{A})}{2\pi(1.0 \times 10^{-2}\ \text{m})} = 3.50 \times 10^{-5}\ \text{T}$.

EVALUATE: The magnetic field due to the wire is much less than the field at the center of the solenoid. For the solenoid, the fields of all the wires add to give a much larger field.

28.47. IDENTIFY and SET UP: The magnetic field near the center of a long solenoid is given by $B = \mu_0 n I$.

EXECUTE: (a) Turns per unit length $n = \dfrac{B}{\mu_0 I} = \dfrac{0.0270\ \text{T}}{(4\pi \times 10^{-7}\ \text{T}\cdot\text{m/A})(12.0\ \text{A})} = 1790\ \text{turns/m}$.

(b) $N = nL = (1790\ \text{turns/m})(0.400\ \text{m}) = 716\ \text{turns}$.

Each turn of radius R has a length $2\pi R$ of wire. The total length of wire required is
$N(2\pi R) = (716)(2\pi)(1.40 \times 10^{-2}\ \text{m}) = 63.0\ \text{m}$.

EVALUATE: A large length of wire is required. Due to the length of wire the solenoid will have appreciable resistance.

28.51. IDENTIFY: Inside an ideal toroidal solenoid, $B = \dfrac{\mu_0 N I}{2\pi r}$.

SET UP: $r = 0.070\ \text{m}$.

EXECUTE: $B = \dfrac{\mu_0 N I}{2\pi r} = \dfrac{\mu_0 (600)(0.650\ \text{A})}{2\pi(0.070\ \text{m})} = 1.11 \times 10^{-3}\ \text{T}$.

28.53. **IDENTIFY:** The magnetic field from the solenoid alone is $B_0 = \mu_0 n I$. The total magnetic field is $B = K_m B_0$. M is given by $\vec{B} = \vec{B}_0 + \mu_0 \vec{M}$.

SET UP: $n = 6000$ turns/m.

EXECUTE: (a) (i) $B_0 = \mu_0 n I = \mu_0 (6000 \text{ m}^{-1})(0.15 \text{ A}) = 1.13 \times 10^{-3}$ T.

(ii) $M = \dfrac{K_m - 1}{\mu_0} B_0 = \dfrac{5199}{\mu_0}(1.13 \times 10^{-3} \text{ T}) = 4.68 \times 10^6$ A/m.

(iii) $B = K_m B_0 = (5200)(1.13 \times 10^{-3} \text{ T}) = 5.88$ T.

(b) The directions of $\vec{B}$, $\vec{B}_0$ and $\vec{M}$ are shown in Figure 28.53. Silicon steel is paramagnetic and $\vec{B}_0$ and $\vec{M}$ are in the same direction.

EVALUATE: The total magnetic field is much larger than the field due to the solenoid current alone.

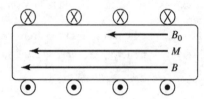

Figure 28.53

28.55. **IDENTIFY:** Moving charges create magnetic fields. The net field is the vector sum of the two fields. A charge moving in an external magnetic field feels a force.

(a) **SET UP:** The magnitude of the magnetic field due to a moving charge is $B = \dfrac{\mu_0}{4\pi} \dfrac{|q|v\sin\phi}{r^2}$. Both fields are into the paper, so their magnitudes add, giving $B_{\text{net}} = B + B' = \dfrac{\mu_0}{4\pi}\left(\dfrac{|q|v\sin\phi}{r^2} + \dfrac{|q'|v'\sin\phi'}{r'^2}\right)$.

EXECUTE: Substituting numbers gives

$$B_{\text{net}} = \dfrac{\mu_0}{4\pi}\left[\dfrac{(8.00 \ \mu\text{C})(9.00\times10^4 \text{ m/s})\sin 90°}{(0.300 \text{ m})^2} + \dfrac{(5.00 \ \mu\text{C})(6.50\times10^4 \text{ m/s})\sin 90°}{(0.400 \text{ m})^2}\right].$$

$B_{\text{net}} = 1.00 \times 10^{-6}$ T $= 1.00 \mu$T, into the paper.

(b) **SET UP:** The magnetic force on a moving charge is $\vec{F} = q\vec{v} \times \vec{B}$, and the magnetic field of charge q' at the location of charge q is into the page. The force on q is

$$\vec{F} = q\vec{v} \times \vec{B}' = (qv)\hat{i} \times \dfrac{\mu_0}{4\pi}\dfrac{q'\vec{v}' \times \hat{r}}{r^2} = (qv)\hat{i} \times \left(\dfrac{\mu_0}{4\pi}\dfrac{qv'\sin\phi}{r^2}\right)(-\hat{k}) = \left(\dfrac{\mu_0}{4\pi}\dfrac{qq'vv'\sin\phi}{r^2}\right)\hat{j}$$

where ϕ is the angle between $\vec{v}'$ and $\hat{r}'$.

EXECUTE: Substituting numbers gives

$$\vec{F} = \dfrac{\mu_0}{4\pi}\left[\dfrac{(8.00\times10^{-6} \text{ C})(5.00\times10^{-6} \text{ C})(9.00\times10^4 \text{ m/s})(6.50\times10^4 \text{ m/s})}{(0.500 \text{ m})^2}\left(\dfrac{0.400}{0.500}\right)\right]\hat{j}.$$

$$\vec{F} = (7.49\times10^{-8} \text{ N})\hat{j}.$$

EVALUATE: These are small fields and small forces, but if the charge has small mass, the force can affect its motion.

28.57. **IDENTIFY:** Use $B = \dfrac{\mu_0 I}{2\pi r}$ and the right-hand rule to determine points where the fields of the two wires cancel.

(a) SET UP: The only place where the magnetic fields of the two wires are in opposite directions is between the wires, in the plane of the wires. Consider a point a distance x from the wire carrying $I_2 = 75.0$ A. B_{tot} will be zero where $B_1 = B_2$.

EXECUTE: $\dfrac{\mu_0 I_1}{2\pi(0.400 \text{ m} - x)} = \dfrac{\mu_0 I_2}{2\pi x}$.

$I_2(0.400 \text{ m} - x) = I_1 x$; $I_1 = 25.0$ A, $I_2 = 75.0$ A.

$x = 0.300$ m; $B_{\text{tot}} = 0$ along a line 0.300 m from the wire carrying 75.0 A and 0.100 m from the wire carrying current 25.0 A.

(b) SET UP: Let the wire with $I_1 = 25.0$ A be 0.400 m above the wire with $I_2 = 75.0$ A. The magnetic fields of the two wires are in opposite directions in the plane of the wires and at points above both wires or below both wires. But to have $B_1 = B_2$ must be closer to wire #1 since $I_1 < I_2$, so can have $B_{\text{tot}} = 0$ only at points above both wires. Consider a point a distance x from the wire carrying $I_1 = 25.0$ A. B_{tot} will be zero where $B_1 = B_2$.

EXECUTE: $\dfrac{\mu_0 I_1}{2\pi x} = \dfrac{\mu_0 I_2}{2\pi(0.400 \text{ m} + x)}$.

$I_2 x = I_1(0.400 \text{ m} + x)$; $x = 0.200$ m.

$B_{\text{tot}} = 0$ along a line 0.200 m from the wire carrying current 25.0 A and 0.600 m from the wire carrying current $I_2 = 75.0$ A.

EVALUATE: For parts (a) and (b) the locations of zero field are in different regions. In each case the points of zero field are closer to the wire that has the smaller current.

28.59. **IDENTIFY:** Find the force that the magnetic field of the wire exerts on the electron.

SET UP: The force on a moving charge has magnitude $F = |q|vB\sin\phi$ and direction given by the right-hand rule. For a long straight wire, $B = \dfrac{\mu_0 I}{2\pi r}$ and the direction of $\vec{B}$ is given by the right-hand rule.

EXECUTE: (a) $a = \dfrac{F}{m} = \dfrac{|q|vB\sin\phi}{m} = \dfrac{ev}{m}\left(\dfrac{\mu_0 I}{2\pi r}\right)$. Substituting numbers gives

$a = \dfrac{(1.6\times 10^{-19}\text{ C})(2.50\times 10^5 \text{ m/s})(4\pi\times 10^{-7}\text{ T}\cdot\text{m/A})(13.0\text{ A})}{(9.11\times 10^{-31}\text{ kg})(2\pi)(0.0200\text{ m})} = 5.7\times 10^{12}$ m/s^2, away from the wire.

(b) The electric force must balance the magnetic force. $eE = evB$, and

$E = vB = v\dfrac{\mu_0 I}{2\pi r} = \dfrac{(250{,}000\text{ m/s})(4\pi\times 10^{-7}\text{ T}\cdot\text{m/A})(13.0\text{ A})}{2\pi(0.0200\text{ m})} = 32.5$ N/C. The magnetic force is directed away from the wire so the force from the electric field must be toward the wire. Since the charge of the electron is negative, the electric field must be directed away from the wire to produce a force in the desired direction.

EVALUATE: (c) $mg = (9.11\times 10^{-31}\text{ kg})(9.8\text{ m/s}^2) \approx 10^{-29}$ N.

$F_{\text{el}} = eE = (1.6\times 10^{-19}\text{ C})(32.5\text{ N/C}) \approx 5\times 10^{-18}$ N. $F_{\text{el}} \approx 5\times 10^{11} F_{\text{grav}}$, so we can neglect gravity.

28.61. **IDENTIFY and SET UP:** The power input of the motor is 65 hp. We know that 1 hp = 746 W. The relation between power, voltage, and current is $P = VI$. The attractive force between two parallel wires is $F = \dfrac{\mu_0 L I_1 I_2}{2\pi r}$.

EXECUTE: (a) We find the current from $I = \dfrac{P}{V} = \dfrac{(65\text{ hp})(746\text{ W/hp})}{600\text{ V}} = 80.8$ A, which rounds to 81 A.

(b) The attractive force between the wires per unit length is

$$F/L = \frac{(4\pi \times 10^{-7}\ \text{T}\cdot\text{m/A})(80.8\ \text{A})^2}{2\pi(0.55\ \text{m})} = 2.4 \times 10^{-3}\ \text{N/m}.$$

EVALUATE: If the current from the cables is in the same direction, the force will be attractive; however, if the current runs in opposite directions the force will be repulsive.

28.65. IDENTIFY: Apply $\sum \vec{F} = 0$ to one of the wires. The force one wire exerts on the other depends on I so $\sum \vec{F} = 0$ gives two equations for the two unknowns T and I.

SET UP: The force diagram for one of the wires is given in Figure 28.65.

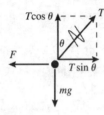

The force one wire exerts on the other is $F = \left(\dfrac{\mu_0 I^2}{2\pi r}\right) L$, where $r = 2(0.040\ \text{m})\sin\theta = 8.362 \times 10^{-3}$ m is the distance between the two wires.

Figure 28.65

EXECUTE: $\sum F_y = 0$ gives $T\cos\theta = mg$ and $T = mg/\cos\theta$.

$\sum F_x = 0$ gives $F = T\sin\theta = (mg/\cos\theta)\sin\theta = mg\tan\theta$.

And $m = \lambda L$, so $F = \lambda L g \tan\theta$.

$$\left(\frac{\mu_0 I^2}{2\pi r}\right) L = \lambda L g \tan\theta.$$

$$I = \sqrt{\frac{\lambda g r \tan\theta}{(\mu_0/2\pi)}}.$$

$$I = \sqrt{\frac{(0.0125\ \text{kg/m})(9.80\ \text{m/s}^2)(\tan 6.00°)(8.362 \times 10^{-3}\ \text{m})}{2 \times 10^{-7}\ \text{T}\cdot\text{m/A}}} = 23.2\ \text{A}.$$

EVALUATE: Since the currents are in opposite directions the wires repel. When I is increased, the angle θ from the vertical increases; a large current is required even for the small displacement specified in this problem.

28.69. (a) IDENTIFY: Consider current density J for a small concentric ring and integrate to find the total current in terms of α and R.

SET UP: We can't say $I = JA = J\pi R^2$, since J varies across the cross section.

To integrate J over the cross section of the wire, divide the wire cross section up into thin concentric rings of radius r and width dr, as shown in Figure 28.69.

Figure 28.69

EXECUTE: The area of such a ring is dA, and the current through it is $dI = J\,dA$; $dA = 2\pi r\,dr$ and $dI = J\,dA = \alpha r(2\pi r\,dr) = 2\pi \alpha r^2 dr$.

$$I = \int dI = 2\pi\alpha \int_0^R r^2 dr = 2\pi\alpha(R^3/3) \text{ so } \alpha = \frac{3I}{2\pi R^3}.$$

(b) IDENTIFY and SET UP: (i) $r \leq R$.

Apply Ampere's law to a circle of radius $r < R$. Use the method of part (a) to find the current enclosed by Ampere's law path.

EXECUTE: $\oint \vec{B} \cdot d\vec{l} = \oint B\,dl = B\oint dl = B(2\pi r)$, by the symmetry and direction of $\vec{B}$. The current passing through the path is $I_{encl} = \int dl$, where the integration is from 0 to r.

$I_{encl} = 2\pi\alpha \int_0^r r^2\,dr = \frac{2\pi\alpha r^3}{3} = \frac{2\pi}{3}\left(\frac{3I}{2\pi R^3}\right)r^3 = \frac{Ir^3}{R^3}$. Thus $\oint \vec{B} \cdot d\vec{l} = \mu_0 I_{encl}$ gives

$B(2\pi r) = \mu_0\left(\frac{Ir^3}{R^3}\right)$ and $B = \frac{\mu_0 Ir^2}{2\pi R^3}$.

(ii) **IDENTIFY** and **SET UP:** $r \geq R$.

Apply Ampere's law to a circle of radius $r > R$.

EXECUTE: $\oint \vec{B} \cdot d\vec{l} = \oint B\,dl = B\oint dl = B(2\pi r)$.

$I_{encl} = I$; all the current in the wire passes through this path. Thus $\oint \vec{B} \cdot d\vec{l} = \mu_0 I_{encl}$ gives $B(2\pi r) = \mu_0 I$

and $B = \frac{\mu_0 I}{2\pi r}$.

EVALUATE: Note that at $r = R$ the expression in (i) (for $r \leq R$) gives $B = \frac{\mu_0 I}{2\pi R}$. At $r = R$ the

expression in (ii) (for $r \geq R$) gives $B = \frac{\mu_0 I}{2\pi R}$, which is the same.

28.73. IDENTIFY: Use what we know about the magnetic field of a long, straight conductor to deduce the symmetry of the magnetic field. Then apply Ampere's law to calculate the magnetic field at a distance a above and below the current sheet.

SET UP: Do parts (a) and (b) together.

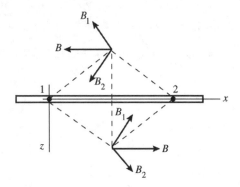

Consider the individual currents in pairs, where the currents in each pair are equidistant on either side of the point where $\vec{B}$ is being calculated. Figure 28.73a shows that for each pair the z-components cancel, and that above the sheet the field is in the $-x$-direction and that below the sheet it is in the $+x$-direction.

Figure 28.73a

Also, by symmetry the magnitude of $\vec{B}$ a distance a above the sheet must equal the magnitude of $\vec{B}$ a distance a below the sheet. Now that we have deduced the symmetry of $\vec{B}$, apply Ampere's law. Use a path that is a rectangle, as shown in Figure 28.73b.

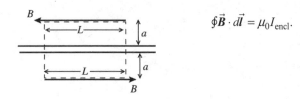

Figure 28.73b

I is directed out of the page, so for I to be positive the integral around the path is taken in the counterclockwise direction.

EXECUTE: Since $\vec{B}$ is parallel to the sheet, on the sides of the rectangle that have length $2a$, $\oint \vec{B} \cdot d\vec{l} = 0$. On the long sides of length L, $\vec{B}$ is parallel to the side, in the direction we are integrating around the path, and has the same magnitude, B, on each side. Thus $\oint \vec{B} \cdot d\vec{l} = 2BL$. n conductors per unit length and current I out of the page in each conductor gives $I_{encl} = InL$. Ampere's law then gives $2BL = \mu_0 InL$ and $B = \frac{1}{2}\mu_0 In$.

EVALUATE: Note that B is independent of the distance a from the sheet. Compare this result to the electric field due to an infinite sheet of charge in Chapter 22.

28.77. **IDENTIFY and SET UP:** The magnitude of the magnetic a distance r from the center of a very long current-carrying wire is $B = \frac{\mu_0 I}{2\pi r}$. In this case, the measured quantity x is the distance from the *surface* of the cable, not from the center.

EXECUTE: **(a)** Multiplying the quantities given in the table in the problem, we get the following values for Bx in units of T·cm, starting with the first pair: 0.812, 1.00, 1.09, 1.13, 1.16. As we can see, these values are not constant. However the last three values are nearly constant. Therefore Bx is not truly constant. The reason for this is that x is the distance from the *surface* of the cable, not from the center. In the formula $B = \frac{\mu_0 I}{2\pi r}$, r is the distance from the center of the cable. In that case, we would expect Br to be constant. For the last three points, it does appear that Bx is nearly constant. The reason for this is that the proper formula for the magnetic field for this cable is $B = \frac{\mu_0 I}{2\pi(R+x)}$, where R is the radius of the cable. As x gets large compared to R, $r \approx x$ and the magnitude approaches $\frac{\mu_0 I}{2\pi r}$.

(b) Using the equation appropriate for the cable and solving for x gives $x = (\mu_0 I/2\pi)\frac{1}{B} - R$. A graph of x versus $1/B$ should have a slope equal to $\mu_0 I/2\pi$ and a y-intercept equal to $-R$. Figure 28.77 shows the graph of x versus $1/B$.

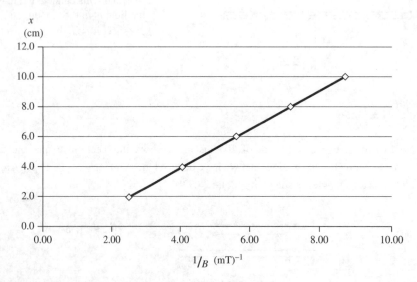

Figure 28.77

(c) The best-fit equation for this graph is $x = (1.2981 \text{ mT} \cdot \text{cm})\frac{1}{B} - 1.1914$ cm. The slope is 1.2981 mT·cm = 1.2981×10^{-5} T·m. Since the slope is equal to $\mu_0 I/2\pi$, we have

$\mu_0 I/2\pi$ = slope, which gives $I = 2\pi(\text{slope})/\mu_0 = 2\pi(1.2981\times10^{-5}\text{ T}\cdot\text{m})/\mu_0 = 64.9$ A, which rounds to 65 A. The y-intercept is $-R$, so $R = -(-1.1914\text{ cm}) = 1.2$ cm.

EVALUATE: As we can see, the field within 2 cm or so of the surface of the cable would vary considerably from the value given by $B = \dfrac{\mu_0 I}{2\pi r}$.

28.83. **IDENTIFY and SET UP:** The enclosure is no longer present to shield the solenoid from the earth's magnetic field of 50 μT, so net field inside is a sum of the solenoid field and the earth's field. Whether the earth's field adds or subtracts from the solenoid's field depends on the orientation of the solenoid. The magnetic field due to the solenoid is 150 μT.

EXECUTE: When the solenoid field is parallel to the earth's field, the net field is 150 μT + 50 μT = 200 μT. When the field's are antiparallel (opposite), the net field is 150 μT – 50 μT = 100 μT. So the field that the bacteria experience is between 100 μT and 200 μT, which is choice (c).

EVALUATE: Since the earth's field is quite appreciable compared to the solenoid's field, it is important to shield the solenoid from external fields, such as that of the earth. The earth's field can make a difference of up to a factor of 2 in the field experienced by the bacteria.

ELECTROMAGNETIC INDUCTION

29.3. **IDENTIFY** and **SET UP:** Use Faraday's law to calculate the average induced emf and apply Ohm's law to the coil to calculate the average induced current and charge that flows.

(a) **EXECUTE:** The magnitude of the average emf induced in the coil is $|\varepsilon_{av}| = N\left|\dfrac{\Delta\Phi_B}{\Delta t}\right|$. Initially, $\Phi_{Bi} = BA\cos\phi = BA$. The final flux is zero, so $|\varepsilon_{av}| = N\dfrac{|\Phi_{Bf} - \Phi_{Bi}|}{\Delta t} = \dfrac{NBA}{\Delta t}$. The average induced current is $I = \dfrac{|\varepsilon_{av}|}{R} = \dfrac{NBA}{R\Delta t}$. The total charge that flows through the coil is $Q = I\Delta t = \left(\dfrac{NBA}{R\Delta t}\right)\Delta t = \dfrac{NBA}{R}$.

EVALUATE: The charge that flows is proportional to the magnetic field but does not depend on the time Δt.

(b) The magnetic stripe consists of a pattern of magnetic fields. The pattern of charges that flow in the reader coil tells the card reader the magnetic field pattern and hence the digital information coded onto the card.

(c) According to the result in part (a) the charge that flows depends only on the change in the magnetic flux and it does not depend on the rate at which this flux changes.

29.5. **IDENTIFY:** Apply Faraday's law.

SET UP: Let $+z$ be the positive direction for $\vec{A}$. Therefore, the initial flux is positive and the final flux is zero.

EXECUTE: (a) and (b) $\varepsilon = -\dfrac{\Delta\Phi_B}{\Delta t} = -\dfrac{0 - (1.5\text{ T})\pi(0.120\text{ m})^2}{2.0\times 10^{-3}\text{ s}} = +34$ V. Since ε is positive and $\vec{A}$ is toward us, the induced current is counterclockwise.

EVALUATE: The shorter the removal time, the larger the average induced emf.

29.9. **IDENTIFY** and **SET UP:** Use Faraday's law to calculate the emf (magnitude and direction). The direction of the induced current is the same as the direction of the emf. The flux changes because the area of the loop is changing; relate dA/dt to dc/dt, where c is the circumference of the loop.

(a) **EXECUTE:** $c = 2\pi r$ and $A = \pi r^2$ so $A = c^2/4\pi$.

$\Phi_B = BA = (B/4\pi)c^2$.

$|\varepsilon| = \left|\dfrac{d\Phi_B}{dt}\right| = \left(\dfrac{B}{2\pi}\right)c\left|\dfrac{dc}{dt}\right|$.

At $t = 9.0$ s, $c = 1.650$ m $-$ (9.0 s)(0.120 m/s) $= 0.570$ m.

$|\varepsilon| = (0.500\text{ T})(1/2\pi)(0.570\text{ m})(0.120\text{ m/s}) = 5.44$ mV.

(b) SET UP: The loop and magnetic field are sketched in Figure 29.9.

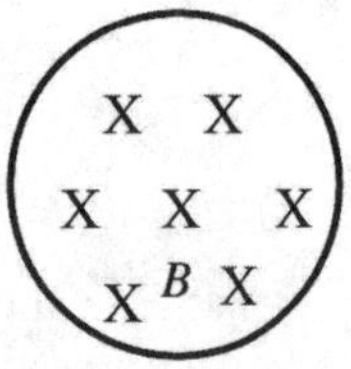

Take into the page to be the positive direction for $\vec{A}$. Then the magnetic flux is positive.

Figure 29.9

EXECUTE: The positive flux is decreasing in magnitude; $d\Phi_B/dt$ is negative and ε is positive. By the right-hand rule, for $\vec{A}$ into the page, positive ε is clockwise.

EVALUATE: Even though the circumference is changing at a constant rate, dA/dt is not constant and $|\varepsilon|$ is not constant. Flux $\otimes$ is decreasing so the flux of the induced current is $\otimes$ and this means that I is clockwise, which checks.

29.11. IDENTIFY: A change in magnetic flux through a coil induces an emf in the coil.
SET UP: The flux through a coil is $\Phi_B = NBA\cos\phi$ and the induced emf is $\varepsilon = -d\Phi_B/dt$.
EXECUTE: (a) $|\varepsilon| = d\Phi_B/dt = d[A(B_0 + bx)]/dt = bA\, dx/dt = bAv$.

(b) Clockwise
(c) Same answers except the current is counterclockwise.
EVALUATE: Even though the coil remains within the magnetic field, the flux through it changes because the strength of the field is changing.

29.13. IDENTIFY: Apply the results of Example 29.3.
SET UP: $\varepsilon_{max} = NBA\omega$.

EXECUTE: $\omega = \dfrac{\varepsilon_{max}}{NBA} = \dfrac{2.40\times 10^{-2}\text{ V}}{(120)(0.0750\text{ T})(0.016\text{ m})^2} = 10.4\text{ rad/s}$.

EVALUATE: We may also express ω as 99.3 rev/min or 1.66 rev/s.

29.23. IDENTIFY: The changing flux through the loop due to the changing magnetic field induces a current in the wire.
SET UP: The magnitude of the induced emf is $|\varepsilon| = \left|\dfrac{d\Phi_B}{dt}\right| = \pi r^2 \left|\dfrac{dB}{dt}\right|$, $I = \varepsilon/R$.

EXECUTE: $\vec{B}$ is into the page and Φ_B is increasing, so the field of the induced current is directed out of the page inside the loop and the induced current is counterclockwise.

$|\varepsilon| = \left|\dfrac{d\Phi_B}{dt}\right| = \pi r^2 \left|\dfrac{dB}{dt}\right| = \pi(0.0250\text{ m})^2(0.380\text{ T/s}^3)(3t^2) = (2.238\times 10^{-3}\text{ V/s}^2)t^2$.

$I = \dfrac{|\varepsilon|}{R} = (5.739\times 10^{-3}\text{ A/s}^2)t^2$. When $B = 1.33$ T, we have $1.33\text{ T} = (0.380\text{ T/s}^3)t^3$, which gives $t = 1.518$ s. At this t, $I = (5.739\times 10^{-3}\text{ A/s}^2)(1.518\text{ s})^2 = 0.0132$ A.

EVALUATE: As the field changes, the current will also change.

29.29. IDENTIFY and SET UP: $\varepsilon = vBL$. Use Lenz's law to determine the direction of the induced current. The force F_{ext} required to maintain constant speed is equal and opposite to the force F_I that the magnetic field exerts on the rod because of the current in the rod.
EXECUTE: (a) $\varepsilon = vBL = (7.50\text{ m/s})(0.800\text{ T})(0.500\text{ m}) = 3.00$ V.

(b) $\vec{B}$ is into the page. The flux increases as the bar moves to the right, so the magnetic field of the induced current is out of the page inside the circuit. To produce magnetic field in this direction the induced current must be counterclockwise, so from b to a in the rod.

Electromagnetic Induction   29-3

(c) $I = \dfrac{\varepsilon}{R} = \dfrac{3.00 \text{ V}}{1.50 \text{ }\Omega} = 2.00 \text{ A}$. $F_I = ILB\sin\phi = (2.00 \text{ A})(0.500 \text{ m})(0.800 \text{ T})\sin 90° = 0.800 \text{ N}$. $\vec{F}_I$ is to the left. To keep the bar moving to the right at constant speed an external force with magnitude $F_{ext} = 0.800 \text{ N}$ and directed to the right must be applied to the bar.

(d) The rate at which work is done by the force F_{ext} is $F_{ext}v = (0.800 \text{ N})(7.50 \text{ m/s}) = 6.00 \text{ W}$. The rate at which thermal energy is developed in the circuit is $I^2R = (2.00 \text{ A})^2(1.50 \text{ }\Omega) = 6.00 \text{ W}$. These two rates are equal, as is required by conservation of energy.

EVALUATE: The force on the rod due to the induced current is directed to oppose the motion of the rod. This agrees with Lenz's law.

29.35.   IDENTIFY: While the circuit is entering and leaving the region of the magnetic field, the flux through it will be changing. This change will induce an emf in the circuit.

SET UP: When the loop is entering or leaving the region of magnetic field the flux through it is changing and there is an induced emf. The magnitude of this induced emf is $\varepsilon = BLv$. The length L is 0.750 m. When the loop is totally within the field the flux through the loop is not changing so there is no induced emf. The induced current has magnitude $I = \dfrac{\varepsilon}{R}$ and direction given by Lenz's law.

EXECUTE: (a) $I = \dfrac{\varepsilon}{R} = \dfrac{BLv}{R} = \dfrac{(1.25 \text{ T})(0.750 \text{ m})(3.0 \text{ m/s})}{12.5 \text{ }\Omega} = 0.225 \text{ A}$. The magnetic field through the loop is directed out of the page and is increasing, so the magnetic field of the induced current is into the page inside the loop and the induced current is clockwise.

(b) The flux is not changing so ε and I are zero.

(c) $I = \dfrac{\varepsilon}{R} = 0.225 \text{ A}$. The magnetic field through the loop is directed out of the page and is decreasing, so the magnetic field of the induced current is out of the page inside the loop and the induced current is counterclockwise.

(d) Let clockwise currents be positive. At $t = 0$ the loop is entering the field. It is totally in the field at time t_a and beginning to move out of the field at time t_b. The graph of the induced current as a function of time is sketched in Figure 29.35.

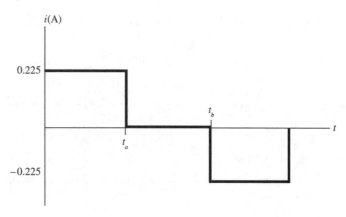

Figure 29.35

EVALUATE: Even though the circuit is moving throughout all parts of this problem, an emf is induced in it only when the flux through it is changing. While the coil is entirely within the field, the flux is constant, so no emf is induced.

29.37.   IDENTIFY: Apply $\varepsilon = \oint \vec{E} \cdot d\vec{l} = -\dfrac{d\Phi_B}{dt}$.

SET UP: Evaluate the integral for a path which is a circle of radius r and concentric with the solenoid. The magnetic field of the solenoid is confined to the region inside the solenoid, so $B(r) = 0$ for $r > R$.

EXECUTE: (a) $\dfrac{d\Phi_B}{dt} = A\dfrac{dB}{dt} = \pi r_1^2 \dfrac{dB}{dt}$.

(b) $E = \dfrac{1}{2\pi r_1}\dfrac{d\Phi_B}{dt} = \dfrac{\pi r_1^2}{2\pi r_1}\dfrac{dB}{dt} = \dfrac{r_1}{2}\dfrac{dB}{dt}$. The direction of $\vec{E}$ is shown in Figure 29.37a.

(c) All the flux is within $r < R$, so outside the solenoid $E = \dfrac{1}{2\pi r_2}\dfrac{d\Phi_B}{dt} = \dfrac{\pi R^2}{2\pi r_2}\dfrac{dB}{dt} = \dfrac{R^2}{2r_2}\dfrac{dB}{dt}$.

(d) The graph is sketched in Figure 29.37b.

(e) At $r = R/2$, $|\varepsilon| = \dfrac{d\Phi_B}{dt} = \pi(R/2)^2 \dfrac{dB}{dt} = \dfrac{\pi R^2}{4}\dfrac{dB}{dt}$.

(f) At $r = R$, $|\varepsilon| = \dfrac{d\Phi_B}{dt} = \pi R^2 \dfrac{dB}{dt}$.

(g) At $r = 2R$, $|\varepsilon| = \dfrac{d\Phi_B}{dt} = \pi R^2 \dfrac{dB}{dt}$.

EVALUATE: The emf is independent of the distance from the center of the cylinder at all points outside it. Even though the magnetic field is zero for $r > R$, the induced electric field is nonzero outside the solenoid and a nonzero emf is induced in a circular turn that has $r > R$.

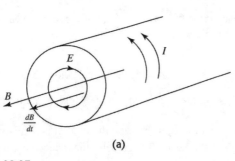

Figure 29.37

29.39. IDENTIFY: Apply $E = \dfrac{1}{2\pi r}\left|\dfrac{d\Phi_B}{dt}\right|$ with $\Phi_B = \mu_0 n i A$.

SET UP: $A = \pi r^2$, where $r = 0.0110$ m. In $E = \dfrac{1}{2\pi r}\left|\dfrac{d\Phi_B}{dt}\right|$, $r = 0.0350$ m.

EXECUTE: $|\varepsilon| = \left|\dfrac{d\Phi_B}{dt}\right| = \left|\dfrac{d}{dt}(BA)\right| = \left|\dfrac{d}{dt}(\mu_0 n i A)\right| = \mu_0 n A \left|\dfrac{di}{dt}\right|$ and $|\varepsilon| = E(2\pi r)$. Therefore, $\left|\dfrac{di}{dt}\right| = \dfrac{E 2\pi r}{\mu_0 n A}$.

$\left|\dfrac{di}{dt}\right| = \dfrac{(8.00\times10^{-6}\text{ V/m}) 2\pi(0.0350\text{ m})}{\mu_0(400\text{ m}^{-1})\pi(0.0110\text{ m})^2} = 9.21$ A/s.

EVALUATE: Outside the solenoid the induced electric field decreases with increasing distance from the axis of the solenoid.

29.43. IDENTIFY: $q = CV$. For a parallel-plate capacitor, $C = \dfrac{\varepsilon A}{d}$, where $\varepsilon = K\varepsilon_0$. $i_C = dq/dt$. $j_D = \varepsilon\dfrac{dE}{dt}$.

SET UP: $E = q/\varepsilon A$ so $dE/dt = i_C/\varepsilon A$.

EXECUTE: (a) $q = CV = \left(\dfrac{\varepsilon A}{d}\right)V = \dfrac{(4.70)\varepsilon_0(3.00\times10^{-4}\text{ m}^2)(120\text{ V})}{2.50\times10^{-3}\text{ m}} = 5.99\times10^{-10}$ C.

(b) $\dfrac{dq}{dt} = i_C = 6.00\times10^{-3}$ A.

(c) $j_D = \varepsilon \dfrac{dE}{dt} = K\varepsilon_0 \dfrac{i_C}{K\varepsilon_0 A} = \dfrac{i_C}{A} = j_C$, so $i_D = i_C = 6.00 \times 10^{-3}$ A.

EVALUATE: $i_D = i_C$, so Kirchhoff's junction rule is satisfied where the wire connects to each capacitor plate.

29.45. IDENTIFY: Apply $\vec{B} = \vec{B}_0 + \mu_0 \vec{M}$.

SET UP: For magnetic fields less than the critical field, there is no internal magnetic field. For fields greater than the critical field, $\vec{B}$ is very nearly equal to $\vec{B}_0$.

EXECUTE: **(a)** The external field is less than the critical field, so inside the superconductor $\vec{B} = 0$ and $\vec{M} = -\dfrac{\vec{B}_0}{\mu_0} = -\dfrac{(0.130\text{ T})\hat{\imath}}{\mu_0} = -(1.03 \times 10^5\text{ A/m})\hat{\imath}$. Outside the superconductor, $\vec{B} = \vec{B}_0 = (0.130\text{ T})\hat{\imath}$ and $\vec{M} = 0$.

(b) The field is greater than the critical field and $\vec{B} = \vec{B}_0 = (0.260\text{ T})\hat{\imath}$, both inside and outside the superconductor.

EVALUATE: Below the critical field the external field is expelled from the superconducting material.

29.47. IDENTIFY: Apply Faraday's law and Lenz's law.

SET UP: For a discharging RC circuit, $i(t) = \dfrac{V_0}{R} e^{-t/RC}$, where V_0 is the initial voltage across the capacitor. The resistance of the small loop is $(25)(0.600\text{ m})(1.0\text{ }\Omega/\text{m}) = 15.0\text{ }\Omega$.

EXECUTE: **(a)** The large circuit is an RC circuit with a time constant of $\tau = RC = (10\text{ }\Omega)(20 \times 10^{-6}\text{ F}) = 200\text{ }\mu\text{s}$. Thus, the current as a function of time is $i = ((100\text{ V})/(10\text{ }\Omega))\, e^{-t/200\,\mu\text{s}}$. At $t = 200\text{ }\mu\text{s}$, we obtain $i = (10\text{ A})(e^{-1}) = 3.7$ A.

(b) Assuming that only the long wire nearest the small loop produces an appreciable magnetic flux through the small loop and referring to the solution of Exercise 29.7 we obtain $\Phi_B = \int_c^{c+a} \dfrac{\mu_0 i b}{2\pi r}\, dr = \dfrac{\mu_0 i b}{2\pi} \ln\!\left(1 + \dfrac{a}{c}\right)$.

Therefore, the emf induced in the small loop at $t = 200\text{ }\mu\text{s}$ is $\varepsilon = -N\dfrac{d\Phi_B}{dt} = -\dfrac{N\mu_0 b}{2\pi} \ln\!\left(1 + \dfrac{a}{c}\right) \dfrac{di}{dt}$.

$\varepsilon = -\dfrac{(25)(4\pi \times 10^{-7}\text{ Wb/A}\cdot\text{m}^2)(0.200\text{ m})}{2\pi} \ln(3.0) \left(-\dfrac{3.7\text{ A}}{200 \times 10^{-6}\text{ s}}\right) = +20.0$ mV. Thus, the induced current in the small loop is $i' = \dfrac{\varepsilon}{R} = \dfrac{20.0\text{ mV}}{15.0\text{ }\Omega} = 1.33$ mA.

(c) The magnetic field from the large loop is directed out of the page within the small loop. The induced current will act to oppose the decrease in flux from the large loop. Thus, the induced current flows counterclockwise.

EVALUATE: **(d)** Three of the wires in the large loop are too far away to make a significant contribution to the flux in the small loop—as can be seen by comparing the distance c to the dimensions of the large loop.

29.49. IDENTIFY: The changing current in the solenoid will cause a changing magnetic field (and hence changing flux) through the secondary winding, which will induce an emf in the secondary coil.

SET UP: The magnetic field of the solenoid is $B = \mu_0 n i$, and the induced emf is $|\varepsilon| = N\left|\dfrac{d\Phi_B}{dt}\right|$.

EXECUTE: $B = \mu_0 n i = (4\pi \times 10^{-7}\text{ T}\cdot\text{m/A})(90.0 \times 10^2\text{ m}^{-1})(0.160\text{ A/s}^2)t^2 = (1.810 \times 10^{-3}\text{ T/s}^2)t^2$. The total flux through secondary winding is $(5.0)B(2.00 \times 10^{-4}\text{ m}^2) = (1.810 \times 10^{-6}\text{ Wb/s}^2)t^2$.

$|\varepsilon| = N\left|\dfrac{d\Phi_B}{dt}\right| = (3.619 \times 10^{-6}\text{ V/s})t$. $i = 3.20$ A says $3.20\text{ A} = (0.160\text{ A/s}^2)t^2$ and $t = 4.472$ s. This gives $|\varepsilon| = (3.619 \times 10^{-6}\text{ V/s})(4.472\text{ s}) = 1.62 \times 10^{-5}$ V.

EVALUATE: This a very small voltage, about 16 μV.

29.59. **IDENTIFY:** Use the expression for motional emf to calculate the emf induced in the rod.
SET UP: (a) The rotating rod is shown in Figure 29.59a.

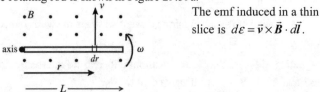

Figure 29.59a

The emf induced in a thin slice is $d\varepsilon = \vec{v} \times \vec{B} \cdot d\vec{l}$.

EXECUTE: Assume that $\vec{B}$ is directed out of the page. Then $\vec{v} \times \vec{B}$ is directed radially outward and $dl = dr$, so $\vec{v} \times \vec{B} \cdot d\vec{l} = vB\,dr$.
$v = r\omega$ so $d\varepsilon = \omega B r\,dr$.
The $d\varepsilon$ for all the thin slices that make up the rod are in series so they add:
$\varepsilon = \int d\varepsilon = \int_0^L \omega B r\,dr = \tfrac{1}{2}\omega B L^2 = \tfrac{1}{2}(8.80 \text{ rad/s})(0.650 \text{ T})(0.240 \text{ m})^2 = 0.165$ V.

EVALUATE: ε increases with $\omega, B,$ or L^2.
(b) SET UP and EXECUTE: No current flows so there is no IR drop in potential. Thus the potential difference between the ends equals the emf of 0.165 V calculated in part (a).
(c) SET UP: The rotating rod is shown in Figure 29.59b.

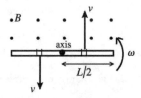

Figure 29.59b

EXECUTE: The emf between the center of the rod and each end is
$\varepsilon = \tfrac{1}{2}\omega B (L/2)^2 = \tfrac{1}{4}(0.165 \text{ V}) = 0.0412$ V, with the direction of the emf from the center of the rod toward each end. The emfs in each half of the rod thus oppose each other and there is no net emf between the ends of the rod.
EVALUATE: ω and B are the same as in part (a) but L of each half is $\tfrac{1}{2}L$ for the whole rod. ε is proportional to L^2, so is smaller by a factor of $\tfrac{1}{4}$.

29.61. (a) **IDENTIFY:** Use Faraday's law to calculate the induced emf, Ohm's law to calculate I, and $\vec{F} = I\vec{l} \times \vec{B}$ to calculate the force on the rod due to the induced current.
SET UP: The force on the wire is shown in Figure 29.61.

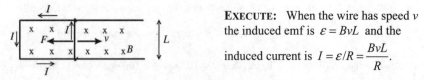

Figure 29.61

EXECUTE: When the wire has speed v the induced emf is $\varepsilon = BvL$ and the induced current is $I = \varepsilon/R = \dfrac{BvL}{R}$.

The induced current flows upward in the wire as shown, so the force $\vec{F} = I\vec{l} \times \vec{B}$ exerted by the magnetic field on the induced current is to the left. $\vec{F}$ opposes the motion of the wire, as it must by Lenz's law. The magnitude of the force is $F = ILB = B^2L^2v/R$.

(b) IDENTIFY and SET UP: Apply $\sum \vec{F} = m\vec{a}$ to the wire. Take $+x$ to be toward the right and let the origin be at the location of the wire at $t = 0$, so $x_0 = 0$.

EXECUTE: $\sum F_x = ma_x$ says $-F = ma_x$.

$a_x = -\dfrac{F}{m} = -\dfrac{B^2L^2v}{mR}$.

Use this expression to solve for $v(t)$:

$a_x = \dfrac{dv}{dt} = -\dfrac{B^2L^2v}{mR}$ and $\dfrac{dv}{v} = -\dfrac{B^2L^2}{mR}dt$.

$\int_{v_0}^{v} \dfrac{dv'}{v'} = -\dfrac{B^2L^2}{mR} \int_0^t dt'$.

$\ln(v) - \ln(v_0) = -\dfrac{B^2L^2 t}{mR}$.

$\ln\left(\dfrac{v}{v_0}\right) = -\dfrac{B^2L^2 t}{mR}$ and $v = v_0 e^{-B^2L^2 t/mR}$.

Note: At $t = 0$, $v = v_0$ and $v \to 0$ when $t \to \infty$.

Now solve for $x(t)$:

$v = \dfrac{dx}{dt} = v_0 e^{-B^2L^2 t/mR}$ so $dx = v_0 e^{-B^2L^2 t/mR} dt$.

$\int_0^x dx' = \int_0^t v_0 e^{-B^2L^2 t'/mR} dt'$.

$x = v_0\left(-\dfrac{mR}{B^2L^2}\right)\left[e^{-B^2L^2 t'/mR}\right]_0^t = \dfrac{mRv_0}{B^2L^2}(1 - e^{-B^2L^2 t/mR})$.

Comes to rest implies $v = 0$. This happens when $t \to \infty$.

$t \to \infty$ gives $x = \dfrac{mRv_0}{B^2L^2}$. Thus this is the distance the wire travels before coming to rest.

EVALUATE: The motion of the slide wire causes an induced emf and current. The magnetic force on the induced current opposes the motion of the wire and eventually brings it to rest. The force and acceleration depend on v and are constant. If the acceleration were constant, not changing from its initial value of $a_x = -B^2L^2 v_0/mR$, then the stopping distance would be $x = -v_0^2/2a_x = mRv_0/2B^2L^2$. The actual stopping distance is twice this.

29.65. IDENTIFY: Apply $i_D = \varepsilon \dfrac{d\Phi_E}{dt}$.

SET UP: $\varepsilon = 3.5 \times 10^{-11}$ F/m.

EXECUTE: $i_D = \varepsilon \dfrac{d\Phi_E}{dt} = (3.5 \times 10^{-11} \text{ F/m})(24.0 \times 10^3 \text{ V} \cdot \text{m/s}^3)t^2$. $i_D = 21 \times 10^{-6}$ A gives $t = 5.0$ s.

EVALUATE: i_D depends on the rate at which Φ_E is changing.

29.67. IDENTIFY: An emf is induced across the moving metal bar, which causes current to flow in the circuit. The magnetic field exerts a force on the moving bar due to the current in it, which causes acceleration of the bar. Newton's second law applies to the accelerating bar. Ohm's law applies to the resistor in the circuit.

SET UP: The induced potential across the moving bar is $\varepsilon = vBL$, the magnetic force on the bar is $F_{mag} = ILB$, and Ohm's law is $\varepsilon = IR$. Newton's second law is $\sum \vec{F} = m\vec{a}$, and $a_x = dv_x/dt$. The flux through the loop is increasing, so the induced current is counterclockwise. Alternatively, the magnetic force $\vec{F} = q\vec{v} \times \vec{B}$ on positive charge in the moving bar is upward, by the right-hand rule, which also gives a counterclockwise current. So the magnetic force on the bar is to the left, opposite to the velocity of the bar.

EXECUTE: (a) Combining the equations discussed in the set up, the magnetic force on the moving bar is $F_{mag} = ILB = (\varepsilon/R)LB = (vBL/R)LB = v(BL)^2/R$.

Newton's second law gives
$F_{mag} = ma$.
$ma = v(BL)^2/R$.
$a = \dfrac{(BL)^2}{mR} v$.

A graph of a versus v should be a straight line having slope equal to $(BL)^2/mR$. The graph of a versus v is shown in Figure 29.67. The best-fit slope of this graph is 0.3071 s^{-1}.

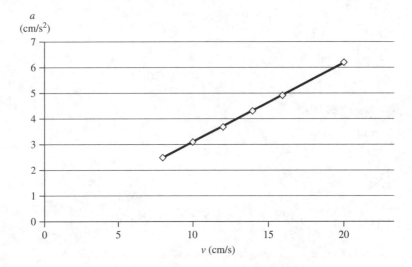

Figure 29.67

(b) $(BL)^2/mR$ = slope, so $B = \sqrt{\dfrac{(\text{slope})mR}{L^2}} = \sqrt{\dfrac{(0.3071 \text{ s}^{-1})(0.200 \text{ kg})(0.800 \text{ }\Omega)}{(0.0600 \text{ m})^2}} = 3.69$ T.

(c) The current flows in a counterclockwise direction in the circuit. Therefore the charges lose potential energy as they pass through the resistor R from a to b, which makes point a at a higher potential than b.

(d) We know that $a_x = dv_x/dt$, and in part (a) we found that the magnitude of the acceleration is $a = \dfrac{(BL)^2}{mR}v$. We also saw that a is opposite to v, so $a_x = -\dfrac{(BL)^2}{mR}v$. Therefore $\dfrac{dv}{dt} = -\dfrac{(BL)^2}{mR}v$. Separating variables and integrating gives

$\int_{20.0 \text{ cm/s}}^{10.0 \text{ cm/s}} \dfrac{dv}{v} = -\int_0^t \dfrac{(BL)^2}{mR} dt'$.

$\ln\left(\dfrac{10}{20}\right) = -\dfrac{(BL)^2}{mR} t$.

$t = -\dfrac{mR}{(BL)^2} \ln(1/2) = -(0.200 \text{ kg})(0.800 \text{ }\Omega)(\ln \tfrac{1}{2})/[(3.69 \text{ T})(0.0600 \text{ m})]^2 = 2.26$ s.

EVALUATE: We cannot use the standard kinematics formulas because the acceleration is not constant.

29.73. **IDENTIFY and SET UP:** Faraday's law gives $\varepsilon = \left|\dfrac{d\Phi_B}{dt}\right| = \dfrac{d(B_{av}A)}{dt} = A\dfrac{dB_{av}}{dt}$. $d(B_{av}A)/dt = A\, dB_{av}/dt$. The quantity dB_{av}/dt is the slope in a B-versus-t graph, so the induced emf is greatest when the slope is steepest. Ohm's law gives $\varepsilon = IR$, so the current will be greatest when ε is the greatest, which is where the slope of the B-versus-t graph is the greatest.

EXECUTE: We need to compare the slopes of graphs A and B with the slope of the graph in part (b) of the introduction to this set of passage problems. The graph in part (b) rises to 4 T in about 0.15 ms. In Figure P29.73, graph A rises to 4 T in less than 0.1 ms, and graph B also reaches 4 T in less than 0.1 ms. Therefore both graphs A and B have steeper slopes than the graph in part (b), so both of them would achieve a larger current than the process shown by the graph in part (b). This makes choice (c) correct.

EVALUATE: It is not the magnitude of the magnetic field that induces potential, but rather the *rate* at which the field changes.

30

INDUCTANCE

30.3. **IDENTIFY:** A coil is wound around a solenoid, so magnetic flux from the solenoid passes through the coil.
SET UP: Example 30.1 shows that the mutual inductance for this configuration of coils is
$M = \dfrac{\mu_0 N_1 N_2 A}{l}$, where l is the length of coil 1.
EXECUTE: Using the formula for M gives
$$M = \dfrac{(4\pi \times 10^{-7} \text{ Wb/m} \cdot \text{A})(800)(50)\pi(0.200 \times 10^{-2} \text{ m})^2}{0.100 \text{ m}} = 6.32 \times 10^{-6} \text{ H} = 6.32 \text{ } \mu\text{H}.$$
EVALUATE: This result is a physically reasonable mutual inductance.

30.7. **IDENTIFY:** We can relate the known self-inductance of the toroidal solenoid to its geometry to calculate the number of coils it has. Knowing the induced emf, we can find the rate of change of the current.
SET UP: Example 30.3 shows that the self-inductance of a toroidal solenoid is $L = \dfrac{\mu_0 N^2 A}{2\pi r}$. The voltage across the coil is related to the rate at which the current in it is changing by $\varepsilon = L \left| \dfrac{di}{dt} \right|$.
EXECUTE: (a) Solving $L = \dfrac{\mu_0 N^2 A}{2\pi r}$ for N gives
$$N = \sqrt{\dfrac{2\pi r L}{\mu_0 A}} = \sqrt{\dfrac{2\pi (0.0600 \text{ m})(2.50 \times 10^{-3} \text{ H})}{(4\pi \times 10^{-7} \text{ T} \cdot \text{m/A})(2.00 \times 10^{-4} \text{ m}^2)}} = 1940 \text{ turns}.$$
(b) $\left|\dfrac{di}{dt}\right| = \dfrac{\varepsilon}{L} = \dfrac{2.00 \text{ V}}{2.50 \times 10^{-3} \text{ H}} = 800 \text{ A/s}.$
EVALUATE: The inductance is determined solely by how the coil is constructed. The induced emf depends on the rate at which the current through the coil is changing.

30.17. **IDENTIFY and SET UP:** Use $U_L = \tfrac{1}{2} L I^2$ to relate the energy stored to the inductance. Example 30.3 gives the inductance of a toroidal solenoid to be $L = \dfrac{\mu_0 N^2 A}{2\pi r}$, so once we know L we can solve for N.
EXECUTE: $U = \tfrac{1}{2} L I^2$ so $L = \dfrac{2U}{I^2} = \dfrac{2(0.390 \text{ J})}{(12.0 \text{ A})^2} = 5.417 \times 10^{-3}$ H.
$$N = \sqrt{\dfrac{2\pi r L}{\mu_0 A}} = \sqrt{\dfrac{2\pi (0.150 \text{ m})(5.417 \times 10^{-3} \text{ H})}{(4\pi \times 10^{-7} \text{ T} \cdot \text{m/A})(5.00 \times 10^{-4} \text{ m}^2)}} = 2850.$$
EVALUATE: L and hence U increase according to the square of N.

30.19. **IDENTIFY:** A current-carrying inductor has a magnetic field inside of itself and hence stores magnetic energy.
(a) **SET UP:** The magnetic field inside a solenoid is $B = \mu_0 n I$.
EXECUTE: $B = \dfrac{(4\pi \times 10^{-7} \text{ T} \cdot \text{m/A})(400)(80.0 \text{ A})}{0.250 \text{ m}} = 0.161$ T.

(b) SET UP: The energy density in a magnetic field is $u = \dfrac{B^2}{2\mu_0}$.

EXECUTE: $u = \dfrac{(0.161\ \text{T})^2}{2(4\pi \times 10^{-7}\ \text{T}\cdot\text{m/A})} = 1.03 \times 10^4\ \text{J/m}^3$.

(c) SET UP: The total stored energy is $U = uV$.

EXECUTE: $U = uV = u(lA) = (1.03 \times 10^4\ \text{J/m}^3)(0.250\ \text{m})(0.500 \times 10^{-4}\ \text{m}^2) = 0.129\ \text{J}$.

(d) SET UP: The energy stored in an inductor is $U = \tfrac{1}{2}LI^2$.

EXECUTE: Solving for L and putting in the numbers gives
$$L = \dfrac{2U}{I^2} = \dfrac{2(0.129\ \text{J})}{(80.0\ \text{A})^2} = 4.02 \times 10^{-5}\ \text{H}.$$

EVALUATE: An inductor stores its energy in the magnetic field inside of it.

30.23. IDENTIFY: Apply Kirchhoff's loop rule to the circuit. $i(t)$ is given by $i = \dfrac{\varepsilon}{R}(1 - e^{-(R/L)t})$.

SET UP: The circuit is sketched in Figure 30.23.

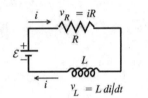

$\dfrac{di}{dt}$ is positive as the current increases from its initial value of zero.

Figure 30.23

EXECUTE: $\varepsilon - v_R - v_L = 0$.

$\varepsilon - iR - L\dfrac{di}{dt} = 0$ so $i = \dfrac{\varepsilon}{R}(1 - e^{-(R/L)t})$.

(a) Initially $(t = 0)$, $i = 0$ so $\varepsilon - L\dfrac{di}{dt} = 0$.

$\dfrac{di}{dt} = \dfrac{\varepsilon}{L} = \dfrac{6.00\ \text{V}}{2.50\ \text{H}} = 2.40\ \text{A/s}$.

(b) $\varepsilon - iR - L\dfrac{di}{dt} = 0$. (Use this equation rather than $\dfrac{di}{dt} = \dfrac{\varepsilon}{L}e^{-(R/L)t}$ since i rather than t is given.)

Thus $\dfrac{di}{dt} = \dfrac{\varepsilon - iR}{L} = \dfrac{6.00\ \text{V} - (0.500\ \text{A})(8.00\ \Omega)}{2.50\ \text{H}} = 0.800\ \text{A/s}$.

(c) $i = \dfrac{\varepsilon}{R}(1 - e^{-(R/L)t}) = \left(\dfrac{6.00\ \text{V}}{8.00\ \Omega}\right)(1 - e^{-(8.00\ \Omega/2.50\ \text{H})(0.250\ \text{s})}) = 0.750\ \text{A}(1 - e^{-0.800}) = 0.413\ \text{A}$.

(d) Final steady state means $t \to \infty$ and $\dfrac{di}{dt} \to 0$, so $\varepsilon - iR = 0$.

$i = \dfrac{\varepsilon}{R} = \dfrac{6.00\ \text{V}}{8.00\ \Omega} = 0.750\ \text{A}$.

EVALUATE: Our results agree with Figure 30.12 in the textbook. The current is initially zero and increases to its final value of ε/R. The slope of the current in the figure, which is di/dt, decreases with t.

30.25. IDENTIFY: $i = \varepsilon/R(1 - e^{-t/\tau})$, with $\tau = L/R$. The energy stored in the inductor is $U = \tfrac{1}{2}Li^2$.

SET UP: The maximum current occurs after a long time and is equal to ε/R.

EXECUTE: (a) $i_{max} = \varepsilon/R$ so $i = i_{max}/2$ when $(1 - e^{-t/\tau}) = \tfrac{1}{2}$ and $e^{-t/\tau} = \tfrac{1}{2}$. $-t/\tau = \ln(\tfrac{1}{2})$.

$t = \dfrac{L\ln 2}{R} = \dfrac{(\ln 2)(1.25 \times 10^{-3}\text{ H})}{50.0\ \Omega} = 17.3\ \mu\text{s}$.

(b) $U = \tfrac{1}{2}U_{max}$ when $i = i_{max}/\sqrt{2}$. $1 - e^{-t/\tau} = 1/\sqrt{2}$, so $e^{-t/\tau} = 1 - 1/\sqrt{2} = 0.2929$.

$t = -L\ln(0.2929)/R = 30.7\ \mu\text{s}$.

EVALUATE: $\tau = L/R = 2.50 \times 10^{-5}$ s $= 25.0\ \mu$s. The time in part (a) is 0.692τ and the time in part (b) is 1.23τ.

30.27. IDENTIFY: Apply the concepts of current decay in an R-L circuit. Apply the loop rule to the circuit. $i(t)$ is given by $i = I_0 e^{-(R/L)t}$. The voltage across the resistor depends on i and the voltage across the inductor depends on di/dt.

SET UP: The circuit with S_1 closed and S_2 open is sketched in Figure 30.27a.

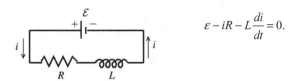

$\varepsilon - iR - L\dfrac{di}{dt} = 0$.

Figure 30.27a

Constant current established means $\dfrac{di}{dt} = 0$.

$i = \dfrac{\varepsilon}{R} = \dfrac{60.0\text{ V}}{240\ \Omega} = 0.250$ A.

EXECUTE: (a) The circuit with S_2 closed and S_1 open is shown in Figure 30.27b.

$i = I_0 e^{-(R/L)t}$.

At $t = 0$, $i = I_0 = 0.250$ A.

Figure 30.27b

The inductor prevents an instantaneous change in the current; the current in the inductor just after S_2 is closed and S_1 is opened equals the current in the inductor just before this is done.

(b) $i = I_0 e^{-(R/L)t} = (0.250\text{ A})e^{-(240\ \Omega/0.160\text{ H})(4.00 \times 10^{-4}\text{ s})} = (0.250\text{ A})e^{-0.600} = 0.137$ A.

(c) See Figure 30.27c.

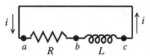

Figure 30.27c

If we trace around the loop in the direction of the current the potential falls as we travel through the resistor so it must rise as we pass through the inductor: $v_{ab} > 0$ and $v_{bc} < 0$. So point c is at a higher potential than point b.

$v_{ab} + v_{bc} = 0$ and $v_{bc} = -v_{ab}$.

Or, $v_{cb} = v_{ab} = iR = (0.137\text{ A})(240\ \Omega) = 32.9$ V.

(d) $i = I_0 e^{-(R/L)t}$.

$i = \tfrac{1}{2}I_0$ says $\tfrac{1}{2}I_0 = I_0 e^{-(R/L)t}$ and $\tfrac{1}{2} = e^{-(R/L)t}$.

Taking natural logs of both sides of this equation gives $\ln(\tfrac{1}{2}) = -Rt/L$.

$t = \left(\dfrac{0.160 \text{ H}}{240 \text{ }\Omega}\right)\ln 2 = 4.62 \times 10^{-4}$ s.

EVALUATE: The current decays, as shown in Figure 30.13 in the textbook. The time constant is $\tau = L/R = 6.67 \times 10^{-4}$ s. The values of t in the problem are less than one time constant. At any instant the potential drop across the resistor (in the direction of the current) equals the potential rise across the inductor.

30.29. IDENTIFY: With S_1 closed and S_2 open, the current builds up to a steady value.

SET UP: Applying Kirchhoff's loop rule gives $\varepsilon - iR - L\dfrac{di}{dt} = 0$.

EXECUTE: $v_R = \varepsilon - L\dfrac{di}{dt} = 18.0 \text{ V} - (0.380 \text{ H})(7.20 \text{ A/s}) = 15.3$ V.

EVALUATE: The rest of the 18.0 V of the emf is across the inductor.

30.31. IDENTIFY: Apply $\tfrac{1}{2}Li^2 + \dfrac{q^2}{2C} = \dfrac{Q^2}{2C}$.

SET UP: $q = Q$ when $i = 0$. $i = i_{max}$ when $q = 0$. $1/\sqrt{LC} = 1917$ s^{-1}.

EXECUTE: **(a)** $\tfrac{1}{2}Li_{max}^2 = \dfrac{Q^2}{2C}$.

$Q = i_{max}\sqrt{LC} = (0.850 \times 10^{-3} \text{ A})\sqrt{(0.0850 \text{ H})(3.20 \times 10^{-6} \text{ F})} = 4.43 \times 10^{-7}$ C.

(b) $q = \sqrt{Q^2 - LCi^2} = \sqrt{(4.43 \times 10^{-7} \text{ C})^2 - \left(\dfrac{5.00 \times 10^{-4} \text{ A}}{1917 \text{ s}^{-1}}\right)^2} = 3.58 \times 10^{-7}$ C.

EVALUATE: The value of q calculated in part (b) is less than the maximum value Q calculated in part (a).

30.35. IDENTIFY and SET UP: The angular frequency is given by $\omega = \dfrac{1}{\sqrt{LC}}$. $q(t)$ and $i(t)$ are given by $q = Q\cos(\omega t + \phi)$ and $i = -\omega Q\sin(\omega t + \phi)$. The energy stored in the capacitor is $U_C = \tfrac{1}{2}CV^2 = q^2/2C$. The energy stored in the inductor is $U_L = \tfrac{1}{2}Li^2$.

EXECUTE: **(a)** $\omega = \dfrac{1}{\sqrt{LC}} = \dfrac{1}{\sqrt{(1.50 \text{ H})(6.00 \times 10^{-5} \text{ F})}} = 105.4$ rad/s, which rounds to 105 rad/s. The

period is given by $T = \dfrac{2\pi}{\omega} = \dfrac{2\pi}{105.4 \text{ rad/s}} = 0.0596$ s.

(b) The circuit containing the battery and capacitor is sketched in Figure 30.35.

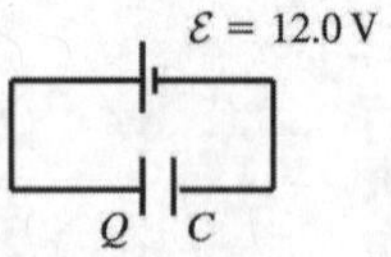

$\varepsilon - \dfrac{Q}{C} = 0$.

$Q = \varepsilon C = (12.0 \text{ V})(6.00 \times 10^{-5} \text{ F}) = 7.20 \times 10^{-4}$ C.

Figure 30.35

(c) $U = \tfrac{1}{2}CV^2 = \tfrac{1}{2}(6.00 \times 10^{-5} \text{ F})(12.0 \text{ V})^2 = 4.32 \times 10^{-3}$ J.

(d) $q = Q\cos(\omega t + \phi)$ (Eq. 30.21).

$q = Q$ at $t = 0$ so $\phi = 0$.

$q = Q\cos\omega t = (7.20\times10^{-4}\text{ C})\cos([105.4\text{ rad/s}][0.0230\text{ s}]) = -5.42\times10^{-4}\text{ C}.$

The minus sign means that the capacitor has discharged fully and then partially charged again by the current maintained by the inductor; the plate that initially had positive charge now has negative charge and the plate that initially had negative charge now has positive charge.

(e) The current is $i = -\omega Q \sin(\omega t + \phi)$

$i = -(105\text{ rad/s})(7.20\times10^{-4}\text{ C})\sin[(105.4\text{ rad/s})(0.0230\text{ s})] = -0.050\text{ A}.$

The negative sign means the current is counterclockwise in Figure 30.15 in the textbook.

or

$\frac{1}{2}Li^2 + \frac{q^2}{2C} = \frac{Q^2}{2C}$ gives $i = \pm\sqrt{\frac{1}{LC}}\sqrt{Q^2 - q^2}$ (Eq. 30.26).

$i = \pm(105\text{ rad/s})\sqrt{(7.20\times10^{-4}\text{ C})^2 - (-5.42\times10^{-4}\text{ C})^2} = \pm 0.050\text{ A}$, which checks.

(f) $U_C = \frac{q^2}{2C} = \frac{(-5.42\times10^{-4}\text{ C})^2}{2(6.00\times10^{-5}\text{ F})} = 2.45\times10^{-3}\text{ J}.$

$U_L = \frac{1}{2}Li^2 = \frac{1}{2}(1.50\text{ H})(0.050\text{ A})^2 = 1.87\times10^{-3}\text{ J}.$

EVALUATE: Note that $U_C + U_L = 2.45\times10^{-3}\text{ J} + 1.87\times10^{-3}\text{ J} = 4.32\times10^{-3}\text{ J}.$

This agrees with the total energy initially stored in the capacitor,

$U = \frac{Q^2}{2C} = \frac{(7.20\times10^{-4}\text{ C})^2}{2(6.00\times10^{-5}\text{ F})} = 4.32\times10^{-3}\text{ J}.$

Energy is conserved. At some times there is energy stored in both the capacitor and the inductor. When $i = 0$ all the energy is stored in the capacitor and when $q = 0$ all the energy is stored in the inductor. But at all times the total energy stored is the same.

30.39. **IDENTIFY:** Evaluate $\omega' = \sqrt{\frac{1}{LC} - \frac{R^2}{4L^2}}.$

SET UP: The angular frequency of the circuit is ω'.

EXECUTE: **(a)** When $R = 0$, $\omega_0 = \frac{1}{\sqrt{LC}} = \frac{1}{\sqrt{(0.450\text{ H})(2.50\times10^{-5}\text{ F})}} = 298\text{ rad/s}.$

(b) We want $\frac{\omega'}{\omega_0} = 0.95$, so $\frac{(1/LC - R^2/4L^2)}{1/LC} = 1 - \frac{R^2C}{4L} = (0.95)^2.$ This gives

$R = \sqrt{\frac{4L}{C}(1 - (0.95)^2)} = \sqrt{\frac{4(0.450\text{ H})(0.0975)}{(2.50\times10^{-5}\text{ F})}} = 83.8\text{ Ω}.$

EVALUATE: When R increases, the angular frequency decreases and approaches zero as $R \to 2\sqrt{L/C}$.

30.41. **IDENTIFY:** The presence of resistance in an L-R-C circuit affects the frequency of oscillation and causes the amplitude of the oscillations to decrease over time.

(a) **SET UP:** The frequency of damped oscillations is $\omega' = \sqrt{\frac{1}{LC} - \frac{R^2}{4L^2}}.$

EXECUTE: $\omega' = \sqrt{\frac{1}{(22\times10^{-3}\text{ H})(15.0\times10^{-9}\text{ F})} - \frac{(75.0\text{ Ω})^2}{4(22\times10^{-3}\text{ H})^2}} = 5.5\times10^4\text{ rad/s}.$

The frequency f is $f = \frac{\omega}{2\pi} = \frac{5.50\times10^4\text{ rad/s}}{2\pi} = 8.76\times10^3\text{ Hz} = 8.76\text{ kHz}.$

(b) **SET UP:** The amplitude decreases as $A(t) = A_0 e^{-(R/2L)t}.$

EXECUTE: Solving for t and putting in the numbers gives:

$t = \frac{-2L\ln(A/A_0)}{R} = \frac{-2(22.0\times10^{-3}\text{ H})\ln(0.100)}{75.0\text{ Ω}} = 1.35\times10^{-3}\text{ s} = 1.35\text{ ms}.$

(c) SET UP: At critical damping, $R = \sqrt{4L/C}$.

EXECUTE: $R = \sqrt{\dfrac{4(22.0 \times 10^{-3}\text{ H})}{15.0 \times 10^{-9}\text{ F}}} = 2420\ \Omega$.

EVALUATE: The frequency with damping is almost the same as the resonance frequency of this circuit ($1/\sqrt{LC}$), which is plausible because the 75-Ω resistance is considerably less than the 2420 Ω required for critical damping.

30.45. IDENTIFY: Set $U_B = K$, where $K = \tfrac{1}{2}mv^2$.

SET UP: The energy density in the magnetic field is $u_B = B^2/2\mu_0$. Consider volume $V = 1\text{ m}^3$ of sunspot material.

EXECUTE: The energy density in the sunspot is $u_B = B^2/2\mu_0 = 6.366 \times 10^4\text{ J/m}^3$. The total energy stored in volume V of the sunspot is $U_B = u_B V$. The mass of the material in volume V of the sunspot is $m = \rho V$. $K = U_B$ so $\tfrac{1}{2}mv^2 = U_B$. $\tfrac{1}{2}\rho V v^2 = u_B V$. The volume divides out, and $v = \sqrt{2u_B/\rho} = 2 \times 10^4$ m/s.

EVALUATE: The speed we calculated is about 30 times smaller than the escape speed.

30.47. (a) IDENTIFY and SET UP: An end view is shown in Figure 30.47.

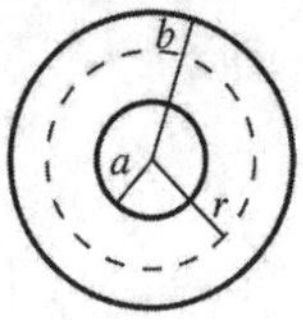

Apply Ampere's law to a circular path of radius r.
$\oint \vec{B} \cdot d\vec{l} = \mu_0 I_{\text{encl}}$.

Figure 30.47

EXECUTE: $\oint \vec{B} \cdot d\vec{l} = B(2\pi r)$.

$I_{\text{encl}} = i$, the current in the inner conductor.

Thus $B(2\pi r) = \mu_0 i$ and $B = \dfrac{\mu_0 i}{2\pi r}$.

(b) IDENTIFY and SET UP: Follow the procedure specified in the problem.

EXECUTE: $u = \dfrac{B^2}{2\mu_0}$.

$dU = u\, dV$, where $dV = 2\pi r l\, dr$.

$dU = \dfrac{1}{2\mu_0}\left(\dfrac{\mu_0 i}{2\pi r}\right)^2 (2\pi r l)\, dr = \dfrac{\mu_0 i^2 l}{4\pi r}\, dr$.

(c) $U = \int dU = \dfrac{\mu_0 i^2 l}{4\pi}\int_a^b \dfrac{dr}{r} = \dfrac{\mu_0 i^2 l}{4\pi}[\ln r]_a^b$.

$U = \dfrac{\mu_0 i^2 l}{4\pi}(\ln b - \ln a) = \dfrac{\mu_0 i^2 l}{4\pi}\ln\!\left(\dfrac{b}{a}\right)$.

(d) Eq. (30.9): $U = \tfrac{1}{2}Li^2$.

Part (c): $U = \dfrac{\mu_0 i^2 l}{4\pi}\ln\!\left(\dfrac{b}{a}\right)$.

$\tfrac{1}{2}Li^2 = \dfrac{\mu_0 i^2 l}{4\pi}\ln\!\left(\dfrac{b}{a}\right)$.

$L = \dfrac{\mu_0 l}{2\pi}\ln\!\left(\dfrac{b}{a}\right)$.

EVALUATE: The value of L we obtain from these energy considerations agrees with L calculated in part (d) of Problem 30.46 by considering flux and $L = \dfrac{N\Phi_B}{i}$.

30.57. IDENTIFY: The current through an inductor doesn't change abruptly. After a long time the current isn't changing and the voltage across each inductor is zero.
SET UP: For part (c) combine the inductors.
EXECUTE: **(a)** Just after the switch is closed there is no current in the inductors. There is no current in the resistors so there is no voltage drop across either resistor. A reads zero and V reads 20.0 V.
(b) After a long time the currents are no longer changing, there is no voltage across the inductors, and the inductors can be replaced by short-circuits. The circuit becomes equivalent to the circuit shown in Figure 30.57a. $I = (20.0 \text{ V})/(75.0 \, \Omega) = 0.267$ A. The voltage between points a and b is zero, so the voltmeter reads zero.
(c) Combine the inductor network into its equivalent, as shown in Figure 30.57b. $R = 75.0 \, \Omega$ is the equivalent resistance. The current is $i = (\varepsilon/R)(1 - e^{-t/\tau})$ with $\tau = L/R = (10.8 \text{ mH})/(75.0 \, \Omega) = 0.144$ ms. $\varepsilon = 20.0$ V, $R = 75.0 \, \Omega$, $t = 0.115$ ms so $i = 0.147$ A. $V_R = iR = (0.147 \text{ A})(75.0 \, \Omega) = 11.0$ V. $20.0 \text{ V} - V_R - V_L = 0$ and $V_L = 20.0 \text{ V} - V_R = 9.0$ V. The ammeter reads 0.147 A and the voltmeter reads 9.0 V.
EVALUATE: The current through the battery increases from zero to a final value of 0.267 A. The voltage across the inductor network drops from 20.0 V to zero.

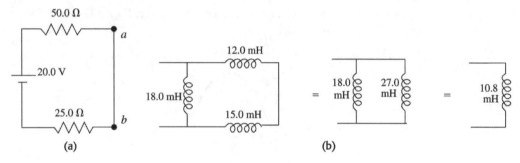

Figure 30.57

30.59. IDENTIFY and SET UP: Just after the switch is closed, the current in each branch containing an inductor is zero and the voltage across any capacitor is zero. The inductors can be treated as breaks in the circuit and the capacitors can be replaced by wires. After a long time there is no voltage across each inductor and no current in any branch containing a capacitor. The inductors can be replaced by wires and the capacitors by breaks in the circuit.
EXECUTE: **(a)** Just after the switch is closed the voltage V_5 across the capacitor is zero and there is also no current through the inductor, so $V_3 = 0$. $V_2 + V_3 = V_4 = V_5$, and since $V_5 = 0$ and $V_3 = 0$, V_4 and V_2 are also zero. $V_4 = 0$ means V_3 reads zero. V_1 then must equal 40.0 V, and this means the current read by A_1 is $(40.0 \text{ V})/(50.0 \, \Omega) = 0.800$ A. $A_2 + A_3 + A_4 = A_1$, but $A_2 = A_3 = 0$ so $A_4 = A_1 = 0.800$ A. $A_1 = A_4 = 0.800$ A; all other ammeters read zero. $V_1 = 40.0$ V and all other voltmeters read zero.
(b) After a long time the capacitor is fully charged so $A_4 = 0$. The current through the inductor isn't changing, so $V_2 = 0$. The currents can be calculated from the equivalent circuit that replaces the inductor by a short circuit, as shown in Figure 30.59a.

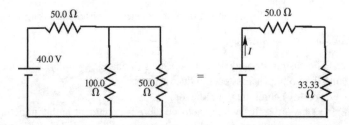

Figure 30.59a

$I = (40.0 \text{ V})/(83.33 \text{ }\Omega) = 0.480 \text{ A}$; A_1 reads 0.480 A.

$V_1 = I(50.0 \text{ }\Omega) = 24.0 \text{ V}$.

The voltage across each parallel branch is $40.0 \text{ V} - 24.0 \text{ V} = 16.0 \text{ V}$.

$V_2 = 0, V_3 = V_4 = V_5 = 16.0 \text{ V}$.

$V_3 = 16.0 \text{ V}$ means A_2 reads 0.160 A. $V_4 = 16.0 \text{ V}$ means A_3 reads 0.320 A. A_4 reads zero. Note that $A_2 + A_3 = A_1$.

(c) $V_5 = 16.0 \text{ V}$ so $Q = CV = (12.0 \text{ }\mu\text{F})(16.0 \text{ V}) = 192 \text{ }\mu\text{C}$.

(d) At $t = 0$ and $t \to \infty$, $V_2 = 0$. As the current in this branch increases from zero to 0.160 A the voltage V_2 reflects the rate of change of the current. The graph is sketched in Figure 30.59b.

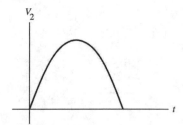

Figure 30.59b

EVALUATE: This reduction of the circuit to resistor networks only apply at $t = 0$ and $t \to \infty$. At intermediate times the analysis is complicated.

30.61. **IDENTIFY:** Apply the loop rule to each parallel branch. The voltage across a resistor is given by iR and the voltage across an inductor is given by $L|di/dt|$. The rate of change of current through the inductor is limited.

SET UP: With S closed the circuit is sketched in Figure 30.61a.

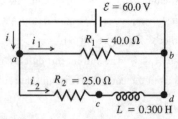

The rate of change of the current through the inductor is limited by the induced emf. Just after the switch is closed the current in the inductor has not had time to increase from zero, so $i_2 = 0$.

Figure 30.61a

EXECUTE : (a) $\varepsilon - v_{ab} = 0$, so $v_{ab} = 60.0 \text{ V}$.

(b) The voltage drops across R, as we travel through the resistor in the direction of the current, so point a is at higher potential.

(c) $i_2 = 0$ so $v_{R_2} = i_2 R_2 = 0.$

$\varepsilon - v_{R_2} - v_L = 0$ so $v_L = \varepsilon = 60.0$ V.

(d) The voltage rises when we go from b to a through the emf, so it must drop when we go from a to b through the inductor. Point c must be at higher potential than point d.

(e) After the switch has been closed a long time, $\dfrac{di_2}{dt} \to 0$ so $v_L = 0$. Then $\varepsilon - v_{R_2} = 0$ and $i_2 R_2 = \varepsilon$

so $i_2 = \dfrac{\varepsilon}{R_2} = \dfrac{60.0 \text{ V}}{25.0 \text{ }\Omega} = 2.40$ A.

SET UP: The rate of change of the current through the inductor is limited by the induced emf. Just after the switch is opened again the current through the inductor hasn't had time to change and is still $i_2 = 2.40$ A. The circuit is sketched in Figure 30.61b.

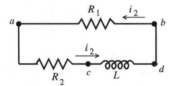

EXECUTE: The current through R_1 is $i_2 = 2.40$ A in the direction b to a.
Thus $v_{ab} = -i_2 R_1 = -(2.40 \text{ A})(40.0 \text{ }\Omega).$
$v_{ab} = -96.0$ V.

Figure 30.61b

(f) Point where current enters resistor is at higher potential; point b is at higher potential.

(g) $v_L - v_{R_1} - v_{R_2} = 0.$

$v_L = v_{R_1} + v_{R_2}.$

$v_{R_1} = -v_{ab} = 96.0$ V; $v_{R_2} = i_2 R_2 = (2.40 \text{ A})(25.0 \text{ }\Omega) = 60.0$ V.

Then $v_L = v_{R_1} + v_{R_2} = 96.0 \text{ V} + 60.0 \text{ V} = 156$ V.

As you travel counterclockwise around the circuit in the direction of the current, the voltage drops across each resistor, so it must rise across the inductor and point d is at higher potential than point c. The current is decreasing, so the induced emf in the inductor is directed in the direction of the current. Thus, $v_{cd} = -156$ V.

(h) Point d is at higher potential.

EVALUATE: The voltage across R_1 is constant once the switch is closed. In the branch containing R_2, just after S is closed the voltage drop is all across L and after a long time it is all across R_2. Just after S is opened the same current flows in the single loop as had been flowing through the inductor and the sum of the voltage across the resistors equals the voltage across the inductor. This voltage dies away, as the energy stored in the inductor is dissipated in the resistors.

30.63. IDENTIFY and SET UP: The circuit is sketched in Figure 30.63a. Apply the loop rule. Just after S_1 is closed, $i = 0$. After a long time i has reached its final value and $di/dt = 0$. The voltage across a resistor depends on i and the voltage across an inductor depends on di/dt.

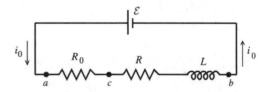

Figure 30.63a

EXECUTE: **(a)** At time $t = 0$, $i_0 = 0$ so $v_{ac} = i_0 R_0 = 0$. By the loop rule $\varepsilon - v_{ac} - v_{cb} = 0$ so $v_{cb} = \varepsilon - v_{ac} = \varepsilon = 36.0$ V. ($i_0 R = 0$ so this potential difference of 36.0 V is across the inductor and is an induced emf produced by the changing current.)

(b) After a long time $\dfrac{di_0}{dt} \to 0$ so the potential $-L\dfrac{di_0}{dt}$ across the inductor becomes zero. The loop rule gives $\varepsilon - i_0(R_0 + R) = 0$.

$i_0 = \dfrac{\varepsilon}{R_0 + R} = \dfrac{36.0 \text{ V}}{50.0\ \Omega + 150\ \Omega} = 0.180$ A.

$v_{ac} = i_0 R_0 = (0.180 \text{ A})(50.0\ \Omega) = 9.0$ V.

Thus $v_{cb} = i_0 R + L\dfrac{di_0}{dt} = (0.180 \text{ A})(150\ \Omega) + 0 = 27.0$ V. (Note that $v_{ac} + v_{cb} = \varepsilon$.)

(c) $\varepsilon - v_{ac} - v_{cb} = 0$.

$\varepsilon - iR_0 - iR - L\dfrac{di}{dt} = 0$.

$L\dfrac{di}{dt} = \varepsilon - i(R_0 + R)$ and $\left(\dfrac{L}{R + R_0}\right)\dfrac{di}{dt} = -i + \dfrac{\varepsilon}{R + R_0}$.

$\dfrac{di}{-i + \varepsilon/(R + R_0)} = \left(\dfrac{R + R_0}{L}\right)dt$.

Integrate from $t = 0$, when $i = 0$, to t, when $i = i_0$:

$\displaystyle\int_0^{i_0} \dfrac{di}{-i + \varepsilon/(R + R_0)} = \dfrac{R + R_0}{L}\int_0^t dt = -\ln\left[-i + \dfrac{\varepsilon}{R + R_0}\right]_0^{i_0} = \left(\dfrac{R + R_0}{L}\right)t$, so

$\ln\left(-i_0 + \dfrac{\varepsilon}{R + R_0}\right) - \ln\left(\dfrac{\varepsilon}{R + R_0}\right) = -\left(\dfrac{R + R_0}{L}\right)t$.

$\ln\left(\dfrac{-i_0 + \varepsilon/(R + R_0)}{\varepsilon/(R + R_0)}\right) = -\left(\dfrac{R + R_0}{L}\right)t$.

Taking exponentials of both sides gives $\dfrac{-i_0 + \varepsilon/(R + R_0)}{\varepsilon/(R + R_0)} = e^{-(R + R_0)t/L}$ and $i_0 = \dfrac{\varepsilon}{R + R_0}(1 - e^{-(R + R_0)t/L})$.

Substituting in the numerical values gives $i_0 = \dfrac{36.0 \text{ V}}{50\ \Omega + 150\ \Omega}(1 - e^{-(200\ \Omega / 4.00 \text{ H})t}) = (0.180 \text{ A})(1 - e^{-t/0.020\text{ s}})$.

At $t \to 0$, $i_0 = (0.180 \text{ A})(1 - 1) = 0$ (agrees with part (a)). At $t \to \infty$, $i_0 = (0.180 \text{ A})(1 - 0) = 0.180$ A (agrees with part (b)).

$v_{ac} = i_0 R_0 = \dfrac{\varepsilon R_0}{R + R_0}(1 - e^{-(R + R_0)t/L}) = 9.0 \text{ V}(1 - e^{-t/0.020\text{ s}})$.

$v_{cb} = \varepsilon - v_{ac} = 36.0 \text{ V} - 9.0 \text{ V}(1 - e^{-t/0.020\text{ s}}) = 9.0 \text{ V}(3.00 + e^{-t/0.020\text{ s}})$.

At $t \to 0$, $v_{ac} = 0$, $v_{cb} = 36.0$ V (agrees with part (a)). At $t \to \infty$, $v_{ac} = 9.0$ V, $v_{cb} = 27.0$ V (agrees with part (b)). The graphs are given in Figure 30.63b.

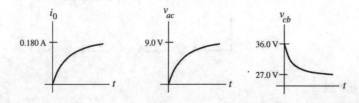

Figure 30.63b

EVALUATE: The expression for $i(t)$ we derived becomes $i = \frac{\varepsilon}{R}(1 - e^{-(R/L)t})$ if the two resistors R_0 and R in series are replaced by a single equivalent resistance $R_0 + R$.

30.67. IDENTIFY and SET UP: Kirchhoff's loop rule applies, the emf across an inductor is $\varepsilon_L = -L\frac{di}{dt}$, the potential across a resistor is $V = Ri$, and the time constant for an L-R circuit is $\tau = L/R$.

EXECUTE: (a) First find the current as a function of time. The inductor has a resistance R_L which is in series with the 10.0-Ω resistor R. Apply Kirchhoff's loop rule to the circuit: $\varepsilon - iR - iR_L - L\frac{di}{dt} = 0$. Now separate variables and integrate.

$$\int_0^t -\frac{R+R_L}{L}dt' = \int_0^i \frac{di'}{i' - \varepsilon/(R+R_L)}.$$

$$-\frac{R+R_L}{L}t = \ln\left(\frac{i - \varepsilon/(R+R_L)}{-\varepsilon/(R+R_L)}\right).$$

$$i = \frac{\varepsilon}{R+R_L}(1 - e^{-(R+R_L)t/L}).$$

The potential across the inductor is the sum of the potential due to the resistance and the potential due to the inductance, so $v_L = iR_L + L\,di/dt$. Using the equation we just found for the current i and taking $L\,di/dt$, we get

$$v_L = iR_L + L\frac{di}{dt} = \left[\frac{\varepsilon}{R+R_L}(1 - e^{-(R+R_L)t/L})\right]R_L + \varepsilon e^{-(R+R_L)t/L}.$$

Collecting terms and taking out common factors, the result is

$$v_L = \frac{\varepsilon}{R+R_L}(R_L + Re^{-(R+R_L)t/L}).$$

(b) Initially there is no current in the circuit due to the inductor, so the potential across the resistance R is zero. Therefore the potential across the inductor is equal to the emf of the battery.

$$v_L(0) = \frac{\varepsilon}{R+R_L}(R_L + R) = \varepsilon = 50.0\text{ V}.$$

(c) As $t \to \infty$, we know that $v_L = 20.0$ V. So $v_R = \varepsilon - v_L = 50.0$ V $-$ 20.0 V $=$ 30.0 V. The current in R is therefore $i = (30.0\text{ V})/(10.0\ \Omega) = 3.00$ A, which is also the current in the circuit.

(d) As $t \to \infty$, the potential across the inductor is due only to its resistance R_L, the potential across it is 20.0 V, and the current through it is 3.00 A. Therefore $R_L = (20.0\text{ V})/(3.00\text{ A}) = 6.67\ \Omega$.

(e) The time constant for this circuit is $\tau = L/(R + R_L)$. Using the equation derived in (a) for v_L, at the end of one time constant v_L is

$$v_L = \frac{\varepsilon}{R+R_L}(R_L + Re^{-1}) = \frac{50.0\text{ V}}{16.67\ \Omega}\left[6.67\ \Omega + (10.0\ \Omega)e^{-1}\right] = 31.0\text{ V}.$$

From the graph shown with the problem in the textbook, we read that $t = 2.4$ ms when $v_L = 31.0$ V. So the time constant is 2.4 ms. Solving $\tau = L/(R + R_L)$ for L gives

$L = \tau(R + R_L) = (2.4\text{ ms})(10.0\ \Omega + 6.67\ \Omega) = 40$ mH.

EVALUATE: In this case, the resistance of the inductor is close to the external resistance in the circuit, so it is significant and cannot be ignored.

30.69. IDENTIFY and SET UP: The current in an R-L circuit is given by $i = i_0 e^{-Rt/L}$, where R is the total resistance. In our measurements, the current is one-half the initial current, so $i = i_0/2$.

EXECUTE: (a) Taking natural logarithms of the current equation, with $R = R_L + R_{\text{ext}}$ and $i = i_0/2$, we get
$\ln(i/i_0) = -Rt/L$.
$\ln(1/2) = -(R_L + R_{\text{ext}})t_{\text{half}}/L$.
$\ln 2 = t_{\text{half}}(R_L + R_{\text{ext}})/L$.

where t_{half} is the time for the current to decrease to half its initial value. Solving for $1/t_{half}$ gives $\dfrac{1}{t_{half}} = \dfrac{R_{ext}}{L\ln 2} + \dfrac{R_L}{L\ln 2}$. Therefore a graph of $1/t_{half}$ versus R_{ext} should be a straight line having a slope equal to $1/(L\ln 2)$ and a y-intercept equal to $R_L/(L\ln 2)$. Figure 30.69 shows this graph.

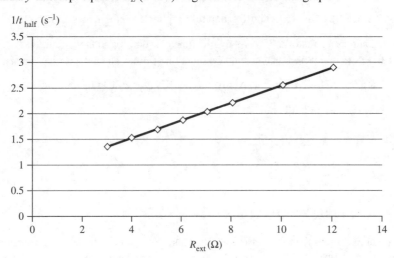

Figure 30.69

(b) The best-fit equation for the line in the graph is $\dfrac{1}{t_{half}} = 0.1692\ (\Omega\cdot s)^{-1} R_{ext} + 0.8524\ s^{-1}$. Using the slope and solving for L gives $L = \dfrac{1}{(\text{slope})\ln 2} = \dfrac{1}{[0.1692\ (\Omega\cdot s)^{-1}]\ln 2} = 8.53$ H, which rounds to 8.5 H. Now use the y-intercept and solve for R_L.

$\dfrac{R_L}{L\ln 2} = y$-intercept, so $R_L = (y\text{-intercept})(L\ln 2) = (0.8524\ s^{-1})(8.53\ H)\ln 2 = 5.04\ \Omega$, which rounds to 5.0 Ω.

(c) $U_L = \tfrac{1}{2}Li^2 = (1/2)(8.53\ H)(20.0\ A)^2 = 1.7\times 10^3$ J = 1.7 kJ.
$P_R = i^2 R = (20.0\ A)^2(5.04\ \Omega) = 2.0\times 10^3$ W = 2.0 kW.
EVALUATE: Whether the 5.0-Ω resistance of this inductor would be significant would depend on the external resistance in the circuit. For the data of this problem, the solenoid resistance would definitely be significant for the external resistances used.

30.75. IDENTIFY: The magnetic energy stored in the magnet is converted into thermal energy which evaporates the liquid helium.
SET UP: The magnetic energy is $U_L = \tfrac{1}{2}Li^2$. The heat Q to evaporate a mass m of liquid is $Q = mL_v$.
EXECUTE: $\tfrac{1}{2}Li^2 = mL_v$. Solving for m gives
$m = Li^2/2L_v = (4.4\ H)(750\ A)^2/[2(20900\ J/kg)] = 59$ kg $\approx$ 60 kg, which is choice (c).
EVALUATE: This is a lot of liquid helium! It is important to avoid quenches!

ALTERNATING CURRENT

31.5. **IDENTIFY:** We want the phase angle for the source voltage relative to the current, and we want the inductance if we know the current amplitude.

SET UP: $X_L = \dfrac{V}{I}$ and $X_L = 2\pi fL$.

EXECUTE: **(a)** $\phi = +90°$. The source voltage leads the current by 90°.

(b) $X_L = \dfrac{V}{I} = \dfrac{45.0 \text{ V}}{3.90 \text{ A}} = 11.54 \text{ }\Omega$. Solving $X_L = 2\pi fL$ for f gives $f = \dfrac{X_L}{2\pi L} = \dfrac{11.54 \text{ }\Omega}{2\pi(9.50 \times 10^{-3} \text{ H})} = 193 \text{ Hz}$.

EVALUATE: The angular frequency is about 1200 rad/s.

31.7. **IDENTIFY** and **SET UP:** Apply $X_C = \dfrac{1}{\omega C}$ and $V_C = IX_C$.

EXECUTE: $V = IX_C$ so $X_C = \dfrac{V}{I} = \dfrac{170 \text{ V}}{0.850 \text{ A}} = 200 \text{ }\Omega$.

$X_C = \dfrac{1}{\omega C}$ gives $C = \dfrac{1}{2\pi f X_C} = \dfrac{1}{2\pi(60.0 \text{ Hz})(200 \text{ }\Omega)} = 1.33 \times 10^{-5} \text{ F} = 13.3 \text{ }\mu\text{F}$.

EVALUATE: The reactance relates the voltage amplitude to the current amplitude and is similar to Ohm's law.

31.11. **IDENTIFY:** In an L-R ac circuit, we want to find out how the voltage across a resistor varies with time if we know how the voltage varies across the inductor.

SET UP: $v_L = -I\omega L \sin\omega t$ and $v_R = V_R \cos(\omega t)$.

EXECUTE: **(a)** $v_L = -I\omega L \sin\omega t$. $\omega = 480$ rad/s. $I\omega L = 12.0$ V.

$I = \dfrac{12.0 \text{ V}}{\omega L} = \dfrac{12.0 \text{ V}}{(480 \text{ rad/s})(0.180 \text{ H})} = 0.1389 \text{ A}$. $V_R = IR = (0.1389 \text{ A})(90.0 \text{ }\Omega) = 12.5 \text{ V}$.

$v_R = V_R \cos(\omega t) = (12.5 \text{ V})\cos[(480 \text{ rad/s})t]$.

(b) $v_R = (12.5 \text{ V})\cos[(480 \text{ rad/s})(2.00 \times 10^{-3} \text{ s})] = 7.17 \text{ V}$.

EVALUATE: The instantaneous voltage (7.17 V) is less than the voltage amplitude (12.5 V).

31.13. **IDENTIFY** and **SET UP:** The voltage and current for a resistor are related by $v_R = iR$. Deduce the frequency of the voltage and use this in $X_L = \omega L$ to calculate the inductive reactance. The equation $v_L = I\omega L \cos(\omega t + 90°)$ gives the voltage across the inductor.

EXECUTE: **(a)** $v_R = (3.80 \text{ V})\cos[(720 \text{ rad/s})t]$.

$v_R = iR$, so $i = \dfrac{v_R}{R} = \left(\dfrac{3.80 \text{ V}}{150 \text{ }\Omega}\right)\cos[(720 \text{ rad/s})t] = (0.0253 \text{ A})\cos[(720 \text{ rad/s})t]$.

(b) $X_L = \omega L$.

$\omega = 720$ rad/s, $L = 0.250$ H, so $X_L = \omega L = (720 \text{ rad/s})(0.250 \text{ H}) = 180 \text{ }\Omega$.

(c) If $i = I\cos\omega t$ then $v_L = V_L \cos(\omega t + 90°)$ (from Eq. 31.10).

$V_L = I\omega L = IX_L = (0.02533 \text{ A})(180 \text{ }\Omega) = 4.56 \text{ V}$.

$v_L = (4.56 \text{ V})\cos[(720 \text{ rad/s})t + 90°]$.

But $\cos(a + 90°) = -\sin a$ (Appendix B), so $v_L = -(4.56 \text{ V})\sin[(720 \text{ rad/s})t]$.

EVALUATE: The current is the same in the resistor and inductor and the voltages are 90° out of phase, with the voltage across the inductor leading.

31.15. IDENTIFY: Apply the equations in Section 31.3.

SET UP: $\omega = 250$ rad/s, $R = 200 \, \Omega$, $L = 0.400$ H, $C = 6.00 \, \mu\text{F}$, and $V = 30.0$ V.

EXECUTE: (a) $Z = \sqrt{R^2 + (\omega L - 1/\omega C)^2}$.

$Z = \sqrt{(200 \, \Omega)^2 + ((250 \text{ rad/s})(0.400 \text{ H}) - 1/((250 \text{ rad/s})(6.00 \times 10^{-6} \text{ F})))^2} = 601 \, \Omega$.

(b) $I = \dfrac{V}{Z} = \dfrac{30 \text{ V}}{601 \, \Omega} = 0.0499$ A.

(c) $\phi = \arctan\left(\dfrac{\omega L - 1/\omega C}{R}\right) = \arctan\left(\dfrac{100 \, \Omega - 667 \, \Omega}{200 \, \Omega}\right) = -70.6°$, and the voltage lags the current.

(d) $V_R = IR = (0.0499 \text{ A})(200 \, \Omega) = 9.98$ V; $V_L = I\omega L = (0.0499 \text{ A})(250 \text{ rad/s})(0.400 \text{ H}) = 4.99$ V;

$V_C = \dfrac{I}{\omega C} = \dfrac{(0.0499 \text{ A})}{(250 \text{ rad/s})(6.00 \times 10^{-6} \text{ F})} = 33.3$ V.

EVALUATE: (e) At any instant, $v = v_R + v_C + v_L$. But v_C and v_L are 180° out of phase, so v_C can be larger than v at a value of t, if $v_L + v_R$ is negative at that t.

31.23. IDENTIFY and SET UP: Use the equations of Section 31.3 to calculate ϕ, Z, and V_{rms}. The average power delivered by the source is given by $P_{\text{av}} = I_{\text{rms}}V_{\text{rms}}\cos\phi$ and the average power dissipated in the resistor is $I_{\text{rms}}^2 R$.

EXECUTE: (a) $X_L = \omega L = 2\pi f L = 2\pi(400 \text{ Hz})(0.120 \text{ H}) = 301.6 \, \Omega$.

$X_C = \dfrac{1}{\omega C} = \dfrac{1}{2\pi f C} = \dfrac{1}{2\pi(400 \text{ Hz})(7.3 \times 10^{-6} \text{ F})} = 54.51 \, \Omega$.

$\tan\phi = \dfrac{X_L - X_C}{R} = \dfrac{301.6 \, \Omega - 54.41 \, \Omega}{240 \, \Omega}$, so $\phi = +45.8°$. The power factor is $\cos\phi = +0.697$.

(b) $Z = \sqrt{R^2 + (X_L - X_C)^2} = \sqrt{(240 \, \Omega)^2 + (301.6 \, \Omega - 54.51 \, \Omega)^2} = 344 \, \Omega$.

(c) $V_{\text{rms}} = I_{\text{rms}}Z = (0.450 \text{ A})(344 \, \Omega) = 155$ V.

(d) $P_{\text{av}} = I_{\text{rms}}V_{\text{rms}}\cos\phi = (0.450 \text{ A})(155 \text{ V})(0.697) = 48.6$ W.

(e) $P_{\text{av}} = I_{\text{rms}}^2 R = (0.450 \text{ A})^2(240 \, \Omega) = 48.6$ W.

EVALUATE: The average electrical power delivered by the source equals the average electrical power consumed in the resistor.

(f) All the energy stored in the capacitor during one cycle of the current is released back to the circuit in another part of the cycle. There is no net dissipation of energy in the capacitor.

(g) The answer is the same as for the capacitor. Energy is repeatedly being stored and released in the inductor, but no net energy is dissipated there.

31.25. IDENTIFY: The angular frequency and the capacitance can be used to calculate the reactance X_C of the capacitor. The angular frequency and the inductance can be used to calculate the reactance X_L of the inductor. Calculate the phase angle ϕ and then the power factor is $\cos\phi$. Calculate the impedance of the circuit and then the rms current in the circuit. The average power is $P_{\text{av}} = V_{\text{rms}}I_{\text{rms}}\cos\phi$. On the average no power is consumed in the capacitor or the inductor, it is all consumed in the resistor.

SET UP: The source has rms voltage $V_{\text{rms}} = \dfrac{V}{\sqrt{2}} = \dfrac{45 \text{ V}}{\sqrt{2}} = 31.8$ V.

EXECUTE: (a) $X_L = \omega L = (360 \text{ rad/s})(15 \times 10^{-3} \text{ H}) = 5.4 \text{ }\Omega$.

$X_C = \dfrac{1}{\omega C} = \dfrac{1}{(360 \text{ rad/s})(3.5 \times 10^{-6} \text{ F})} = 794 \text{ }\Omega$. $\tan\phi = \dfrac{X_L - X_C}{R} = \dfrac{5.4 \text{ }\Omega - 794 \text{ }\Omega}{250 \text{ }\Omega}$ and $\phi = -72.4°$.

The power factor is $\cos\phi = 0.302$.

(b) $Z = \sqrt{R^2 + (X_L - X_C)^2} = \sqrt{(250 \text{ }\Omega)^2 + (5.4 \text{ }\Omega - 794 \text{ }\Omega)^2} = 827 \text{ }\Omega$. $I_{rms} = \dfrac{V_{rms}}{Z} = \dfrac{31.8 \text{ V}}{827 \text{ }\Omega} = 0.0385 \text{ A}$.

$P_{av} = V_{rms} I_{rms} \cos\phi = (31.8 \text{ V})(0.0385 \text{ A})(0.302) = 0.370 \text{ W}$.

(c) The average power delivered to the resistor is $P_{av} = I_{rms}^2 R = (0.0385 \text{ A})^2 (250 \text{ }\Omega) = 0.370 \text{ W}$. The average power delivered to the capacitor and to the inductor is zero.

EVALUATE: On average the power delivered to the circuit equals the power consumed in the resistor. The capacitor and inductor store electrical energy during part of the current oscillation but each return the energy to the circuit during another part of the current cycle.

31.27. IDENTIFY and SET UP: The current is largest at the resonance frequency. At resonance, $X_L = X_C$ and $Z = R$. For part (b), calculate Z and use $I = V/Z$.

EXECUTE: (a) $f_0 = \dfrac{1}{2\pi\sqrt{LC}} = 113 \text{ Hz}$. $I = V/R = 15.0 \text{ mA}$.

(b) $X_C = 1/\omega C = 500 \text{ }\Omega$. $X_L = \omega L = 160 \text{ }\Omega$.

$Z = \sqrt{R^2 + (X_L - X_C)^2} = \sqrt{(200 \text{ }\Omega)^2 + (160 \text{ }\Omega - 500 \text{ }\Omega)^2} = 394.5 \text{ }\Omega$. $I = V/Z = 7.61 \text{ mA}$. $X_C > X_L$ so the source voltage lags the current.

EVALUATE: $\omega_0 = 2\pi f_0 = 710 \text{ rad/s}$. $\omega = 400 \text{ rad/s}$ and is less than ω_0. When $\omega < \omega_0$, $X_C > X_L$. Note that I in part (b) is less than I in part (a).

31.33. IDENTIFY: At resonance $Z = R$ and $X_L = X_C$.

SET UP: $\omega_0 = \dfrac{1}{\sqrt{LC}}$. $V = IZ$. $V_R = IR$, $V_L = IX_L$ and $V_C = V_L$.

EXECUTE: (a) $\omega_0 = \dfrac{1}{\sqrt{LC}} = \dfrac{1}{\sqrt{(0.280 \text{ H})(4.00 \times 10^{-6} \text{ F})}} = 945 \text{ rad/s}$.

(b) $I = 1.70 \text{ A}$ at resonance, so $R = Z = \dfrac{V}{I} = \dfrac{120 \text{ V}}{1.70 \text{ A}} = 70.6 \text{ }\Omega$.

(c) At resonance, $V_R = 120 \text{ V}$, $V_L = V_C = I\omega L = (1.70 \text{ A})(945 \text{ rad/s})(0.280 \text{ H}) = 450 \text{ V}$.

EVALUATE: At resonance, $V_R = V$ and $V_L - V_C = 0$.

31.35. IDENTIFY and SET UP: The equation $\dfrac{V_2}{V_1} = \dfrac{N_2}{N_1}$ relates the primary and secondary voltages to the number of turns in each. $I = V/R$ and the power consumed in the resistive load is $I_{rms}^2 = V_{rms}^2 / R$. Let I_1, V_1 and I_2, V_2 be rms values for the primary and secondary.

EXECUTE: (a) $\dfrac{V_2}{V_1} = \dfrac{N_2}{N_1}$ so $\dfrac{N_1}{N_2} = \dfrac{V_1}{V_2} = \dfrac{120 \text{ V}}{12.0 \text{ V}} = 10$.

(b) $I_2 = \dfrac{V_2}{R} = \dfrac{12.0 \text{ V}}{5.00 \text{ }\Omega} = 2.40 \text{ A}$.

(c) $P_{av} = I_2^2 R = (2.40 \text{ A})^2 (5.00 \text{ }\Omega) = 28.8 \text{ W}$.

(d) The power drawn from the line by the transformer is the 28.8 W that is delivered by the load.

$P_{av} = \dfrac{V_1^2}{R}$ so $R = \dfrac{V_1^2}{P_{av}} = \dfrac{(120 \text{ V})^2}{28.8 \text{ W}} = 500 \text{ }\Omega$.

And $\left(\dfrac{N_1}{N_2}\right)^2 (5.00\ \Omega) = (10)^2 (5.00\ \Omega) = 500\ \Omega$, as was to be shown.

EVALUATE: The resistance is "transformed." A load of resistance R connected to the secondary draws the same power as a resistance $(N_1/N_2)^2 R$ connected directly to the supply line, without using the transformer.

31.41. IDENTIFY: We can use geometry to calculate the capacitance and inductance, and then use these results to calculate the resonance angular frequency.

SET UP: The capacitance of an air-filled parallel plate capacitor is $C = \dfrac{\varepsilon_0 A}{d}$. The inductance of a long solenoid is $L = \dfrac{\mu_0 A N^2}{l}$. The inductor has $N = (125\text{ coils/cm})(9.00\text{ cm}) = 1125$ coils. The resonance frequency is $f_0 = \dfrac{1}{2\pi\sqrt{LC}}$. $\varepsilon_0 = 8.85\times 10^{-12}\ \text{C}^2/\text{N}\cdot\text{m}^2$. $\mu_0 = 4\pi\times 10^{-7}\ \text{T}\cdot\text{m/A}$.

EXECUTE: $C = \dfrac{\varepsilon_0 A}{d} = \dfrac{(8.85\times 10^{-12}\ \text{C}^2/\text{N}\cdot\text{m}^2)(4.50\times 10^{-2}\ \text{m})^2}{8.00\times 10^{-3}\ \text{m}} = 2.24\times 10^{-12}\ \text{F}$.

$L = \dfrac{\mu_0 A N^2}{l} = \dfrac{(4\pi\times 10^{-7}\ \text{T}\cdot\text{m/A})\pi(0.250\times 10^{-2}\ \text{m})^2 (1125)^2}{9.00\times 10^{-2}\ \text{m}} = 3.47\times 10^{-4}\ \text{H}$.

$\omega_0 = \dfrac{1}{\sqrt{(3.47\times 10^{-4}\ \text{H})(2.24\times 10^{-12}\ \text{F})}} = 3.59\times 10^{7}\ \text{rad/s}$.

EVALUATE: The result is a rather high angular frequency.

31.43. IDENTIFY and SET UP: Source voltage lags current so it must be that $X_C > X_L$.

EXECUTE: **(a)** We must add an inductor in series with the circuit. When $X_C = X_L$ the power factor has its maximum value of unity, so calculate the additional L needed to raise X_L to equal X_C.

(b) Power factor $\cos\phi$ equals 1 so $\phi = 0$ and $X_C = X_L$. Calculate the present value of $X_C - X_L$ to see how much more X_L is needed: $R = Z\cos\phi = (60.0\ \Omega)(0.720) = 43.2\ \Omega$

$\tan\phi = \dfrac{X_L - X_C}{R}$ so $X_L - X_C = R\tan\phi$.

$\cos\phi = 0.720$ gives $\phi = -43.95°$ (ϕ is negative since the voltage lags the current).

Then $X_L - X_C = R\tan\phi = (43.2\ \Omega)\tan(-43.95°) = -41.64\ \Omega$.

Therefore need to add $41.64\ \Omega$ of X_L.

$X_L = \omega L = 2\pi f L$ and $L = \dfrac{X_L}{2\pi f} = \dfrac{41.64\ \Omega}{2\pi(50.0\ \text{Hz})} = 0.133\ \text{H}$, amount of inductance to add.

EVALUATE: From the information given we can't calculate the original value of L in the circuit, just how much to add. When this L is added the current in the circuit will increase.

31.47. IDENTIFY and SET UP: Express Z and I in terms of ω, L, C, and R. The voltages across the resistor and the inductor are $90°$ out of phase, so $V_{\text{out}} = \sqrt{V_R^2 + V_L^2}$.

EXECUTE: The circuit is sketched in Figure 31.47.

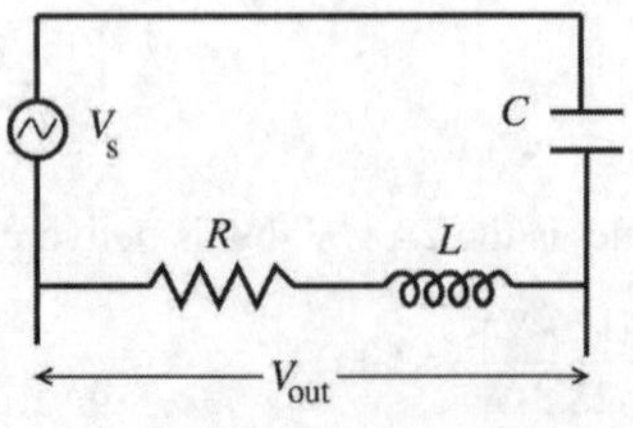

$X_L = \omega L,\ X_C = \dfrac{1}{\omega C}$

$Z = \sqrt{R^2 + \left(\omega L - \dfrac{1}{\omega C}\right)^2}$

$I = \dfrac{V_s}{Z} = \dfrac{V_s}{\sqrt{R^2 + \left(\omega L - \dfrac{1}{\omega C}\right)^2}}$

Figure 31.47

$$V_{out} = I\sqrt{R^2 + X_L^2} = I\sqrt{R^2 + \omega^2 L^2} = V_s \sqrt{\dfrac{R^2 + \omega^2 L^2}{R^2 + \left(\omega L - \dfrac{1}{\omega C}\right)^2}}$$

$$\dfrac{V_{out}}{V_s} = \sqrt{\dfrac{R^2 + \omega^2 L^2}{R^2 + \left(\omega L - \dfrac{1}{\omega C}\right)^2}}$$

ω small:

As ω gets small, $R^2 + \left(\omega L - \dfrac{1}{\omega C}\right)^2 \to \dfrac{1}{\omega^2 C^2}$, $R^2 + \omega^2 L^2 \to R^2$.

Therefore $\dfrac{V_{out}}{V_s} \to \sqrt{\dfrac{R^2}{(1/\omega^2 C^2)}} = \omega RC$ as ω becomes small.

ω large:

As ω gets large, $R^2 + \left(\omega L - \dfrac{1}{\omega C}\right)^2 \to R^2 + \omega^2 L^2 \to \omega^2 L^2$, $R^2 + \omega^2 L^2 \to \omega^2 L^2$.

Therefore, $\dfrac{V_{out}}{V_s} \to \sqrt{\dfrac{\omega^2 L^2}{\omega^2 L^2}} = 1$ as ω becomes large.

EVALUATE: $V_{out}/V_s \to 0$ as ω becomes small, so there is V_{out} only when the frequency ω of V_s is large. If the source voltage contains a number of frequency components, only the high frequency ones are passed by this filter.

31.51. IDENTIFY: We know R, X_C, and ϕ so $\tan\phi = \dfrac{X_L - X_C}{R}$ tells us X_L. Use $P_{av} = I_{rms}^2 R$ to calculate I_{rms}. Then calculate Z and use $V_{rms} = I_{rms} Z$ to calculate V_{rms} for the source.

SET UP: Source voltage lags current so $\phi = -54.0°$. $X_C = 350\,\Omega$, $R = 180\,\Omega$, $P_{av} = 140$ W.

EXECUTE: (a) $\tan\phi = \dfrac{X_L - X_C}{R}$.

$X_L = R\tan\phi + X_C = (180\,\Omega)\tan(-54.0°) + 350\,\Omega = -248\,\Omega + 350\,\Omega = 102\,\Omega$.

(b) $P_{av} = V_{rms} I_{rms} \cos\phi = I_{rms}^2 R$ (Exercise 31.22). $I_{rms} = \sqrt{\dfrac{P_{av}}{R}} = \sqrt{\dfrac{140\text{ W}}{180\,\Omega}} = 0.882$ A.

(c) $Z = \sqrt{R^2 + (X_L - X_C)^2} = \sqrt{(180\,\Omega)^2 + (102\,\Omega - 350\,\Omega)^2} = 306\,\Omega$.

$V_{rms} = I_{rms} Z = (0.882\text{ A})(306\,\Omega) = 270$ V.

EVALUATE: We could also use $P_{av} = V_{rms} I_{rms} \cos\phi$.

$V_{rms} = \dfrac{P_{av}}{I_{rms}\cos\phi} = \dfrac{140\text{ W}}{(0.882\text{ A})\cos(-54.0°)} = 270$ V, which agrees. The source voltage lags the current when $X_C > X_L$, and this agrees with what we found.

31.53. IDENTIFY and SET UP: Calculate Z and $I = V/Z$.

EXECUTE: (a) For $\omega = 800$ rad/s:

$Z = \sqrt{R^2 + (\omega L - 1/\omega C)^2} = \sqrt{(500\,\Omega)^2 + \{(800\text{ rad/s})(2.0\text{ H}) - 1/[(800\text{ rad/s})(5.0\times 10^{-7}\text{ F})]\}^2}$. $Z = 1030\,\Omega$.

$I = \dfrac{V}{Z} = \dfrac{100\text{ V}}{1030\,\Omega} = 0.0971$ A. $V_R = IR = (0.0971\text{ A})(500\,\Omega) = 48.6$ V,

$V_C = IX_C = \dfrac{I}{\omega C} = \dfrac{0.0971\text{ A}}{(800\text{ rad/s})(5.0\times 10^{-7}\text{ F})} = 243$ V and

$V_L = I\omega L = (0.0971 \text{ A})(800 \text{ rad/s})(2.00 \text{ H}) = 155 \text{ V}$. $\phi = \arctan\left(\dfrac{\omega L - 1/(\omega C)}{R}\right) = -60.9°$. The graph of each voltage versus time is given in Figure 31.53a.

(b) Repeating exactly the same calculations as above for $\omega = 1000$ rad/s:

$Z = R = 500 \text{ }\Omega$; $\phi = 0$; $I = 0.200$ A; $V_R = V = 100$ V; $V_C = V_L = 400$ V. The graph of each voltage versus time is given in Figure 31.53b.

(c) Repeating exactly the same calculations as part (a) for $\omega = 1250$ rad/s:

$Z = 1030 \text{ }\Omega$; $\phi = +60.9°$; $I = 0.0971$ A; $V_R = 48.6$ V; $V_C = 155$ V; $V_L = 243$ V. The graph of each voltage versus time is given in Figure 31.53c.

EVALUATE: The resonance frequency is $\omega_0 = \dfrac{1}{\sqrt{LC}} = \dfrac{1}{\sqrt{(2.00 \text{ H})(0.500 \text{ }\mu\text{F})}} = 1000$ rad/s. For $\omega < \omega_0$ the phase angle is negative and for $\omega > \omega_0$ the phase angle is positive.

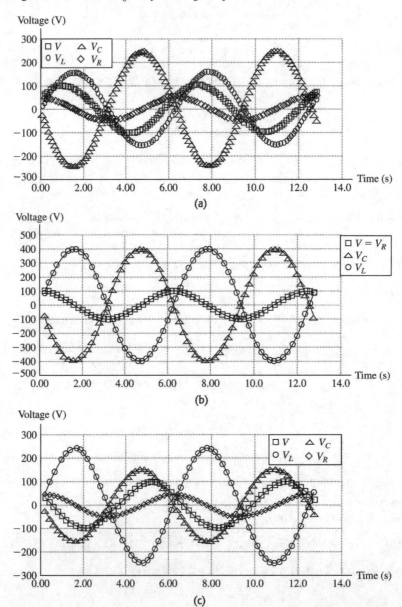

Figure 31.53

31.57. **IDENTIFY:** The average power depends on the phase angle ϕ.

SET UP: The average power is $P_{av} = V_{rms}I_{rms}\cos\phi$, and the impedance is $Z = \sqrt{R^2 + \left(\omega L - \dfrac{1}{\omega C}\right)^2}$.

EXECUTE: (a) $P_{av} = V_{rms}I_{rms}\cos\phi = \tfrac{1}{2}(V_{rms}I_{rms})$, which gives $\cos\phi = \tfrac{1}{2}$, so $\phi = \pi/3 = 60°$. $\tan\phi = (X_L - X_C)/R$, which gives $\tan 60° = (\omega L - 1/\omega C)/R$. Using $R = 75.0\,\Omega$, $L = 5.00$ mH and $C = 2.50\,\mu$F and solving for ω we get $\omega = 28760$ rad/s $= 28,800$ rad/s.

(b) $Z = \sqrt{R^2 + (X_L - X_C)^2}$, where $X_L = \omega L = (28,760$ rad/s$)(5.00$ mH$) = 144\,\Omega$ and $X_C = 1/\omega C = 1/[(28,760$ rad/s$)(2.50\,\mu$F$)] = 13.9\,\Omega$, giving $Z = \sqrt{(75\,\Omega)^2 + (144\,\Omega - 13.9\,\Omega)^2} = 150\,\Omega$; $I = V/Z = (15.0$ V$)/(150\,\Omega) = 0.100$ A and $P_{av} = \tfrac{1}{2}VI\cos\phi = \tfrac{1}{2}(15.0$ V$)(0.100$ A$)(1/2) = 0.375$ W.

EVALUATE: All this power is dissipated in the resistor because the average power delivered to the inductor and capacitor is zero.

31.63. **IDENTIFY:** $P_{av} = V_{rms}I_{rms}\cos\phi$ and $I_{rms} = \dfrac{V_{rms}}{Z}$. Calculate Z. $R = Z\cos\phi$.

SET UP: $f = 50.0$ Hz and $\omega = 2\pi f$. The power factor is $\cos\phi$.

EXECUTE: (a) $P_{av} = \dfrac{V_{rms}^2}{Z}\cos\phi$. $Z = \dfrac{V_{rms}^2 \cos\phi}{P_{av}} = \dfrac{(120\text{ V})^2(0.560)}{(220\text{ W})} = 36.7\,\Omega$. $R = Z\cos\phi = (36.7\,\Omega)(0.560) = 20.6\,\Omega$.

(b) $Z = \sqrt{R^2 + X_L^2} \cdot X_L = \sqrt{Z^2 - R^2} = \sqrt{(36.7\,\Omega)^2 - (20.6\,\Omega)^2} = 30.4\,\Omega$. But $\phi = 0$ is at resonance, so the inductive and capacitive reactances equal each other. Therefore we need to add $X_C = 30.4\,\Omega$. $X_C = \dfrac{1}{\omega C}$ therefore gives $C = \dfrac{1}{\omega X_C} = \dfrac{1}{2\pi f X_C} = \dfrac{1}{2\pi(50.0\text{ Hz})(30.4\,\Omega)} = 1.05\times10^{-4}$ F.

(c) At resonance, $P_{av} = \dfrac{V^2}{R} = \dfrac{(120\text{ V})^2}{20.6\,\Omega} = 699$ W.

EVALUATE: $P_{av} = I_{rms}^2 R$ and I_{rms} is maximum at resonance, so the power drawn from the line is maximum at resonance.

31.65. **IDENTIFY and SET UP:** For an L-R-C series circuit, the maximum current occurs at resonance, and the resonance angular frequency is $\omega_{res} = \dfrac{1}{\sqrt{LC}}$.

EXECUTE: At resonance, the angular frequency is $\omega_{res} = \dfrac{1}{\sqrt{LC}}$. Squaring gives $\omega_{res}^2 = \dfrac{1}{L}\cdot\dfrac{1}{C}$, so a graph of ω_{res}^2 versus $1/C$ should be a straight line with a slope equal to $1/L$. Using two convenient points on the graph, we find the slope to be $\dfrac{(25.0 - 1.00)\times10^4 \text{ rad}^2/\text{s}^2}{(4.50 - 1.75)\times10^3 \text{ F}^{-1}} = 5.455$ F/s^2. Solving for L gives $L = (\text{slope})^{-1} = (5.455$ F/s$^2)^{-1} = 0.183$ H, which rounds to 0.18 H, since we cannot determine the slope of the graph in the text with anything better than 2 significant figures. To find R, we realize that at resonance $Z = R$, so $R = V/I = (90.0$ V$)/(4.50$ A$) = 20.0\,\Omega$.

EVALUATE: These are reasonable values for L and R for a laboratory solenoid.

31.69. **IDENTIFY** and **SET UP:** We are told that the platinum electrode behaves like an ideal capacitor in series with the resistance of the fluid. The impedance of an R-C circuit is $Z = \sqrt{R^2 + X_C^2}$, where $X_C = \dfrac{1}{\omega C}$.

EXECUTE: For a dc signal we have $\omega = 2\pi f = 0$. Using $X_C = \dfrac{1}{\omega C}$ we see that as $\omega \to 0$ we have $X_C \to \infty$, and so $Z \to \infty$. The correct choice is (b).

EVALUATE: The oscillation period of such a circuit is $T = 1/f$, so $T \to \infty$ as $\omega \to 0$.

31.71. **IDENTIFY** and **SET UP:** We know that $V_{rms} = \dfrac{V}{\sqrt{2}}$, where V is the amplitude (peak value) of the voltage. According to the problem, the peak-to-peak voltage V_{pp} is the difference between the two extreme values of voltage.

EXECUTE: Since the voltage oscillates between $+V$ and $-V$ the peak-to-peak voltage is $V_{pp} = V - (-V) = 2V = 2\sqrt{2} V_{rms}$. Thus, the correct answer is (d).

EVALUATE: The voltage amplitude is half the peak-to-peak voltage.

ELECTROMAGNETIC WAVES

32.3. **IDENTIFY:** $E_{max} = cB_{max}$. $\vec{E} \times \vec{B}$ is in the direction of propagation.

SET UP: $c = 3.00 \times 10^8$ m/s. $E_{max} = 4.00$ V/m.

EXECUTE: $B_{max} = E_{max}/c = 1.33 \times 10^{-8}$ T. For $\vec{E}$ in the $+x$-direction, $\vec{E} \times \vec{B}$ is in the $+z$-direction when $\vec{B}$ is in the $+y$-direction.

EVALUATE: $\vec{E}$, $\vec{B}$, and the direction of propagation are all mutually perpendicular.

32.7. **IDENTIFY:** $c = f\lambda$. $E_{max} = cB_{max}$. $k = 2\pi/\lambda$. $\omega = 2\pi f$.

SET UP: Since the wave is traveling in empty space, its wave speed is $c = 3.00 \times 10^8$ m/s.

EXECUTE: **(a)** $f = \dfrac{c}{\lambda} = \dfrac{3.00 \times 10^8 \text{ m/s}}{432 \times 10^{-9} \text{ m}} = 6.94 \times 10^{14}$ Hz.

(b) $E_{max} = cB_{max} = (3.00 \times 10^8 \text{ m/s})(1.25 \times 10^{-6} \text{ T}) = 375$ V/m.

(c) $k = \dfrac{2\pi}{\lambda} = \dfrac{2\pi \text{ rad}}{432 \times 10^{-9} \text{ m}} = 1.45 \times 10^7$ rad/m. $\omega = (2\pi \text{ rad})(6.94 \times 10^{14} \text{ Hz}) = 4.36 \times 10^{15}$ rad/s.

$E = E_{max} \cos(kx - \omega t) = (375 \text{ V/m}) \cos[(1.45 \times 10^7 \text{ rad/m})x - (4.36 \times 10^{15} \text{ rad/s})t]$.

$B = B_{max} \cos(kx - \omega t) = (1.25 \times 10^{-6} \text{ T}) \cos[(1.45 \times 10^7 \text{ rad/m})x - (4.36 \times 10^{15} \text{ rad/s})t]$.

EVALUATE: The $\cos(kx - \omega t)$ factor is common to both the electric and magnetic field expressions, since these two fields are in phase.

32.11. **IDENTIFY and SET UP:** Compare the $\vec{E}(y,t)$ given in the problem to the general form given by Eq. (32.17). Use the direction of propagation and of $\vec{E}$ to find the direction of $\vec{B}$.

EXECUTE: **(a)** The equation for the electric field contains the factor $\cos(ky - \omega t)$ so the wave is traveling in the $+y$-direction.

(b) $\vec{E}(y,t) = (3.10 \times 10^5 \text{ V/m})\hat{k} \cos[ky - (12.65 \times 10^{12} \text{ rad/s})t]$.

Comparing to Eq. (32.17) gives $\omega = 12.65 \times 10^{12}$ rad/s

$\omega = 2\pi f = \dfrac{2\pi c}{\lambda}$ so $\lambda = \dfrac{2\pi c}{\omega} = \dfrac{2\pi(2.998 \times 10^8 \text{ m/s})}{(12.65 \times 10^{12} \text{ rad/s})} = 1.49 \times 10^{-4}$ m.

(c)

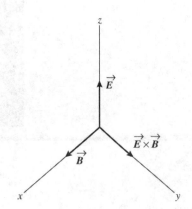

$\vec{E} \times \vec{B}$ must be in the $+y$-direction (the direction in which the wave is traveling). When $\vec{E}$ is in the $+z$-direction then $\vec{B}$ must be in the $+x$-direction, as shown in Figure 32.11.

Figure 32.11

$k = \dfrac{2\pi}{\lambda} = \dfrac{\omega}{c} = \dfrac{12.65 \times 10^{12} \text{ rad/s}}{2.998 \times 10^8 \text{ m/s}} = 4.22 \times 10^4 \text{ rad/m}.$

$E_{max} = 3.10 \times 10^5 \text{ V/m}.$

Then $B_{max} = \dfrac{E_{max}}{c} = \dfrac{3.10 \times 10^5 \text{ V/m}}{2.998 \times 10^8 \text{ m/s}} = 1.03 \times 10^{-3} \text{ T}.$

Using Eq. (32.17) and the fact that $\vec{B}$ is in the $+\hat{i}$-direction when $\vec{E}$ is in the $+\hat{k}$-direction,

$\vec{B} = +(1.03 \times 10^{-3} \text{ T})\hat{i} \cos[(4.22 \times 10^4 \text{ rad/m})y - (12.65 \times 10^{12} \text{ rad/s})t].$

EVALUATE: $\vec{E}$ and $\vec{B}$ are perpendicular and oscillate in phase.

32.15. IDENTIFY and SET UP: $v = f\lambda$ relates frequency and wavelength to the speed of the wave. Use $n = \sqrt{KK_m} \approx \sqrt{K}$ to calculate n and K.

EXECUTE: (a) $\lambda = \dfrac{v}{f} = \dfrac{2.17 \times 10^8 \text{ m/s}}{5.70 \times 10^{14} \text{ Hz}} = 3.81 \times 10^{-7}$ m.

(b) $\lambda = \dfrac{c}{f} = \dfrac{2.998 \times 10^8 \text{ m/s}}{5.70 \times 10^{14} \text{ Hz}} = 5.26 \times 10^{-7}$ m.

(c) $n = \dfrac{c}{v} = \dfrac{2.998 \times 10^8 \text{ m/s}}{2.17 \times 10^8 \text{ m/s}} = 1.38.$

(d) $n = \sqrt{KK_m} \approx \sqrt{K}$ so $K = n^2 = (1.38)^2 = 1.90.$

EVALUATE: In the material $v < c$ and f is the same, so λ is less in the material than in air. $v < c$ always, so n is always greater than unity.

32.17. IDENTIFY: $I = P/A$. $I = \tfrac{1}{2}\varepsilon_0 c E_{max}^2$. $E_{max} = cB_{max}$.

SET UP: The surface area of a sphere of radius r is $A = 4\pi r^2$. $\varepsilon_0 = 8.85 \times 10^{-12}$ C^2/N·m^2.

EXECUTE: (a) $I = \dfrac{P}{A} = \dfrac{(0.05)(75 \text{ W})}{4\pi(3.0 \times 10^{-2} \text{ m})^2} = 330$ W/m^2.

(b) $E_{max} = \sqrt{\dfrac{2I}{\varepsilon_0 c}} = \sqrt{\dfrac{2(330 \text{ W/m}^2)}{(8.85 \times 10^{-12} \text{ C}^2/\text{N} \cdot \text{m}^2)(3.00 \times 10^8 \text{ m/s})}} = 500$ V/m.

$B_{max} = \dfrac{E_{max}}{c} = 1.7 \times 10^{-6}$ T $= 1.7$ μT.

EVALUATE: At the surface of the bulb the power radiated by the filament is spread over the surface of the bulb. Our calculation approximates the filament as a point source that radiates uniformly in all directions.

32.23. IDENTIFY: $P_{av} = IA$ and $I = E_{max}^2/2\mu_0 c$

SET UP: The surface area of a sphere is $A = 4\pi r^2$.

EXECUTE: $P_{av} = S_{av}A = \left(\dfrac{E_{max}^2}{2c\mu_0}\right)(4\pi r^2)$. $E_{max} = \sqrt{\dfrac{P_{av}c\mu_0}{2\pi r^2}} = \sqrt{\dfrac{(60.0 \text{ W})(3.00\times 10^8 \text{ m/s})\mu_0}{2\pi(5.00 \text{ m})^2}} = 12.0 \text{ V/m}$.

$B_{max} = \dfrac{E_{max}}{c} = \dfrac{12.0 \text{ V/m}}{3.00\times 10^8 \text{ m/s}} = 4.00\times 10^{-8}$ T.

EVALUATE: E_{max} and B_{max} are both inversely proportional to the distance from the source.

32.25. IDENTIFY: Use the radiation pressure to find the intensity, and then $P_{av} = I(4\pi r^2)$.

SET UP: For a perfectly absorbing surface, $p_{rad} = \dfrac{I}{c}$.

EXECUTE: $p_{rad} = I/c$ so $I = cp_{rad} = 2.70\times 10^3$ W/m^2. Then $P_{av} = I(4\pi r^2) = (2.70\times 10^3 \text{ W/m}^2)(4\pi)(5.0 \text{ m})^2 = 8.5\times 10^5$ W.

EVALUATE: Even though the source is very intense the radiation pressure 5.0 m from the surface is very small.

32.27. IDENTIFY: We know the greatest intensity that the eye can safely receive.

SET UP: $I = \dfrac{P}{A}$. $I = \tfrac{1}{2}\varepsilon_0 c E_{max}^2$. $E_{max} = cB_{max}$.

EXECUTE: (a) $P = IA = (1.0\times 10^2 \text{ W/m}^2)\pi(0.75\times 10^{-3} \text{ m})^2 = 1.8\times 10^{-4}$ W $= 0.18$ mW.

(b) $E = \sqrt{\dfrac{2I}{\varepsilon_0 c}} = \sqrt{\dfrac{2(1.0\times 10^2 \text{ W/m}^2)}{(8.85\times 10^{-12} \text{ C}^2/\text{N}\cdot\text{m}^2)(3.00\times 10^8 \text{ m/s})}} = 274$ V/m. $B_{max} = \dfrac{E_{max}}{c} = 9.13\times 10^{-7}$ T.

(c) $P = 0.18$ mW $= 0.18$ mJ/s.

(d) $I = (1.0\times 10^2 \text{ W/m}^2)\left(\dfrac{1 \text{ m}}{10^2 \text{ cm}}\right)^2 = 0.010$ W/cm^2.

EVALUATE: Both the electric and magnetic fields are quite weak compared to normal laboratory fields.

32.31. IDENTIFY: The nodal and antinodal planes are each spaced one-half wavelength apart.

SET UP: $2\tfrac{1}{2}$ wavelengths fit in the oven, so $(2\tfrac{1}{2})\lambda = L$, and the frequency of these waves obeys the equation $f\lambda = c$.

EXECUTE: (a) Since $(2\tfrac{1}{2})\lambda = L$, we have $L = (5/2)(12.2 \text{ cm}) = 30.5$ cm.

(b) Solving for the frequency gives $f = c/\lambda = (3.00\times 10^8 \text{ m/s})/(0.122 \text{ m}) = 2.46\times 10^9$ Hz.

(c) $L = 35.5$ cm in this case. $(2\tfrac{1}{2})\lambda = L$, so $\lambda = 2L/5 = 2(35.5 \text{ cm})/5 = 14.2$ cm.

$f = c/\lambda = (3.00\times 10^8 \text{ m/s})/(0.142 \text{ m}) = 2.11\times 10^9$ Hz.

EVALUATE: Since microwaves have a reasonably large wavelength, microwave ovens can have a convenient size for household kitchens. Ovens using radiowaves would need to be far too large, while ovens using visible light would have to be microscopic.

32.33. IDENTIFY: We know the wavelength and power of a laser beam as well as the area over which it acts and the duration of a pulse.

SET UP: The energy is $U = Pt$. For absorption the radiation pressure is $\dfrac{I}{c}$, where $I = \dfrac{P}{A}$. The wavelength in the eye is $\lambda = \dfrac{\lambda_0}{n}$. $I = \tfrac{1}{2}\varepsilon_0 c E_{max}^2$ and $E_{max} = cB_{max}$.

EXECUTE: (a) $U = Pt = (250\times 10^{-3} \text{ W})(1.50\times 10^{-3} \text{ s}) = 3.75\times 10^{-4}$ J $= 0.375$ mJ.

(b) $I = \dfrac{P}{A} = \dfrac{250 \times 10^{-3} \text{ W}}{\pi(255 \times 10^{-6} \text{ m})^2} = 1.22 \times 10^6 \text{ W/m}^2$. The average pressure is

$\dfrac{I}{c} = \dfrac{1.22 \times 10^6 \text{ W/m}^2}{3.00 \times 10^8 \text{ m/s}} = 4.08 \times 10^{-3} \text{ Pa}$.

(c) $\lambda = \dfrac{\lambda_0}{n} = \dfrac{810 \text{ nm}}{1.34} = 604 \text{ nm}$. $f = \dfrac{v}{\lambda} = \dfrac{c}{\lambda_0} = \dfrac{3.00 \times 10^8 \text{ m/s}}{810 \times 10^{-9} \text{ m}} = 3.70 \times 10^{14} \text{ Hz}$; f is the same in the air and in the vitreous humor.

(d) $E_{max} = \sqrt{\dfrac{2I}{\varepsilon_0 c}} = \sqrt{\dfrac{2(1.22 \times 10^6 \text{ W/m}^2)}{(8.85 \times 10^{-12} \text{ C}^2/\text{N} \cdot \text{m}^2)(3.00 \times 10^8 \text{ m/s})}} = 3.03 \times 10^4 \text{ V/m}$.

$B_{max} = \dfrac{E_{max}}{c} = 1.01 \times 10^{-4} \text{ T}$.

EVALUATE: The intensity of the beam is high, as it must be to weld tissue, but the pressure it exerts on the retina is only around 10^{-8} that of atmospheric pressure. The magnetic field in the beam is about twice that of the earth's magnetic field.

32.39. IDENTIFY: The same intensity light falls on both reflectors, but the force on the reflecting surface will be twice as great as the force on the absorbing surface. Therefore there will be a net torque about the rotation axis.

SET UP: For a totally absorbing surface, $F = p_{rad}A = (I/c)A$, while for a totally reflecting surface the force will be twice as great. The intensity of the wave is $I = \tfrac{1}{2}\varepsilon_0 c E_{max}^2$. Once we have the torque, we can use the rotational form of Newton's second law, $\tau_{net} = I\alpha$, to find the angular acceleration.

EXECUTE: The force on the absorbing reflector is $F_{abs} = p_{rad}A = (I/c)A = \dfrac{\tfrac{1}{2}\varepsilon_0 c E_{max}^2 A}{c} = \tfrac{1}{2}\varepsilon_0 A E_{max}^2$.

For a totally reflecting surface, the force will be twice as great, which is $\varepsilon_0 c E_{max}^2$. The net torque is therefore $\tau_{net} = F_{refl}(L/2) - F_{abs}(L/2) = \varepsilon_0 A E_{max}^2 L/4$.

Newton's second law for rotation gives $\tau_{net} = I\alpha$. $\varepsilon_0 A E_{max}^2 L/4 = 2m(L/2)^2 \alpha$.
Solving for α gives

$\alpha = \varepsilon_0 A E_{max}^2/(2mL) = \dfrac{(8.85 \times 10^{-12} \text{ C}^2/\text{N} \cdot \text{m}^2)(0.0150 \text{ m})^2(1.25 \text{ N/C})^2}{(2)(0.00400 \text{ kg})(1.00 \text{ m})} = 3.89 \times 10^{-13} \text{ rad/s}^2$.

EVALUATE: This is an extremely small angular acceleration. To achieve a larger value, we would have to greatly increase the intensity of the light wave or decrease the mass of the reflectors.

32.41. IDENTIFY and SET UP: In the wire the electric field is related to the current density by $\vec{E} = \rho \vec{J}$. Use Ampere's law to calculate $\vec{B}$. The Poynting vector is given by $\vec{S} = \dfrac{1}{\mu_0}\vec{E} \times \vec{B}$ and $P = \oint \vec{S} \cdot d\vec{A}$ relates the energy flow through a surface to $\vec{S}$.

EXECUTE: **(a)** The direction of $\vec{E}$ is parallel to the axis of the cylinder, in the direction of the current. $E = \rho J = \rho I/\pi a^2$. ($E$ is uniform across the cross section of the conductor.)

(b) A cross-sectional view of the conductor is given in Figure 32.41a; take the current to be coming out of the page.

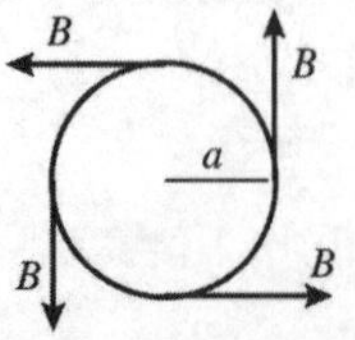

Apply Ampere's law to a circle of radius a.
$\oint \vec{B} \cdot d\vec{l} = B(2\pi a)$
$I_{encl} = I$.

Figure 32.41a

$\oint \vec{B} \cdot d\vec{l} = \mu_0 I_{\text{encl}}$ gives $B(2\pi a) = \mu_0 I$ and $B = \dfrac{\mu_0 I}{2\pi a}$.

The direction of $\vec{B}$ is counterclockwise around the circle.

(c) The directions of $\vec{E}$ and $\vec{B}$ are shown in Figure 32.41b.

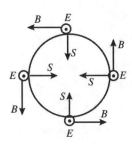

The direction of $\vec{S} = \dfrac{1}{\mu_0} \vec{E} \times \vec{B}$.

is radially inward.

$S = \dfrac{1}{\mu_0} EB = \dfrac{1}{\mu_0} \left(\dfrac{\rho I}{\pi a^2}\right)\left(\dfrac{\mu_0 I}{2\pi a}\right)$.

$S = \dfrac{\rho I^2}{2\pi^2 a^3}$.

Figure 32.41b

EVALUATE: **(d)** Since S is constant over the surface of the conductor, the rate of energy flow P is given by S times the surface of a length l of the conductor: $P = SA = S(2\pi al) = \dfrac{\rho I^2}{2\pi^2 a^3}(2\pi al) = \dfrac{\rho l I^2}{\pi a^2}$. But $R = \dfrac{\rho l}{\pi a^2}$, so the result from the Poynting vector is $P = RI^2$. This agrees with $P_R = I^2 R$, the rate at which electrical energy is being dissipated by the resistance of the wire. Since $\vec{S}$ is radially inward at the surface of the wire and has magnitude equal to the rate at which electrical energy is being dissipated in the wire, this energy can be thought of as entering through the cylindrical sides of the conductor.

32.43. **IDENTIFY:** The nodal planes are one-half wavelength apart.

SET UP: The nodal planes of B are at $x = \lambda/4, 3\lambda/4, 5\lambda/4, \ldots$, which are $\lambda/2$ apart.

EXECUTE: **(a)** The wavelength is $\lambda = c/f = (2.998 \times 10^8 \text{ m/s})/(110.0 \times 10^6 \text{ Hz}) = 2.725$ m. So the nodal planes are at $(2.725 \text{ m})/2 = 1.363$ m apart.

(b) For the nodal planes of E, we have $\lambda_n = 2L/n$, so $L = n\lambda/2 = (8)(2.725 \text{ m})/2 = 10.90$ m.

EVALUATE: Because radiowaves have long wavelengths, the distances involved are easily measurable using ordinary metersticks.

32.45. **IDENTIFY:** The orbiting satellite obeys Newton's second law of motion. The intensity of the electromagnetic waves it transmits obeys the inverse-square distance law, and the intensity of the waves depends on the amplitude of the electric and magnetic fields.

SET UP: Newton's second law applied to the satellite gives $mv^2/r = GmM/r^2$, where M is the mass of the earth and m is the mass of the satellite. The intensity I of the wave is $I = S_{\text{av}} = \tfrac{1}{2}\varepsilon_0 c E_{\text{max}}^2$, and by definition, $I = P_{\text{av}}/A$.

EXECUTE: **(a)** The period of the orbit is 12 hr. Applying Newton's second law to the satellite gives $mv^2/r = GmM/r^2$, which gives $\dfrac{m(2\pi r/T)^2}{r} = \dfrac{GmM}{r^2}$. Solving for r, we get

$r = \left(\dfrac{GMT^2}{4\pi^2}\right)^{1/3} = \left[\dfrac{(6.67 \times 10^{-11} \text{ N} \cdot \text{m}^2/\text{kg}^2)(5.97 \times 10^{24} \text{ kg})(12 \times 3600 \text{ s})^2}{4\pi^2}\right]^{1/3} = 2.66 \times 10^7$ m.

The height above the surface is $h = 2.66 \times 10^7$ m $- 6.37 \times 10^6$ m $= 2.02 \times 10^7$ m. The satellite only radiates its energy to the lower hemisphere, so the area is 1/2 that of a sphere. Thus, from the definition of intensity, the intensity at the ground is

$I = P_{\text{av}}/A = P_{\text{av}}/(2\pi h^2) = (25.0 \text{ W})/[2\pi(2.02 \times 10^7 \text{ m})^2] = 9.75 \times 10^{-15}$ W/m^2

(b) $I = S_{av} = \frac{1}{2}\epsilon_0 c E_{max}^2$, so $E_{max} = \sqrt{\dfrac{2I}{\epsilon_0 c}} = \sqrt{\dfrac{2(9.75\times 10^{-15}\text{ W/m}^2)}{(8.85\times 10^{-12}\text{ C}^2/\text{N}\cdot\text{m}^2)(3.00\times 10^8\text{ m/s})}} = 2.71\times 10^{-6}$ N/C.

$B_{max} = E_{max}/c = (2.71\times 10^{-6}\text{ N/C})/(3.00\times 10^8\text{ m/s}) = 9.03\times 10^{-15}$ T.

$t = d/c = (2.02\times 10^7\text{ m})/(3.00\times 10^8\text{ m/s}) = 0.0673$ s.

(c) $p_{rad} = I/c = (9.75\times 10^{-15}\text{ W/m}^2)/(3.00\times 10^8\text{ m/s}) = 3.25\times 10^{-23}$ Pa.

(d) $\lambda = c/f = (3.00\times 10^8\text{ m/s})/(1575.42\times 10^6\text{ Hz}) = 0.190$ m.

EVALUATE: The fields and pressures due to these waves are very small compared to typical laboratory quantities.

32.49. IDENTIFY and SET UP: The intensity of the light beam is $I = \frac{1}{2}\epsilon_0 c E_{max}^2$.

EXECUTE: **(a)** A graph of I versus E_{max}^2 should be a straight line having slope equal to $\frac{1}{2}\epsilon_0 c$.

(b) Using the slope of the graph given with the problem, we have $\frac{1}{2}\epsilon_0 c = 1.33\times 10^{-3}$ J/(V$^2\cdot$s). Solving for c gives $c = 2[1.33\times 10^{-3}\text{ J/(V}^2\cdot\text{s)}]/(8.854\times 10^{-12}\text{ C}^2/\text{N}\cdot\text{m}^2) = 3.00\times 10^8$ m/s.

EVALUATE: This result is nearly identical to the speed of light in vacuum.

THE NATURE AND PROPAGATION OF LIGHT

33.7. **IDENTIFY:** Apply the law of reflection and Snell's law to calculate θ_r and θ_b. The angles in these equations are measured with respect to the normal, not the surface.
SET UP: The incident, reflected and refracted rays are shown in Figure 33.7. The law of reflection is $\theta_r = \theta_a$, and Snell's law is $n_a \sin\theta_a = n_b \sin\theta_b$.

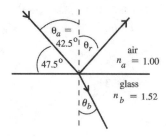

EXECUTE: (a) $\theta_r = \theta_a = 42.5°$ The reflected ray makes an angle of $90.0° - \theta_r = 47.5°$ with the surface of the glass.

Figure 33.7

(b) $n_a \sin\theta_a = n_b \sin\theta_b$, where the angles are measured from the normal to the interface.
$\sin\theta_b = \dfrac{n_a \sin\theta_a}{n_b} = \dfrac{(1.00)(\sin 42.5°)}{1.66} = 0.4070$.
$\theta_b = 24.0°$.
The refracted ray makes an angle of $90.0° - \theta_b = 66.0°$ with the surface of the glass.
EVALUATE: The light is bent toward the normal when the light enters the material of larger refractive index.

33.11. **IDENTIFY:** The figure shows the angle of incidence and angle of refraction for light going from the water into material X. Snell's law applies at the air-water and water-X boundaries.
SET UP: Snell's law says $n_a \sin\theta_a = n_b \sin\theta_b$. Apply Snell's law to the refraction from material X into the water and then from the water into the air.
EXECUTE: (a) Material X to water: $n_a = n_X$, $n_b = n_w = 1.333$. $\theta_a = 25°$ and $\theta_b = 48°$.
$n_a = n_b \left(\dfrac{\sin\theta_b}{\sin\theta_a}\right) = (1.333)\left(\dfrac{\sin 48°}{\sin 25°}\right) = 2.34$.
(b) Water to air: As Figure 33.11 shows, $\theta_a = 48°$. $n_a = 1.333$ and $n_b = 1.00$.
$\sin\theta_b = \left(\dfrac{n_a}{n_b}\right)\sin\theta_a = (1.333)\sin 48° = 82°$.
EVALUATE: $n > 1$ for material X, as it must be.

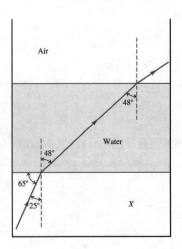

Figure 33.11

33.15. **IDENTIFY:** The critical angle for total internal reflection is θ_a that gives $\theta_b = 90°$ in Snell's law.
SET UP: In Figure 33.15 the angle of incidence θ_a is related to angle θ by $\theta_a + \theta = 90°$.
EXECUTE: **(a)** Calculate θ_a that gives $\theta_b = 90°$. $n_a = 1.60$, $n_b = 1.00$ so $n_a \sin\theta_a = n_b \sin\theta_b$ gives $(1.60)\sin\theta_a = (1.00)\sin 90°$. $\sin\theta_a = \dfrac{1.00}{1.60}$ and $\theta_a = 38.7°$. $\theta = 90° - \theta_a = 51.3°$.

(b) $n_a = 1.60$, $n_b = 1.333$. $(1.60)\sin\theta_a = (1.333)\sin 90°$. $\sin\theta_a = \dfrac{1.333}{1.60}$ and $\theta_a = 56.4°$. $\theta = 90° - \theta_a = 33.6°$.

EVALUATE: The critical angle increases when the ratio $\dfrac{n_a}{n_b}$ decreases.

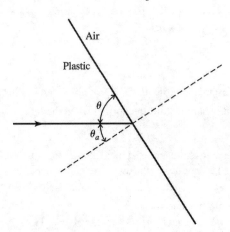

Figure 33.15

33.17. **IDENTIFY:** Use the critical angle to find the index of refraction of the liquid.
SET UP: Total internal reflection requires that the light be incident on the material with the larger n, in this case the liquid. Apply $n_a \sin\theta_a = n_b \sin\theta_b$ with a = liquid and b = air, so $n_a = n_{\text{liq}}$ and $n_b = 1.0$.
EXECUTE: $\theta_a = \theta_{\text{crit}}$ when $\theta_b = 90°$, so $n_{\text{liq}} \sin\theta_{\text{crit}} = (1.0)\sin 90°$.
$n_{\text{liq}} = \dfrac{1}{\sin\theta_{\text{crit}}} = \dfrac{1}{\sin 42.5°} = 1.48$.

(a) $n_a \sin\theta_a = n_b \sin\theta_b$ (a = liquid, b = air).

$\sin\theta_b = \dfrac{n_a \sin\theta_a}{n_b} = \dfrac{(1.48)\sin 35.0°}{1.0} = 0.8489$ and $\theta_b = 58.1°$.

(b) Now $n_a \sin\theta_a = n_b \sin\theta_b$ with a = air, b = liquid.

$\sin\theta_b = \dfrac{n_a \sin\theta_a}{n_b} = \dfrac{(1.0)\sin 35.0°}{1.48} = 0.3876$ and $\theta_b = 22.8°$.

EVALUATE: Light traveling from liquid to air is bent away from the normal. Light traveling from air to liquid is bent toward the normal.

33.21. IDENTIFY: If no light refracts out of the glass at the glass to air interface, then the incident angle at that interface is θ_{crit}.

SET UP: The ray has an angle of incidence of $0°$ at the first surface of the glass, so enters the glass without being bent, as shown in Figure 33.21. The figure shows that $\alpha + \theta_{crit} = 90°$.

EXECUTE: (a) For the glass-air interface $\theta_a = \theta_{crit}$, $n_a = 1.52$, $n_b = 1.00$, and $\theta_b = 90°$.

$n_a \sin\theta_a = n_b \sin\theta_b$ gives $\sin\theta_{crit} = \dfrac{(1.00)(\sin 90°)}{1.52}$ and $\theta_{crit} = 41.1°$. $\alpha = 90° - \theta_{crit} = 48.9°$.

(b) Now the second interface is glass → water and $n_b = 1.333$. $n_a \sin\theta_a = n_b \sin\theta_b$ gives

$\sin\theta_{crit} = \dfrac{(1.333)(\sin 90°)}{1.52}$ and $\theta_{crit} = 61.3°$. $\alpha = 90° - \theta_{crit} = 28.7°$.

EVALUATE: The critical angle increases when the air is replaced by water.

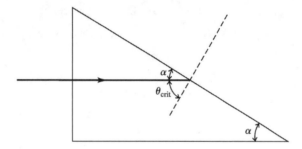

Figure 33.21

33.23. IDENTIFY: The index of refraction depends on the wavelength of light, so the light from the red and violet ends of the spectrum will be bent through different angles as it passes into the glass. Snell's law applies at the surface.

SET UP: $n_a \sin\theta_a = n_b \sin\theta_b$. From the graph in Figure 33.17 in the textbook, for $\lambda = 400$ nm (the violet end of the visible spectrum), $n = 1.67$ and for $\lambda = 700$ nm (the red end of the visible spectrum), $n = 1.62$. The path of a ray with a single wavelength is sketched in Figure 33.23.

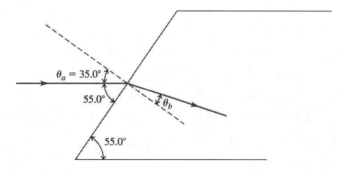

Figure 33.23

EXECUTE: For $\lambda = 400$ nm, $\sin\theta_b = \dfrac{n_a}{n_b}\sin\theta_a = \dfrac{1.00}{1.67}\sin 35.0°$, so $\theta_b = 20.1°$. For $\lambda = 700$ nm, $\sin\theta_b = \dfrac{1.00}{1.62}\sin 35.0°$, so $\theta_b = 20.7°$. $\Delta\theta$ is about $0.6°$.

EVALUATE: This angle is small, but the separation of the beams could be fairly large if the light travels through a fairly large slab.

33.27. IDENTIFY: When unpolarized light passes through a polarizer the intensity is reduced by a factor of $\tfrac{1}{2}$ and the transmitted light is polarized along the axis of the polarizer. When polarized light of intensity I_{max} is incident on a polarizer, the transmitted intensity is $I = I_{max}\cos^2\phi$, where ϕ is the angle between the polarization direction of the incident light and the axis of the filter.

SET UP: For the second polarizer $\phi = 60°$. For the third polarizer, $\phi = 90° - 60° = 30°$.

EXECUTE: (a) At point A the intensity is $I_0/2$ and the light is polarized along the vertical direction. At point B the intensity is $(I_0/2)(\cos 60°)^2 = 0.125I_0$, and the light is polarized along the axis of the second polarizer. At point C the intensity is $(0.125I_0)(\cos 30°)^2 = 0.0938I_0$.

(b) Now for the last filter $\phi = 90°$ and $I = 0$.

EVALUATE: Adding the middle filter increases the transmitted intensity.

33.29. IDENTIFY and SET UP: Reflected beam completely linearly polarized implies that the angle of incidence equals the polarizing angle, so $\theta_p = 54.5°$. Use Brewster's law, $\tan\theta_p = \dfrac{n_b}{n_a}$, to calculate the refractive index of the glass. Then use Snell's law to calculate the angle of refraction. See Figure 33.29.

EXECUTE: (a) $\tan\theta_p = \dfrac{n_b}{n_a}$ gives $n_{glass} = n_{air}\tan\theta_p = (1.00)\tan 54.5° = 1.40$.

(b) $n_a\sin\theta_a = n_b\sin\theta_b$.

$\sin\theta_b = \dfrac{n_a\sin\theta_a}{n_b} = \dfrac{(1.00)\sin 54.5°}{1.40} = 0.5815$ and $\theta_b = 35.5°$.

EVALUATE:

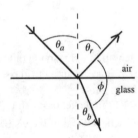

Note: $\phi = 180.0° - \theta_r - \theta_b$ and $\theta_r = \theta_a$. Thus $\phi = 180.0° - 54.5° - 35.5° = 90.0°$; the reflected ray and the refracted ray are perpendicular to each other. This agrees with Figure 33.28 in the textbook.

Figure 33.29

33.33. IDENTIFY: When unpolarized light of intensity I_0 is incident on a polarizing filter, the transmitted light has intensity $\tfrac{1}{2}I_0$ and is polarized along the filter axis. When polarized light of intensity I_0 is incident on a polarizing filter the transmitted light has intensity $I_0\cos^2\phi$.

SET UP: For the second filter, $\phi = 62.0° - 25.0° = 37.0°$.

EXECUTE: After the first filter the intensity is $\tfrac{1}{2}I_0 = 10.0$ W/cm^2 and the light is polarized along the axis of the first filter. The intensity after the second filter is $I = I_0\cos^2\phi$, where $I_0 = 10.0$ W/cm^2 and $\phi = 37.0°$. This gives $I = 6.38$ W/cm^2.

EVALUATE: The transmitted intensity depends on the angle between the axes of the two filters.

33.35. IDENTIFY: The shorter the wavelength of light, the more it is scattered. The intensity is inversely proportional to the fourth power of the wavelength.

SET UP: The intensity of the scattered light is proportional to $1/\lambda^4$; we can write it as $I = (\text{constant})/\lambda^4$.

EXECUTE: **(a)** Since I is proportional to $1/\lambda^4$, we have $I = (\text{constant})/\lambda^4$. Taking the ratio of the intensity of the red light to that of the green light gives

$$\frac{I_R}{I} = \frac{(\text{constant})/\lambda_R^4}{(\text{constant})/\lambda_G^4} = \left(\frac{\lambda_G}{\lambda_R}\right)^4 = \left(\frac{532 \text{ nm}}{685 \text{ nm}}\right)^4 = 0.364, \text{ so } I_R = 0.364I.$$

(b) Following the same procedure as in part (a) gives $\dfrac{I_V}{I} = \left(\dfrac{\lambda_G}{\lambda_V}\right)^4 = \left(\dfrac{532 \text{ nm}}{415 \text{ nm}}\right)^4 = 2.70$, so $I_V = 2.70I$.

EVALUATE: In the scattered light, the intensity of the short-wavelength violet light is about 7 times as great as that of the red light, so this scattered light will have a blue-violet color.

33.37. IDENTIFY: Snell's law applies to the sound waves in the heart.

SET UP: $n_a \sin\theta_a = n_b \sin\theta_b$. If θ_a is the critical angle then $\theta_b = 90°$. For air, $n_\text{air} = 1.00$. For heart muscle, $n_\text{mus} = \dfrac{344 \text{ m/s}}{1480 \text{ m/s}} = 0.2324$.

EXECUTE: **(a)** $n_a \sin\theta_a = n_b \sin\theta_b$ gives $(1.00)\sin(9.73°) = (0.2324)\sin\theta_b$. $\sin\theta_b = \dfrac{\sin(9.73°)}{0.2324}$ so $\theta_b = 46.7°$.

(b) $(1.00)\sin\theta_\text{crit} = (0.2324)\sin 90°$ gives $\theta_\text{crit} = 13.4°$.

EVALUATE: To interpret a sonogram, it should be important to know the true direction of travel of the sound waves within muscle. This would require knowledge of the refractive index of the muscle.

33.41. IDENTIFY: For total internal reflection, the angle of incidence must be at least as large as the critical angle.

SET UP: The angle of incidence for the glass-oil interface must be the critical angle, so $\theta_b = 90°$.

$n_a \sin\theta_a = n_b \sin\theta_b$.

EXECUTE: $n_a \sin\theta_a = n_b \sin\theta_b$ gives $(1.52)\sin 57.2° = n_\text{oil} \sin 90°$. $n_\text{oil} = (1.52)\sin 57.2° = 1.28$.

EVALUATE: $n_\text{oil} > 1$, which it must be, and 1.28 is a reasonable value for an oil.

33.47. IDENTIFY: Apply Snell's law to the water $\to$ ice and ice $\to$ air interfaces.

(a) SET UP: Consider the ray shown in Figure 33.47.

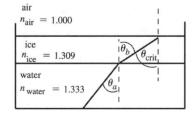

We want to find the incident angle θ_a at the water-ice interface that causes the incident angle at the ice-air interface to be the critical angle.

Figure 33.47

EXECUTE: ice-air interface: $n_\text{ice} \sin\theta_\text{crit} = 1.0 \sin 90°$.

$n_\text{ice} \sin\theta_\text{crit} = 1.0$ so $\sin\theta_\text{crit} = \dfrac{1}{n_\text{ice}}$.

But from the diagram we see that $\theta_b = \theta_\text{crit}$, so $\sin\theta_b = \dfrac{1}{n_\text{ice}}$.

water-ice interface: $n_w \sin\theta_a = n_\text{ice} \sin\theta_b$.

But $\sin\theta_b = \dfrac{1}{n_\text{ice}}$ so $n_w \sin\theta_a = 1.0$. $\sin\theta_a = \dfrac{1}{n_w} = \dfrac{1}{1.333} = 0.7502$ and $\theta_a = 48.6°$.

33.49. **IDENTIFY:** Apply Snell's law to the refraction of each ray as it emerges from the glass. The angle of incidence equals the angle $A = 25.0°$.

SET UP: The paths of the two rays are sketched in Figure 33.49.

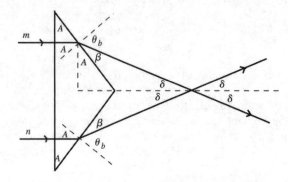

Figure 33.49

EXECUTE: $n_a \sin\theta_a = n_b \sin\theta_b$.

$n_{\text{glass}} \sin 25.0° = 1.00 \sin\theta_b$.

$\sin\theta_b = n_{\text{glass}} \sin 25.0°$.

$\sin\theta_b = 1.66 \sin 25.0° = 0.7015$.

$\theta_b = 44.55°$.

$\beta = 90.0° - \theta_b = 45.45°$.

Then $\delta = 90.0° - A - \beta = 90.0° - 25.0° - 45.45° = 19.55°$. The angle between the two rays is $2\delta = 39.1°$.

EVALUATE: The light is incident normally on the front face of the prism so the light is not bent as it enters the prism.

33.51. **IDENTIFY:** Apply Snell's law to the refraction of the light as it enters the atmosphere.

SET UP: The path of a ray from the sun is sketched in Figure 33.51.

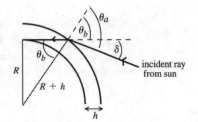

$\delta = \theta_a - \theta_b$.

From the diagram $\sin\theta_b = \dfrac{R}{R+h}$.

$\theta_b = \arcsin\left(\dfrac{R}{R+h}\right)$.

Figure 33.51

EXECUTE: **(a)** Apply Snell's law to the refraction that occurs at the top of the atmosphere:

$n_a \sin\theta_a = n_b \sin\theta_b$

(a = vacuum of space, refractive index 1.0; b = atmosphere, refractive index n).

$\sin\theta_a = n\sin\theta_b = n\left(\dfrac{R}{R+h}\right)$ so $\theta_a = \arcsin\left(\dfrac{nR}{R+h}\right)$.

$\delta = \theta_a - \theta_b = \arcsin\left(\dfrac{nR}{R+h}\right) - \arcsin\left(\dfrac{R}{R+h}\right)$.

(b) $\dfrac{R}{R+h} = \dfrac{6.38 \times 10^6 \text{ m}}{6.38 \times 10^6 \text{ m} + 20 \times 10^3 \text{ m}} = 0.99688$.

$\dfrac{nR}{R+h} = 1.0003(0.99688) = 0.99718$.

$\theta_b = \arcsin\left(\dfrac{R}{R+h}\right) = 85.47°$.

$\theta_a = \arcsin\left(\dfrac{nR}{R+h}\right) = 85.70°$.

$\delta = \theta_a - \theta_b = 85.70° - 85.47° = 0.23°$.

EVALUATE: The calculated δ is about the same as the angular radius of the sun.

33.53. IDENTIFY: Apply Snell's law to the two refractions of the ray.
SET UP: Refer to the figure that accompanies the problem.
EXECUTE: **(a)** $n_a \sin\theta_a = n_b \sin\theta_b$ gives $\sin\theta_a = n_b \sin\dfrac{A}{2}$. But $\theta_a = \dfrac{A}{2} + \alpha$, so

$\sin\left(\dfrac{A}{2} + \alpha\right) = \sin\dfrac{A + 2\alpha}{2} = n\sin\dfrac{A}{2}$. At each face of the prism the deviation is α, so $2\alpha = \delta$ and

$\sin\dfrac{A + \delta}{2} = n\sin\dfrac{A}{2}$.

(b) From part (a), $\delta = 2\arcsin\left(n\sin\dfrac{A}{2}\right) - A$. $\delta = 2\arcsin\left((1.52)\sin\dfrac{60.0°}{2}\right) - 60.0° = 38.9°$.

(c) If two colors have different indices of refraction for the glass, then the deflection angles for them will differ:

$\delta_\text{red} = 2\arcsin\left((1.61)\sin\dfrac{60.0°}{2}\right) - 60.0° = 47.2°$.

$\delta_\text{violet} = 2\arcsin\left((1.66)\sin\dfrac{60.0°}{2}\right) - 60.0° = 52.2° \Rightarrow \Delta\delta = 52.2° - 47.2° = 5.0°$.

EVALUATE: The violet light has a greater refractive index and therefore the angle of deviation is greater for the violet light.

33.57. IDENTIFY: Apply Snell's law in part (a). In part (b), we know from Chapter 32 that in a dielectric, $n^2 = KK_\text{m}$. In this case, we are told that K_m is very close to 1, so $n^2 \approx K$, where K is the dielectric constant of the material.
SET UP: Use $n_a \sin\theta_a = n_b \sin\theta_b$ (Snell's law) in (a) and $n^2 \approx K$ in (b). Use $n = 1.00$ for air. $f\lambda = c$.
EXECUTE: **(a)** For each liquid, apply $n_a \sin\theta_a = n_b \sin\theta_b$ using the data in the table with the problem. The angle of incidence is $60.0°$ in each case.

$\sin(60.0°) = n \sin\theta_b$, which gives $n = \dfrac{\sin(60.0°)}{\sin\theta_b}$. Apply this formula for each liquid and then use the information in Table 33.1 to identify the liquids.

Liquid A: $n_A = \dfrac{\sin(60.0°)}{\sin(36.4°)} = 1.46$ (carbon tetrachloride).

Liquid B: $n_B = \dfrac{\sin(60.0°)}{\sin(40.5°)} = 1.33$ (water).

Liquid C: $n_C = \dfrac{\sin(60.0°)}{\sin(32.1°)} = 1.63$ (carbon disulfide).

Liquid D: $n_D = \dfrac{\sin(60.0°)}{\sin(35.2°)} = 1.50$ (benzene).

(b) Use $K = n^2$ for each liquid.
$K_A = (1.46)^2 = 2.13$.
$K_B = (1.33)^2 = 1.77$.
$K_C = (1.63)^2 = 2.66$.
$K_D = (1.50)^2 = 2.25$.
(c) Use $f\lambda = c$: $f = c/\lambda = (3.00 \times 10^8 \text{ m/s})/(589 \times 10^{-9} \text{ m}) = 5.09 \times 10^{14}$ Hz. This is the frequency in air and also in each liquid.
EVALUATE: The indexes of refraction are accurate, but the dielectric constants are less so because $n^2 \approx K$ is an approximation.

33.59. **IDENTIFY and SET UP:** The polarizer passes $\frac{1}{2}$ of the intensity of the unpolarized component, independent of α. Malus's law tells us that out of the intensity I_p of the polarized component, the polarizer passes intensity $I_p \cos^2(\alpha - \theta)$, where $\alpha - \theta$ is the angle between the plane of polarization and the axis of the polarizer.
EXECUTE: **(a)** Use the angle where the transmitted intensity is maximum or minimum to find θ. See Figure 33.59.

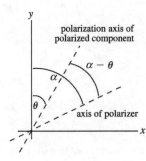

Figure 33.59

The total transmitted intensity is $I = \frac{1}{2}I_0 + I_p \cos^2(\alpha - \theta)$. This is maximum when $\theta = \alpha$, and from the graph in the problem this occurs when α is approximately 35°, so $\theta = 35°$. Alternatively, the total transmitted intensity is minimum when $\alpha - \theta = 90°$ and from the graph this occurs for $\alpha = 125°$. Thus, $\theta = \alpha - 90° = 125° - 90° = 35°$, which is in agreement with what we just found.

(b) For the equation, $I = \frac{1}{2}I_0 + I_p \cos^2(\alpha - \theta)$, we use data at two values of α to determine I_0 and I_p. It is easiest to use data where I is a maximum and a minimum. From the graph, we see that these extremes are 25 W/m² at $\alpha = 35°$ and 5.0 W/m² at $\alpha = 125°$.
At $\alpha = 125°$ the net intensity is 5.0 W/m², so we have
$5.0 \text{ W/m}^2 = \frac{1}{2}I_0 + I_p \cos^2(125° - 35°) = \frac{1}{2}I_0 + I_p \cos^2(90°) = \frac{1}{2}I_0 \rightarrow I_0 = 10 \text{ W/m}^2$.
At $\alpha = 35°$ the net intensity is 25 W/m², so we have
$25 \text{ W/m}^2 = \frac{1}{2}I_0 + I_p \cos^2 0° = \frac{1}{2}I_0 + I_p = 5 \text{ W/m}^2 + I_p \rightarrow I_p = 20 \text{ W/m}^2$.

EVALUATE: Now that we have $I_0, I_p,$ and θ we can verify that $I = \frac{1}{2}I_0 + I_p \cos^2(\phi - \theta)$ describes the data in the graph.

GEOMETRIC OPTICS

34.7. **IDENTIFY:** $\frac{1}{s}+\frac{1}{s'}=\frac{1}{f}$. $m=-\frac{s'}{s}$. $|m|=\frac{|y'|}{y}$. Find m and calculate y'.

SET UP: $f=+1.75$ m.

EXECUTE: $s \gg f$ so $s'=f=1.75$ m.

$m=-\frac{s'}{s}=-\frac{1.75 \text{ m}}{5.58\times 10^{10} \text{ m}}=-3.14\times 10^{-11}$.

$|y'|=|m||y|=(3.14\times 10^{-11})(6.794\times 10^6 \text{ m})=2.13\times 10^{-4}$ m $=0.213$ mm.

EVALUATE: The image is real and is 1.75 m in front of the mirror.

34.13. **IDENTIFY:** $\frac{1}{s}+\frac{1}{s'}=\frac{1}{f}$ and $m=\frac{y'}{y}=-\frac{s'}{s}$.

SET UP: $m=+2.00$ and $s=1.25$ cm. An erect image must be virtual.

EXECUTE: (a) $s'=\frac{sf}{s-f}$ and $m=-\frac{f}{s-f}$. For a concave mirror, m can be larger than 1.00. For a convex mirror, $|f|=-f$ so $m=+\frac{|f|}{s+|f|}$ and m is always less than 1.00. The mirror must be concave ($f>0$).

(b) $\frac{1}{f}=\frac{s'+s}{ss'}$. $f=\frac{ss'}{s+s'}$. $m=-\frac{s'}{s}=+2.00$ and $s'=-2.00s$. $f=\frac{s(-2.00s)}{s-2.00s}=+2.00s=+2.50$ cm.

$R=2f=+5.00$ cm.

(c) The principal-ray diagram is drawn in Figure 34.13.

EVALUATE: The principal-ray diagram agrees with the description from the equations.

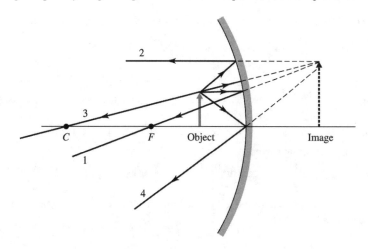

Figure 34.13

34-1

34-2 Chapter 34

34.15. IDENTIFY: In part (a), the shell is a concave mirror, but in (b) it is a convex mirror. The magnitude of its focal length is the same in both cases, but the sign reverses.

SET UP: For the orientation of the shell shown in the figure in the problem, $R = +12.0$ cm. When the glass is reversed, so the seed faces a convex surface, $R = -12.0$ cm. $\dfrac{1}{s} + \dfrac{1}{s'} = \dfrac{2}{R}$ and $m = \dfrac{y'}{y} = -\dfrac{s'}{s}$.

EXECUTE: (a) $R = +12.0$ cm. $\dfrac{1}{s'} = \dfrac{2}{R} - \dfrac{1}{s} = \dfrac{2s-R}{Rs}$ and $s' = \dfrac{Rs}{2s-R} = \dfrac{(12.0\text{ cm})(15.0\text{ cm})}{30.0\text{ cm}-12.0\text{ cm}} = +10.0$ cm.

$m = -\dfrac{s'}{s} = -\dfrac{10.0\text{ cm}}{15.0\text{ cm}} = -0.667$. $y' = my = -2.20$ mm. The image is 10.0 cm to the left of the shell vertex and is 2.20 mm tall.

(b) $R = -12.0$ cm. $s' = \dfrac{(-12.0\text{ cm})(15.0\text{ cm})}{30.0\text{ cm}+12.0\text{ cm}} = -4.29$ cm. $m = -\dfrac{-4.29\text{ cm}}{15.0\text{ cm}} = +0.286$.

$y' = my = 0.944$ mm. The image is 4.29 cm to the right of the shell vertex and is 0.944 mm tall.

EVALUATE: In (a), $s > R/2$ and the mirror is concave, so the image is real. In (b) the image is virtual because a convex mirror always forms a virtual image.

34.17. IDENTIFY: Apply $\dfrac{n_a}{s} + \dfrac{n_b}{s'} = \dfrac{n_b - n_a}{R}$, with $R \to \infty$. $|s'|$ is the apparent depth.

SET UP: The image and object are shown in Figure 34.17.

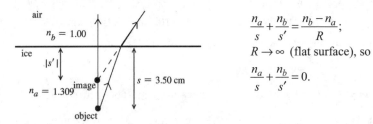

$\dfrac{n_a}{s} + \dfrac{n_b}{s'} = \dfrac{n_b - n_a}{R}$;

$R \to \infty$ (flat surface), so

$\dfrac{n_a}{s} + \dfrac{n_b}{s'} = 0$.

Figure 34.17

EXECUTE: $s' = -\dfrac{n_b s}{n_a} = -\dfrac{(1.00)(3.50\text{ cm})}{1.309} = -2.67$ cm.

The apparent depth is 2.67 cm.

EVALUATE: When the light goes from ice to air (larger to smaller n), it is bent away from the normal and the virtual image is closer to the surface than the object is.

34.19. IDENTIFY: Think of the surface of the water as a section of a sphere having an infinite radius of curvature.

SET UP: $\dfrac{n_a}{s} + \dfrac{n_b}{s'} = 0$. $n_a = 1.00$. $n_b = 1.333$.

EXECUTE: The image is 5.20 m − 0.80 m = 4.40 m above the surface of the water, so $s' = -4.40$ m.

$s = -\dfrac{n_a}{n_b} s' = -\left(\dfrac{1.00}{1.333}\right)(-4.40\text{ m}) = +3.30$ m.

EVALUATE: The diving board is closer to the water than it looks to the swimmer.

34.21. IDENTIFY: $\dfrac{n_a}{s} + \dfrac{n_b}{s'} = \dfrac{n_b - n_a}{R}$. $m = -\dfrac{n_a s'}{n_b s}$. Light comes from the fish to the person's eye.

SET UP: $R = -14.0$ cm. $s = +14.0$ cm. $n_a = 1.333$ (water). $n_b = 1.00$ (air). Figure 34.21 shows the object and the refracting surface.

EXECUTE: (a) $\dfrac{1.333}{14.0\text{ cm}} + \dfrac{1.00}{s'} = \dfrac{1.00-1.333}{-14.0\text{ cm}}$. $s' = -14.0$ cm. $m = -\dfrac{(1.333)(-14.0\text{ cm})}{(1.00)(14.0\text{ cm})} = +1.33$.

The fish's image is 14.0 cm to the left of the bowl surface so is at the center of the bowl and the magnification is 1.33.

(b) The focal point is at the image location when $s \to \infty$. $\dfrac{n_b}{s'} = \dfrac{n_b - n_a}{R}$. $n_a = 1.00$. $n_b = 1.333$.

$R = +14.0$ cm. $\dfrac{1.333}{s'} = \dfrac{1.333 - 1.00}{14.0 \text{ cm}}$. $s' = +56.0$ cm. s' is greater than the diameter of the bowl, so the surface facing the sunlight does not focus the sunlight to a point inside the bowl. The focal point is outside the bowl and there is no danger to the fish.

EVALUATE: In part (b) the rays refract when they exit the bowl back into the air so the image we calculated is not the final image.

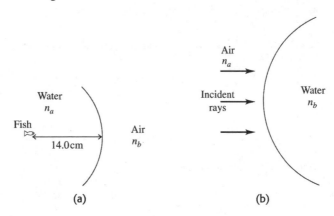

Figure 34.21

34.23. IDENTIFY: The hemispherical glass surface forms an image by refraction. The location of this image depends on the curvature of the surface and the indices of refraction of the glass and oil.

SET UP: The image and object distances are related to the indices of refraction and the radius of curvature by the equation $\dfrac{n_a}{s} + \dfrac{n_b}{s'} = \dfrac{n_b - n_a}{R}$.

EXECUTE: $\dfrac{n_a}{s} + \dfrac{n_b}{s'} = \dfrac{n_b - n_a}{R} \Rightarrow \dfrac{1.45}{s} + \dfrac{1.60}{1.20 \text{ m}} = \dfrac{0.15}{0.0300 \text{ m}} \Rightarrow s = 39.5$ cm.

EVALUATE: The presence of the oil changes the location of the image.

34.29. IDENTIFY: Use the lensmaker's equation and the thin-lens equation.

SET UP: Combine the lensmaker's equation and the thin-lens equation to get $\dfrac{1}{s} + \dfrac{1}{s'} = (n-1)\left(\dfrac{1}{R_1} - \dfrac{1}{R_2}\right)$, and use the fact that the magnification of the lens is $m = -\dfrac{s'}{s}$.

EXECUTE: **(a)** $\dfrac{1}{s} + \dfrac{1}{s'} = (n-1)\left(\dfrac{1}{R_1} - \dfrac{1}{R_2}\right) \Rightarrow \dfrac{1}{24.0 \text{ cm}} + \dfrac{1}{s'} = (1.52 - 1)\left(\dfrac{1}{-7.00 \text{ cm}} - \dfrac{1}{-4.00 \text{ cm}}\right)$

$\Rightarrow s' = 71.2$ cm, to the right of the lens.

(b) $m = -\dfrac{s'}{s} = -\dfrac{71.2 \text{ cm}}{24.0 \text{ cm}} = -2.97$.

EVALUATE: Since the magnification is negative, the image is inverted.

34.31. IDENTIFY: The thin-lens equation applies in this case.

SET UP: The thin-lens equation is $\dfrac{1}{s} + \dfrac{1}{s'} = \dfrac{1}{f}$, and the magnification is $m = -\dfrac{s'}{s} = \dfrac{y'}{y}$.

EXECUTE: $m = \dfrac{y'}{y} = \dfrac{34.0 \text{ mm}}{8.00 \text{ mm}} = 4.25 = -\dfrac{s'}{s} = -\dfrac{-12.0 \text{ cm}}{s} \Rightarrow s = 2.82$ cm. The thin-lens equation gives

$\dfrac{1}{s} + \dfrac{1}{s'} = \dfrac{1}{f} \Rightarrow f = 3.69$ cm.

EVALUATE: Since the focal length is positive, this is a converging lens. The image distance is negative because the object is inside the focal point of the lens.

34.37. IDENTIFY: First use the figure that accompanies the problem to decide if each radius of curvature is positive or negative. Then apply the lensmaker's formula to calculate the focal length of each lens.

SET UP: Use $\frac{1}{f} = (n-1)\left(\frac{1}{R_1} - \frac{1}{R_2}\right)$ to calculate f and then use $\frac{1}{s} + \frac{1}{s'} = \frac{1}{f}$ to locate the image. $s = 18.0$ cm.

EXECUTE: (a) $\frac{1}{f} = (0.5)\left(\frac{1}{10.0 \text{ cm}} - \frac{1}{-15.0 \text{ cm}}\right)$ and $f = +12.0$ cm. $\frac{1}{s'} = \frac{1}{f} - \frac{1}{s} = \frac{s-f}{sf}$.

$s' = \frac{sf}{s-f} = \frac{(18.0 \text{ cm})(12.0 \text{ cm})}{18.0 \text{ cm} - 12.0 \text{ cm}} = +36.0$ cm. The image is 36.0 cm to the right of the lens.

(b) $\frac{1}{f} = (0.5)\left(\frac{1}{10.0 \text{ cm}} - \frac{1}{\infty}\right)$ so $f = +20.0$ cm. $s' = \frac{sf}{s-f} = \frac{(18.0 \text{ cm})(20.0 \text{ cm})}{18.0 \text{ cm} - 20.0 \text{ cm}} = -180$ cm. The image is 180 cm to the left of the lens.

(c) $\frac{1}{f} = (0.5)\left(\frac{1}{-10.0 \text{ cm}} - \frac{1}{15.0 \text{ cm}}\right)$ so $f = -12.0$ cm. $s' = \frac{sf}{s-f} = \frac{(18.0 \text{ cm})(-12.0 \text{ cm})}{18.0 \text{ cm} + 12.0 \text{ cm}} = -7.20$ cm.

The image is 7.20 cm to the left of the lens.

(d) $\frac{1}{f} = (0.5)\left(\frac{1}{-10.0 \text{ cm}} - \frac{1}{-15.0 \text{ cm}}\right)$ so $f = -60.0$ cm. $s' = \frac{sf}{s-f} = \frac{(18.0 \text{ cm})(-60.0 \text{ cm})}{18.0 \text{ cm} + 60.0 \text{ cm}} = -13.8$ cm.

The image is 13.8 cm to the left of the lens.

EVALUATE: The focal length of a lens is determined by *both* of its radii of curvature.

34.39. IDENTIFY: Use $\frac{1}{s} + \frac{1}{s'} = \frac{1}{f}$ to calculate the object distance s. m calculated from $m = -\frac{s'}{s}$ determines the size and orientation of the image.

SET UP: $f = -48.0$ cm. Virtual image 17.0 cm from lens so $s' = -17.0$ cm.

EXECUTE: $\frac{1}{s} + \frac{1}{s'} = \frac{1}{f}$, so $\frac{1}{s} = \frac{1}{f} - \frac{1}{s'} = \frac{s'-f}{s'f}$.

$s = \frac{s'f}{s'-f} = \frac{(-17.0 \text{ cm})(-48.0 \text{ cm})}{-17.0 \text{ cm} - (-48.0 \text{ cm})} = +26.3$ cm.

$m = -\frac{s'}{s} = -\frac{-17.0 \text{ cm}}{+26.3 \text{ cm}} = +0.646$.

$m = \frac{y'}{y}$ so $|y| = \frac{|y'|}{|m|} = \frac{8.00 \text{ mm}}{0.646} = 12.4$ mm.

The principal-ray diagram is sketched in Figure 34.39.

EVALUATE: Virtual image, real object ($s > 0$) so image and object are on same side of lens. $m > 0$ so image is erect with respect to the object. The height of the object is 12.4 mm.

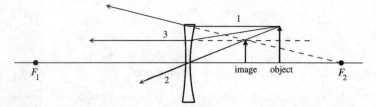

Figure 34.39

34.41. IDENTIFY: The first lens forms an image that is then the object for the second lens.

SET UP: Apply $\frac{1}{s} + \frac{1}{s'} = \frac{1}{f}$ to each lens. $m_1 = \frac{y'_1}{y_1}$ and $m_2 = \frac{y'_2}{y_2}$.

EXECUTE: **(a)** Lens 1: $\frac{1}{s}+\frac{1}{s'}=\frac{1}{f}$ gives $s'_1 = \frac{s_1 f_1}{s_1 - f_1} = \frac{(50.0\text{ cm})(40.0\text{ cm})}{50.0\text{ cm} - 40.0\text{ cm}} = +200\text{ cm}.$

$m_1 = -\frac{s'_1}{s_1} = -\frac{200\text{ cm}}{50\text{ cm}} = -4.00.$ $y'_1 = m_1 y_1 = (-4.00)(1.20\text{ cm}) = -4.80\text{ cm}.$ The image I_1 is 200 cm to the right of lens 1, is 4.80 cm tall and is inverted.

(b) Lens 2: $y_2 = -4.80\text{ cm}.$ The image I_1 is $300\text{ cm} - 200\text{ cm} = 100\text{ cm}$ to the left of lens 2, so $s_2 = +100\text{ cm}.$ $s'_2 = \frac{s_2 f_2}{s_2 - f_2} = \frac{(100\text{ cm})(60.0\text{ cm})}{100\text{ cm} - 60.0\text{ cm}} = +150\text{ cm}.$ $m_2 = -\frac{s'_2}{s_2} = -\frac{150\text{ cm}}{100\text{ cm}} = -1.50.$

$y'_2 = m_2 y_2 = (-1.50)(-4.80\text{ cm}) = +7.20\text{ cm}.$ The image is 150 cm to the right of the second lens, is 7.20 cm tall, and is erect with respect to the original object.

EVALUATE: The overall magnification of the lens combination is $m_{\text{tot}} = m_1 m_2.$

34.47. IDENTIFY: Apply $\frac{1}{s}+\frac{1}{s'}=\frac{1}{f}$ and $m = \frac{y'}{y} = -\frac{s'}{s}.$

SET UP: $s = 3.90$ m. $f = 0.085$ m.

EXECUTE: $\frac{1}{s}+\frac{1}{s'}=\frac{1}{f} \Rightarrow \frac{1}{3.90\text{ m}}+\frac{1}{s'}=\frac{1}{0.085\text{ m}} \Rightarrow s' = 0.0869$ m.

$y' = -\frac{s'}{s}y = -\frac{0.0869}{3.90}1750\text{ mm} = -39.0\text{ mm},$ so it will not fit on the $24\text{-mm}\times 36\text{-mm}$ sensor.

EVALUATE: The image is just outside the focal point and $s' \approx f.$ To have $|y'| = 36$ mm, so that the image will fit on the sensor, $s = -\frac{s'y}{y'} \approx -\frac{(0.085\text{ m})(1.75\text{ m})}{-0.036\text{ m}} = 4.1$ m. The person would need to stand about 4.1 m from the lens.

34.49. IDENTIFY: The f-number of a lens is the ratio of its focal length to its diameter. To maintain the same exposure, the amount of light passing through the lens during the exposure must remain the same.

SET UP: The f-number is $f/D.$

EXECUTE: **(a)** $f\text{-number} = \frac{f}{D} \Rightarrow f\text{-number} = \frac{180.0\text{ mm}}{16.36\text{ mm}} \Rightarrow f\text{-number} = f/11.$ (The f-number is an integer.)

(b) $f/11$ to $f/2.8$ is four steps of 2 in intensity, so one needs $1/16^{\text{th}}$ the exposure. The exposure should be $1/480\text{ s} = 2.1\times 10^{-3}\text{ s} = 2.1$ ms.

EVALUATE: When opening the lens from $f/11$ to $f/2.8$, the area increases by a factor of 16, so 16 times as much light is allowed in. Therefore the exposure time must be decreased by a factor of 1/16 to maintain the same exposure on the film or light receptors of a digital camera.

34.51. (a) IDENTIFY: The purpose of the corrective lens is to take an object 25 cm from the eye and form a virtual image at the eye's near point. Use $\frac{1}{s}+\frac{1}{s'}=\frac{1}{f}$ to solve for the image distance when the object distance is 25 cm.

SET UP: $\frac{1}{f} = +2.75$ diopters means $f = +\frac{1}{2.75}$ m $= +0.3636$ m (converging lens)

$f = 36.36$ cm; $s = 25$ cm; $s' = ?$

EXECUTE: $\frac{1}{s}+\frac{1}{s'}=\frac{1}{f}$ so

$s' = \frac{sf}{s-f} = \frac{(25\text{ cm})(36.36\text{ cm})}{25\text{ cm} - 36.36\text{ cm}} = -80.0$ cm.

The eye's near point is 80.0 cm from the eye.

(b) IDENTIFY: The purpose of the corrective lens is to take an object at infinity and form a virtual image of it at the eye's far point. Use $\frac{1}{s}+\frac{1}{s'}=\frac{1}{f}$ to solve for the image distance when the object is at infinity.

SET UP: $\frac{1}{f} = -1.30$ diopters means $f = -\frac{1}{1.30}$ m $= -0.7692$ m (diverging lens). $f = -76.92$ cm; $s = \infty$; $s' = ?$

EXECUTE: $\frac{1}{s}+\frac{1}{s'}=\frac{1}{f}$ and $s = \infty$ says $\frac{1}{s'}=\frac{1}{f}$ and $s' = f = -76.9$ cm. The eye's far point is 76.9 cm from the eye.

EVALUATE: In each case a virtual image is formed by the lens. The eye views this virtual image instead of the object. The object is at a distance where the eye can't focus on it, but the virtual image is at a distance where the eye can focus.

34.53. IDENTIFY and SET UP: For an object 25.0 cm from the eye, the corrective lens forms a virtual image at the near point of the eye. The distances from the corrective lens are $s = 23.0$ cm and $s' = -43.0$ cm. $\frac{1}{s}+\frac{1}{s'}=\frac{1}{f}$. P(in diopters) $= 1/f$ (in m).

EXECUTE: Solving $\frac{1}{s}+\frac{1}{s'}=\frac{1}{f}$ for f gives $f = \frac{ss'}{s+s'} = \frac{(23.0 \text{ cm})(-43.0 \text{ cm})}{23.0 \text{ cm} - 43.0 \text{ cm}} = +49.4$ cm. The power is $\frac{1}{0.494 \text{ m}} = 2.02$ diopters.

EVALUATE: In Problem 34.52 the contact lenses have power 1.78 diopters. The power of the lenses is different for ordinary glasses versus contact lenses.

34.57. IDENTIFY: Use $\frac{1}{s}+\frac{1}{s'}=\frac{1}{f}$ and $m = \frac{y'}{y} = -\frac{s'}{s}$ to calculate s and y'.

SET UP: $f = 8.00$ cm; $s' = -25.0$ cm; $s = ?$

EXECUTE: (a) $\frac{1}{s}+\frac{1}{s'}=\frac{1}{f}$, so $\frac{1}{s}=\frac{1}{f}-\frac{1}{s'}=\frac{s'-f}{s'f}$.

$s = \frac{s'f}{s'-f} = \frac{(-25.0 \text{ cm})(+8.00 \text{ cm})}{-25.0 \text{ cm} - 8.00 \text{ cm}} = +6.06$ cm.

(b) $m = -\frac{s'}{s} = -\frac{-25.0 \text{ cm}}{6.06 \text{ cm}} = +4.125$.

$|m| = \frac{|y'|}{|y|}$ so $|y'| = |m||y| = (4.125)(1.00 \text{ mm}) = 4.12$ mm.

EVALUATE: The lens allows the object to be much closer to the eye than the near point. The lens allows the eye to view an image at the near point rather than the object.

34.59. (a) IDENTIFY and SET UP:

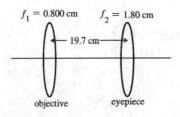

Figure 34.59

Final image is at ∞ so the object for the eyepiece is at its focal point. But the object for the eyepiece is the image of the objective so the image formed by the objective is 19.7 cm − 1.80 cm = 17.9 cm to the right of the lens. Apply $\frac{1}{s}+\frac{1}{s'}=\frac{1}{f}$ to the image formation by the objective, solve for the object distance s.

$f = 0.800$ cm; $s' = 17.9$ cm; $s = ?$

$\frac{1}{s}+\frac{1}{s'}=\frac{1}{f}$, so $\frac{1}{s}=\frac{1}{f}-\frac{1}{s'}=\frac{s'-f}{s'f}$.

EXECUTE: $s=\frac{s'f}{s'-f}=\frac{(17.9\text{ cm})(+0.800\text{ cm})}{17.9\text{ cm}-0.800\text{ cm}}=+8.37$ mm.

(b) SET UP: Use $\frac{1}{s}+\frac{1}{s'}=\frac{1}{f}$.

EXECUTE: $m_1=-\frac{s'}{s}=-\frac{17.9\text{ cm}}{0.837\text{ cm}}=-21.4$.

The magnitude of the linear magnification of the objective is 21.4.

(c) SET UP: Use $M=m_1M_2$.

EXECUTE: $M_2=\frac{25\text{ cm}}{f_2}=\frac{25\text{ cm}}{1.80\text{ cm}}=13.9$.

$M=m_1M_2=(-21.4)(13.9)=-297$.

EVALUATE: M is not accurately given by $(25\text{ cm})s_1'/f_1f_2=311$, because the object is not quite at the focal point of the objective ($s_1=0.837$ cm and $f_1=0.800$ cm).

34.61. **(a) IDENTIFY and SET UP:** Use $M=-\frac{f_1}{f_2}$, with $f_1=95.0$ cm (objective) and $f_2=15.0$ cm (eyepiece).

EXECUTE: $M=-\frac{f_1}{f_2}=-\frac{95.0\text{ cm}}{15.0\text{ cm}}=-6.33$.

(b) IDENTIFY: Use $m=\frac{y'}{y}=-\frac{s'}{s}$ to calculate y'.

SET UP: $s=3.00\times 10^3$ m.

$s'=f_1=95.0$ cm (since s is very large, $s'\approx f$).

EXECUTE: $m=-\frac{s'}{s}=-\frac{0.950\text{ m}}{3.00\times 10^3\text{ m}}=-3.167\times 10^{-4}$.

$|y'|=|m||y|=(3.167\times 10^{-4})(60.0\text{ m})=0.0190\text{ m}=1.90$ cm.

(c) IDENTIFY and SET UP:: Use $M=\frac{\theta'}{\theta}$ and the angular magnification M obtained in part (a) to calculate θ'. The angular size θ of the image formed by the objective (object for the eyepiece) is its height divided by its distance from the objective.

EXECUTE: The angular size of the object for the eyepiece is $\theta=\frac{0.0190\text{ m}}{0.950\text{ m}}=0.0200$ rad.

(Note that this is also the angular size of the object for the objective: $\theta=\frac{60.0\text{ m}}{3.00\times 10^3\text{ m}}=0.0200$ rad. For a thin lens the object and image have the same angular size and the image of the objective is the object for the eyepiece.) $M=\frac{\theta'}{\theta}$, so the angular size of the image is $\theta'=M\theta=-(6.33)(0.0200\text{ rad})=-0.127$ rad.

(The minus sign shows that the final image is inverted.)

EVALUATE: The lateral magnification of the objective is small; the image it forms is much smaller than the object. But the total angular magnification is larger than 1.00; the angular size of the final image viewed by the eye is 6.33 times larger than the angular size of the original object, as viewed by the unaided eye.

34.63. **IDENTIFY:** $f = R/2$ and $M = -\dfrac{f_1}{f_2}$.

SET UP: For object and image both at infinity, $f_1 + f_2$ equals the distance d between the eyepiece and the mirror vertex. $f_2 = 1.10$ cm. $R_1 = 1.30$ m.

EXECUTE: (a) $f_1 = \dfrac{R_1}{2} = 0.650$ m $\Rightarrow d = f_1 + f_2 = 0.661$ m.

(b) $|M| = \dfrac{f_1}{f_2} = \dfrac{0.650 \text{ m}}{0.011 \text{ m}} = 59.1$.

EVALUATE: For a telescope, $f_1 \gg f_2$.

34.67. **IDENTIFY:** We are given the image distance, the image height, and the object height. Use $m = -\dfrac{s'}{s}$ to calculate the object distance s. Then use $\dfrac{1}{s} + \dfrac{1}{s'} = \dfrac{2}{R}$ to calculate R.

SET UP: The image is to be formed on screen so it is a real image; $s' > 0$. The mirror-to-screen distance is 8.00 m, so $s' = +800$ cm. $m = -\dfrac{s'}{s} < 0$ since both s and s' are positive.

EXECUTE: (a) $|m| = \dfrac{|y'|}{|y|} = \dfrac{24.0 \text{ cm}}{0.600 \text{ cm}} = 40.0$, so $m = -40.0$. Then $m = -\dfrac{s'}{s}$ gives

$s = -\dfrac{s'}{m} = -\dfrac{800 \text{ cm}}{-40.0} = +20.0$ cm.

(b) $\dfrac{1}{s} + \dfrac{1}{s'} = \dfrac{2}{R}$, so $\dfrac{2}{R} = \dfrac{s+s'}{ss'}$. $R = 2\left(\dfrac{ss'}{s+s'}\right) = 2\left(\dfrac{(20.0 \text{ cm})(800 \text{ cm})}{20.0 \text{ cm} + 800 \text{ cm}}\right) = 39.0$ cm.

EVALUATE: R is calculated to be positive, which is correct for a concave mirror. Also, in part (a) s is calculated to be positive, as it should be for a real object.

34.69. **IDENTIFY:** Since the truck is moving toward the mirror, its image will also be moving toward the mirror.

SET UP: The equation relating the object and image distances to the focal length of a spherical mirror is $\dfrac{1}{s} + \dfrac{1}{s'} = \dfrac{1}{f}$, where $f = R/2$.

EXECUTE: Since the mirror is convex, $f = R/2 = (-1.50 \text{ m})/2 = -0.75$ m. Applying the equation for a spherical mirror gives $\dfrac{1}{s} + \dfrac{1}{s'} = \dfrac{1}{f} \Rightarrow s' = \dfrac{fs}{s-f}$. Using the chain rule from calculus and the fact that

$v = ds/dt$, we have $v' = \dfrac{ds'}{dt} = \dfrac{ds'}{ds}\dfrac{ds}{dt} = v\dfrac{f^2}{(s-f)^2}$. Solving for v gives

$v = v'\left(\dfrac{s-f}{f}\right)^2 = (1.9 \text{ m/s})\left[\dfrac{2.0 \text{ m} - (-0.75 \text{ m})}{-0.75 \text{ m}}\right]^2 = 25.5$ m/s. This is the velocity of the truck relative to the mirror, so the truck is approaching the mirror at 25.5 m/s. You are traveling at 25 m/s, so the truck must be traveling at 25 m/s + 25.5 m/s = 51 m/s relative to the highway.

EVALUATE: Even though the truck and car are moving at constant speed, the image of the truck is *not* moving at constant speed because its location depends on the distance from the mirror to the truck.

34.71. **IDENTIFY:** Apply $\dfrac{1}{s} + \dfrac{1}{s'} = \dfrac{1}{f}$ to calculate s' and then use $m = -\dfrac{s'}{s} = \dfrac{y'}{y}$ to find the height of the image.

SET UP: For a convex mirror, $R < 0$, so $R = -18.0$ cm and $f = \dfrac{R}{2} = -9.00$ cm.

EXECUTE: (a) $\frac{1}{s} + \frac{1}{s'} = \frac{1}{f}$. $s' = \frac{sf}{s-f} = \frac{(900 \text{ cm})(-9.00 \text{ cm})}{900 \text{ cm} - (-9.00 \text{ cm})} = -8.91 \text{ cm}.$

$m = -\frac{s'}{s} = -\frac{-8.91 \text{ cm}}{900 \text{ cm}} = 9.90 \times 10^{-3}.$ $|y'| = |m|y = (9.90 \times 10^{-3})(1.5 \text{ m}) = 0.0149 \text{ m} = 1.49 \text{ cm}.$

(b) The height of the image is much less than the height of the car, so the car appears to be farther away than its actual distance.

EVALUATE: A plane mirror would form an image the same size as the car. Since the image formed by the convex mirror is smaller than the car, the car appears to be farther away compared to what it would appear using a plane mirror.

34.73. IDENTIFY and SET UP: Rays that pass through the hole are undeflected. All other rays are blocked. $m = -\frac{s'}{s}.$

EXECUTE: (a) The ray diagram is drawn in Figure 34.73. The ray shown is the only ray from the top of the object that reaches the film, so this ray passes through the top of the image. An inverted image is formed on the far side of the box, no matter how far this side is from the pinhole and no matter how far the object is from the pinhole.

(b) $s = 1.5$ m. $s' = 20.0$ cm. $m = -\frac{s'}{s} = -\frac{20.0 \text{ cm}}{150 \text{ cm}} = -0.133.$ $y' = my = (-0.133)(18 \text{ cm}) = -2.4 \text{ cm}.$

The image is 2.4 cm tall.

EVALUATE: A defect of this camera is that not much light energy passes through the small hole each second, so long exposure times are required.

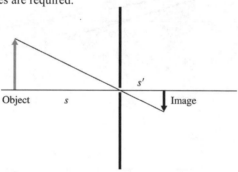

Figure 34.73

34.81. IDENTIFY: $\frac{1}{s} + \frac{1}{s'} = \frac{1}{f}.$ The type of lens determines the sign of f. $m = \frac{y'}{y} = -\frac{s'}{s}.$ The sign of s' depends on whether the image is real or virtual. $s = 16.0$ cm.

SET UP: $s' = -22.0$ cm; s' is negative because the image is on the same side of the lens as the object.

EXECUTE: (a) $\frac{1}{f} = \frac{s+s'}{ss'}$ and $f = \frac{ss'}{s+s'} = \frac{(16.0 \text{ cm})(-22.0 \text{ cm})}{16.0 \text{ cm} - 22.0 \text{ cm}} = +58.7$ cm. f is positive so the lens is converging.

(b) $m = -\frac{s'}{s} = -\frac{-22.0 \text{ cm}}{16.0 \text{ cm}} = 1.38.$ $y' = my = (1.38)(3.25 \text{ mm}) = 4.48$ mm. $s' < 0$ and the image is virtual.

EVALUATE: A converging lens forms a virtual image when the object is closer to the lens than the focal point.

34.89. IDENTIFY: $\frac{1}{s} + \frac{1}{s'} = \frac{1}{f}$ gives $s' = \frac{sf}{s-f}$, for both the mirror and the lens.

SET UP: For the second image, the image formed by the mirror serves as the object for the lens. For the mirror, $f_m = +10.0$ cm. For the lens, $f = 32.0$ cm. The center of curvature of the mirror is $R = 2f_m = 20.0$ cm to the right of the mirror vertex.

EXECUTE: **(a)** The principal-ray diagrams from the two images are sketched in Figure 34.89. In Figure 34.89b, only the image formed by the mirror is shown. This image is at the location of the candle so the principal-ray diagram that shows the image formation when the image of the mirror serves as the object for the lens is analogous to that in Figure 34.89a and is not drawn.
(b) Image formed by the light that passes directly through the lens: The candle is 85.0 cm to the left of the lens. $s' = \dfrac{sf}{s-f} = \dfrac{(85.0 \text{ cm})(32.0 \text{ cm})}{85.0 \text{ cm} - 32.0 \text{ cm}} = +51.3$ cm. $m = -\dfrac{s'}{s} = -\dfrac{51.3 \text{ cm}}{85.0 \text{ cm}} = -0.604$. This image is 51.3 cm to the right of the lens. $s' > 0$ so the image is real. $m < 0$ so the image is inverted. Image formed by the light that first reflects off the mirror: First consider the image formed by the mirror. The candle is 20.0 cm to the right of the mirror, so $s = +20.0$ cm. $s' = \dfrac{sf}{s-f} = \dfrac{(20.0 \text{ cm})(10.0 \text{ cm})}{20.0 \text{ cm} - 10.0 \text{ cm}} = 20.0$ cm.
$m_1 = -\dfrac{s_1'}{s_1} = -\dfrac{20.0 \text{ cm}}{20.0 \text{ cm}} = -1.00$. The image formed by the mirror is at the location of the candle, so $s_2 = +85.0$ cm and $s_2' = 51.3$ cm. $m_2 = -0.604$. $m_{\text{tot}} = m_1 m_2 = (-1.00)(-0.604) = 0.604$. The second image is 51.3 cm to the right of the lens. $s_2' > 0$, so the final image is real. $m_{\text{tot}} > 0$, so the final image is erect.
EVALUATE: The two images are at the same place. They are the same size. One is erect and one is inverted.

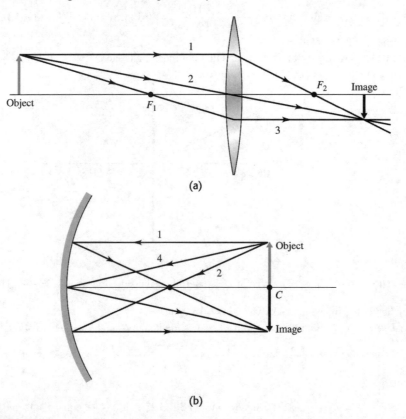

Figure 34.89

34.97. **IDENTIFY:** Apply $\dfrac{1}{s} + \dfrac{1}{s'} = \dfrac{1}{f}$. The near point is at infinity, so that is where the image must be formed for any objects that are close.
SET UP: The power in diopters equals $\dfrac{1}{f}$, with f in meters.

EXECUTE: $\dfrac{1}{f} = \dfrac{1}{s} + \dfrac{1}{s'} = \dfrac{1}{24\text{ cm}} + \dfrac{1}{-\infty} = \dfrac{1}{0.24\text{ m}} = 4.17$ diopters.

EVALUATE: To focus on closer objects, the power must be increased.

34.99. IDENTIFY and **SET UP:** The person's eye cannot focus on anything closer than 85.0 cm. The problem asks us to find the location of an object such that his old lenses produce a virtual image 85.0 cm from his eye. $\dfrac{1}{s} + \dfrac{1}{s'} = \dfrac{1}{f}$. P(in diopters) $= 1/f$ (in m).

EXECUTE: **(a)** $\dfrac{1}{f} = 2.25$ diopters so $f = 44.4$ cm. The image is 85.0 cm from his eye so is 83.0 cm from the eyeglass lens. Solving $\dfrac{1}{s} + \dfrac{1}{s'} = \dfrac{1}{f}$ for s gives $s = \dfrac{s'f}{s' - f} = \dfrac{(-83.0\text{ cm})(44.4\text{ cm})}{-83.0\text{ cm} - 44.4\text{ cm}} = +28.9$ cm. The object is 28.9 cm from the eyeglasses so is 30.9 cm from his eyes.

(b) Now $s' = -85.0$ cm. $s = \dfrac{s'f}{s' - f} = \dfrac{(-85.0\text{ cm})(44.4\text{ cm})}{-85.0\text{ cm} - 44.4\text{ cm}} = +29.2$ cm.

EVALUATE: The old glasses allow him to focus on objects as close as about 30 cm from his eyes. This is much better than a closest distance of 85 cm with no glasses, but his current glasses probably allow him to focus as close as 25 cm.

34.103. IDENTIFY: The thin-lens formula applies. The converging lens forms a real image on its right side. This image acts as the object for the diverging lens. The image formed by the converging lens is on the right side of the diverging lens, so this image acts as a *virtual object* for the diverging lens and its object distance is *negative*.

SET UP: Apply $\dfrac{1}{s} + \dfrac{1}{s'} = \dfrac{1}{f}$ in each case. $m = -\dfrac{s'}{s}$. The total magnification of two lenses is $m_{\text{tot}} = m_1 m_2$.

EXECUTE: **(a)** For the first trial on the diverging lens, we have
$s = 20.0$ cm $- 29.7$ cm $= -9.7$ cm and $s' = 42.8$ cm $- 20.0$ cm $= 22.8$ cm.
$\dfrac{1}{f} = \dfrac{1}{s} + \dfrac{1}{s'} = \dfrac{1}{-9.7\text{ cm}} + \dfrac{1}{22.8\text{ cm}} \rightarrow f = -16.88$ cm.

For the second trial on the diverging lens, we have
$s = 25.0$ cm $- 29.7$ cm $= -4.7$ cm and $s' = 31.6$ cm $- 25.0$ cm $= 6.6$ cm.
$\dfrac{1}{f} = \dfrac{1}{s} + \dfrac{1}{s'} = \dfrac{1}{-4.7\text{ cm}} + \dfrac{1}{6.6\text{ cm}} \rightarrow f = -16.33$ cm.

Taking the average of the focal lengths, we get $f_{\text{av}} = (-16.88\text{ cm} - 16.33\text{ cm})/2 = -16.6$ cm.

(b) The total magnification is $m_{\text{tot}} = m_1 m_2$. The converging lens does not move during the two trials, so m_1 is the same for both of them. But m_2 does change.

At 20.0 cm: $m = -\dfrac{s'}{s} = -(22.8\text{ cm})/(-9.7\text{ cm}) = +2.35$.

At 25.0 cm: $m = -\dfrac{s'}{s} = -(6.6\text{ cm})/(-4.7\text{ cm}) = +1.40$.

The magnification is greater when the lens is at 20.0 cm.

EVALUATE: This is a case where a diverging lens can form a real image, but only when it is used in conjunction with one or more other lenses.

34.109. **IDENTIFY** and **SET UP:** Apply the thin-lens formula $\frac{1}{s}+\frac{1}{s'}=\frac{1}{f}$ to the eye. The lens power in diopters is $1/f$ (in m). For the corrected eye, the image is at infinity, and it would take –6.0 D to correct the frog's vision so it could see at infinity.

EXECUTE: Using $\frac{1}{f}=\frac{1}{s}+\frac{1}{s'}$ gives $-6.0 \text{ m}^{-1} = 1/s + 1/\infty = 1/s$, so $s = 0.17$ m $= 17$ cm, which makes choice (d) the correct one.

EVALUATE: It is reasonable for a fog to see clearly up to only 17 cm since its food consists of insects, which must be fairly close to get caught.

35

INTERFERENCE

35.3. IDENTIFY: Use $c = f\lambda$ to calculate the wavelength of the transmitted waves. Compare the difference in the distance from A to P and from B to P. For constructive interference this path difference is an integer multiple of the wavelength.
SET UP: Consider Figure 35.3.

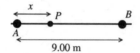

The distance of point P from each coherent source is $r_A = x$ and $r_B = 9.00 \text{ m} - x$.

Figure 35.3

EXECUTE: The path difference is $r_B - r_A = 9.00 \text{ m} - 2x$.
$r_B - r_A = m\lambda$, $m = 0, \pm 1, \pm 2, \ldots$
$\lambda = \dfrac{c}{f} = \dfrac{2.998 \times 10^8 \text{ m/s}}{120 \times 10^6 \text{ Hz}} = 2.50 \text{ m}$.

Thus $9.00 \text{ m} - 2x = m(2.50 \text{ m})$ and $x = \dfrac{9.00 \text{ m} - m(2.50 \text{ m})}{2} = 4.50 \text{ m} - (1.25 \text{ m})m$. x must lie in the range 0 to 9.00 m since P is said to be between the two antennas.
$m = 0$ gives $x = 4.50$ m.
$m = +1$ gives $x = 4.50 \text{ m} - 1.25 \text{ m} = 3.25 \text{ m}$.
$m = +2$ gives $x = 4.50 \text{ m} - 2.50 \text{ m} = 2.00 \text{ m}$.
$m = +3$ gives $x = 4.50 \text{ m} - 3.75 \text{ m} = 0.75 \text{ m}$.
$m = -1$ gives $x = 4.50 \text{ m} + 1.25 \text{ m} = 5.75 \text{ m}$.
$m = -2$ gives $x = 4.50 \text{ m} + 2.50 \text{ m} = 7.00 \text{ m}$.
$m = -3$ gives $x = 4.50 \text{ m} + 3.75 \text{ m} = 8.25 \text{ m}$.
All other values of m give values of x out of the allowed range. Constructive interference will occur for $x = 0.75$ m, 2.00 m, 3.25 m, 4.50 m, 5.75 m, 7.00 m, and 8.25 m.
EVALUATE: Constructive interference occurs at the midpoint between the two sources since that point is the same distance from each source. The other points of constructive interference are symmetrically placed relative to this point.

35.9. IDENTIFY and SET UP: The dark lines correspond to destructive interference and hence are located by
$d \sin\theta = (m + \tfrac{1}{2})\lambda$ so $\sin\theta = \dfrac{(m + \tfrac{1}{2})\lambda}{d}$, $m = 0, \pm 1, \pm 2, \ldots$

Solve for θ that locates the second and third dark lines. Use $y = R\tan\theta$ to find the distance of each of the dark lines from the center of the screen.

EXECUTE: 1st dark line is for $m = 0$.

2nd dark line is for $m = 1$ and $\sin\theta_1 = \dfrac{3\lambda}{2d} = \dfrac{3(500\times 10^{-9}\text{ m})}{2(0.450\times 10^{-3}\text{ m})} = 1.667\times 10^{-3}$ and $\theta_1 = 1.667\times 10^{-3}$ rad.

3rd dark line is for $m = 2$ and $\sin\theta_2 = \dfrac{5\lambda}{2d} = \dfrac{5(500\times 10^{-9}\text{ m})}{2(0.450\times 10^{-3}\text{ m})} = 2.778\times 10^{-3}$ and $\theta_2 = 2.778\times 10^{-3}$ rad.

(Note that θ_1 and θ_2 are small so that the approximation $\theta \approx \sin\theta \approx \tan\theta$ is valid.) The distance of each dark line from the center of the central bright band is given by $y_m = R\tan\theta$, where $R = 0.850$ m is the distance to the screen.
$\tan\theta \approx \theta$ so $y_m = R\theta_m$.

$y_1 = R\theta_1 = (0.750\text{ m})(1.667\times 10^{-3}\text{ rad}) = 1.25\times 10^{-3}$ m.

$y_2 = R\theta_2 = (0.750\text{ m})(2.778\times 10^{-3}\text{ rad}) = 2.08\times 10^{-3}$ m.

$\Delta y = y_2 - y_1 = 2.08\times 10^{-3}\text{ m} - 1.25\times 10^{-3}\text{ m} = 0.83$ mm.

EVALUATE: Since θ_1 and θ_2 are very small we could have used $y_m = R\dfrac{m\lambda}{d}$, generalized to destructive interference: $y_m = R(m + \tfrac{1}{2})\lambda/d$.

35.11. IDENTIFY: Bright fringes are located at angles θ given by $d\sin\theta = m\lambda$.
SET UP: The largest value $\sin\theta$ can have is 1.00.

EXECUTE: (a) $m = \dfrac{d\sin\theta}{\lambda}$. For $\sin\theta = 1$, $m = \dfrac{d}{\lambda} = \dfrac{0.0116\times 10^{-3}\text{ m}}{5.85\times 10^{-7}\text{ m}} = 19.8$. Therefore, the largest m for fringes on the screen is $m = 19$. There are $2(19) + 1 = 39$ bright fringes, the central one and 19 above and 19 below it.

(b) The most distant fringe has $m = \pm 19$. $\sin\theta = m\dfrac{\lambda}{d} = \pm 19\left(\dfrac{5.85\times 10^{-7}\text{ m}}{0.0116\times 10^{-3}\text{ m}}\right) = \pm 0.958$ and $\theta = \pm 73.3°$.

EVALUATE: For small θ the spacing Δy between adjacent fringes is constant but this is no longer the case for larger angles.

35.17. IDENTIFY: Use $I = I_0\cos^2(\phi/2)$ with $\phi = (2\pi/\lambda)(r_2 - r_1)$.
SET UP: ϕ is the phase difference and $(r_2 - r_1)$ is the path difference.
EXECUTE: (a) $I = I_0(\cos 30.0°)^2 = 0.750 I_0$.
(b) $60.0° = (\pi/3)$ rad. $(r_2 - r_1) = (\phi/2\pi)\lambda = [(\pi/3)/2\pi]\lambda = \lambda/6 = 80$ nm.
EVALUATE: $\phi = 360°/6$ and $(r_2 - r_1) = \lambda/6$.

35.21. IDENTIFY: The phase difference ϕ and the path difference $r_1 - r_2$ are related by $\phi = \dfrac{2\pi}{\lambda}(r_1 - r_2)$. The intensity is given by $I = I_0\cos^2\left(\dfrac{\phi}{2}\right)$.

SET UP: $\lambda = \dfrac{c}{f} = \dfrac{3.00\times 10^8\text{ m/s}}{1.20\times 10^8\text{ Hz}} = 2.50$ m. When the receiver measures intensity I_0, $\phi = 0$.

EXECUTE: (a) $\phi = \dfrac{2\pi}{\lambda}(r_1 - r_2) = \dfrac{2\pi}{2.50\text{ m}}(1.8\text{ m}) = 4.52$ rad.

(b) $I = I_0\cos^2\left(\dfrac{\phi}{2}\right) = I_0\cos^2\left(\dfrac{4.52\text{ rad}}{2}\right) = 0.404 I_0$.

EVALUATE: $(r_1 - r_2)$ is greater than $\lambda/2$, so one minimum has been passed as the receiver is moved.

35.25. **IDENTIFY:** The fringes are produced by interference between light reflected from the top and bottom surfaces of the air wedge. The refractive index of glass is greater than that of air, so the waves reflected from the top surface of the air wedge have no reflection phase shift, and the waves reflected from the bottom surface of the air wedge do have a half-cycle reflection phase shift. The condition for constructive interference (bright fringes) is therefore $2t = (m+\frac{1}{2})\lambda$.

SET UP: The geometry of the air wedge is sketched in Figure 35.25. At a distance x from the point of contact of the two plates, the thickness of the air wedge is t.

EXECUTE: $\tan\theta = \dfrac{t}{x}$ so $t = x\tan\theta$. $t_m = (m+\frac{1}{2})\dfrac{\lambda}{2}$. $x_m = (m+\frac{1}{2})\dfrac{\lambda}{2\tan\theta}$ and $x_{m+1} = (m+\frac{3}{2})\dfrac{\lambda}{2\tan\theta}$. The distance along the plate between adjacent fringes is $\Delta x = x_{m+1} - x_m = \dfrac{\lambda}{2\tan\theta}$. 15.0 fringes/cm $= \dfrac{1.00}{\Delta x}$ and

$\Delta x = \dfrac{1.00}{15.0 \text{ fringes/cm}} = 0.0667$ cm. $\tan\theta = \dfrac{\lambda}{2\Delta x} = \dfrac{546\times 10^{-9}\text{ m}}{2(0.0667\times 10^{-2}\text{ m})} = 4.09\times 10^{-4}$. The angle of the wedge is 4.09×10^{-4} rad $= 0.0234°$.

EVALUATE: The fringes are equally spaced; Δx is independent of m.

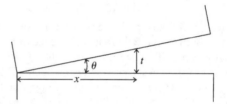

Figure 35.25

35.27. **IDENTIFY:** The light reflected from the top of the TiO_2 film interferes with the light reflected from the top of the glass surface. These waves are out of phase due to the path difference in the film and the phase differences caused by reflection.

SET UP: There is a π phase change at the TiO_2 surface but none at the glass surface, so for destructive interference the path difference must be $m\lambda$ in the film.

EXECUTE: **(a)** Calling T the thickness of the film gives $2T = m\lambda_0/n$, which yields $T = m\lambda_0/(2n)$. Substituting the numbers gives

$T = m$ (520.0 nm)/[2(2.62)] = 99.237 nm.

T must be greater than 1036 nm, so $m = 11$, which gives $T = 1091.6$ nm, since we want to know the minimum thickness to add.

$\Delta T = 1091.6$ nm $- 1036$ nm $= 55.6$ nm.

(b) (i) Path difference $= 2T = 2(1092$ nm$) = 2184$ nm $= 2180$ nm.

(ii) The wavelength in the film is
$\lambda = \lambda_0/n = (520.0$ nm$)/2.62 = 198.5$ nm.
Path difference $= (2180$ nm$)/[(198.5$ nm$)/$wavelength$] = 11.0$ wavelengths.

EVALUATE: Because the path difference in the film is 11.0 wavelengths, the light reflected off the top of the film will be 180° out of phase with the light that traveled through the film and was reflected off the glass due to the phase change at reflection off the top of the film.

35.31. **IDENTIFY:** Require destructive interference between light reflected from the two points on the disc.
SET UP: Both reflections occur for waves in the plastic substrate reflecting from the reflective coating, so they both have the same phase shift upon reflection and the condition for destructive interference (cancellation) is $2t = (m+\frac{1}{2})\lambda$, where t is the depth of the pit. $\lambda = \dfrac{\lambda_0}{n}$. The minimum pit depth is for $m = 0$.

35-4 Chapter 35

EXECUTE: $2t = \dfrac{\lambda}{2}$. $t = \dfrac{\lambda}{4} = \dfrac{\lambda_0}{4n} = \dfrac{790 \text{ nm}}{4(1.8)} = 110 \text{ nm} = 0.11 \, \mu\text{m}$.

EVALUATE: The path difference occurs in the plastic substrate and we must compare the wavelength in the substrate to the path difference.

35.33. IDENTIFY and SET UP: Apply $y = m(\lambda/2)$ and calculate y for $m = 1800$.

EXECUTE: $y = m(\lambda/2) = 1800(633 \times 10^{-9} \text{ m})/2 = 5.70 \times 10^{-4} \text{ m} = 0.570 \text{ mm}$

EVALUATE: A small displacement of the mirror corresponds to many wavelengths and a large number of fringes cross the line.

35.35. IDENTIFY: The two scratches are parallel slits, so the light that passes through them produces an interference pattern. However, the light is traveling through a medium (plastic) that is different from air.

SET UP: The central bright fringe is bordered by a dark fringe on each side of it. At these dark fringes, $d \sin\theta = \frac{1}{2} \lambda/n$, where n is the refractive index of the plastic.

EXECUTE: First use geometry to find the angles at which the two dark fringes occur. At the first dark fringe $\tan\theta = [(5.82 \text{ mm})/2]/(3250 \text{ mm})$, giving $\theta = \pm 0.0513°$.

For destructive interference, we have $d \sin\theta = \frac{1}{2} \lambda/n$ and

$$n = \lambda/(2d \sin\theta) = (632.8 \text{ nm})/[2(0.000225 \text{ m})(\sin 0.0513°)] = 1.57.$$

EVALUATE: The wavelength of the light in the plastic is reduced compared to what it would be in air.

35.39. IDENTIFY: The insertion of the metal foil produces a wedge of air, which is an air film of varying thickness. This film causes a path difference between light reflected off the top and bottom of this film.

SET UP: The two sheets of glass are sketched in Figure 35.39. The thickness of the air wedge at a distance x from the line of contact is $t = x \tan\theta$. Consider rays 1 and 2 that are reflected from the top and bottom surfaces, respectively, of the air film. Ray 1 has no phase change when it reflects and ray 2 has a 180° phase change when it reflects, so the reflections introduce a net 180° phase difference. The path difference is $2t$ and the wavelength in the film is $\lambda = \lambda_{\text{air}}$.

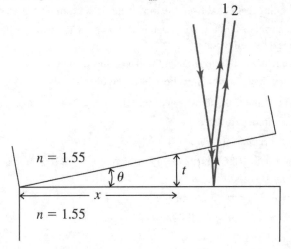

Figure 35.39

EXECUTE: (a) Since there is a 180° phase difference from the reflections, the condition for constructive interference is $2t = (m + \frac{1}{2})\lambda$. The positions of first enhancement correspond to $m = 0$ and $2t = \dfrac{\lambda}{2}$.

$x \tan\theta = \dfrac{\lambda}{4}$. θ is a constant, so $\dfrac{x_1}{\lambda_1} = \dfrac{x_2}{\lambda_2}$. $x_1 = 1.15$ mm, $\lambda_1 = 400.0$ nm. $x_2 = x_1 \left(\dfrac{\lambda_2}{\lambda_1} \right)$. For

$\lambda_2 = 550$ nm (green), $x_2 = (1.15 \text{ mm})\left(\dfrac{550 \text{ nm}}{400 \text{ nm}}\right) = 1.58$ mm. For $\lambda_2 = 600$ nm (orange),

$x_2 = (1.15 \text{ mm})\left(\dfrac{600 \text{ nm}}{400 \text{ nm}}\right) = 1.72$ mm.

(b) The positions of next enhancement correspond to $m=1$ and $2t = \dfrac{3\lambda}{2}$. $x\tan\theta = \dfrac{3\lambda}{4}$. The values of x are 3 times what they are in part (a). Violet: 3.45 mm; green: 4.74 mm; orange: 5.16 mm.

(c) $\tan\theta = \dfrac{\lambda}{4x} = \dfrac{400.0\times 10^{-9} \text{ m}}{4(1.15\times 10^{-3} \text{ m})} = 8.70\times 10^{-5}$. $\tan\theta = \dfrac{t_{\text{foil}}}{11.0 \text{ cm}}$, so $t_{\text{foil}} = 9.57\times 10^{-4}$ cm $= 9.57$ μm.

EVALUATE: The thickness of the foil must be very small to cause these observable interference effects. If it is too thick, the film is no longer a "thin film."

35.41. IDENTIFY: The liquid alters the wavelength of the light and that affects the locations of the interference minima.

SET UP: The interference minima are located by $d\sin\theta = (m+\tfrac{1}{2})\lambda$. For a liquid with refractive index n, $\lambda_{\text{liq}} = \dfrac{\lambda_{\text{air}}}{n}$.

EXECUTE: $\dfrac{\sin\theta}{\lambda} = \dfrac{(m+\tfrac{1}{2})}{d} = $ constant, so $\dfrac{\sin\theta_{\text{air}}}{\lambda_{\text{air}}} = \dfrac{\sin\theta_{\text{liq}}}{\lambda_{\text{liq}}}$. $\dfrac{\sin\theta_{\text{air}}}{\lambda_{\text{air}}} = \dfrac{\sin\theta_{\text{liq}}}{\lambda_{\text{air}}/n}$ and

$n = \dfrac{\sin\theta_{\text{air}}}{\sin\theta_{\text{liq}}} = \dfrac{\sin 35.20°}{\sin 19.46°} = 1.730$.

EVALUATE: In the liquid the wavelength is shorter and $\sin\theta = (m+\tfrac{1}{2})\dfrac{\lambda}{d}$ gives a smaller θ than in air, for the same m.

35.43. IDENTIFY: For destructive interference, $d = r_2 - r_1 = (m+\tfrac{1}{2})\lambda$.

SET UP: $r_2 - r_1 = \sqrt{(200 \text{ m})^2 + x^2} - x$.

EXECUTE: $(200 \text{ m})^2 + x^2 = x^2 + \left[(m+\tfrac{1}{2})\lambda\right]^2 + 2x(m+\tfrac{1}{2})\lambda$.

$x = \dfrac{20{,}000 \text{ m}^2}{(m+\tfrac{1}{2})\lambda} - \tfrac{1}{2}(m+\tfrac{1}{2})\lambda$. The wavelength is calculated by $\lambda = \dfrac{c}{f} = \dfrac{3.00\times 10^8 \text{ m/s}}{5.80\times 10^6 \text{ Hz}} = 51.7$ m.

$m=0: x=761$ m; $m=1: x=219$ m; $m=2: x=90.1$ m; $m=3; x=20.0$ m.

EVALUATE: For $m=3$, $d = 3.5\lambda = 181$ m. The maximum possible path difference is the separation of 200 m between the sources.

35.45. IDENTIFY and SET UP: Consider interference between rays reflected from the upper and lower surfaces of the film to relate the thickness of the film to the wavelengths for which there is destructive interference. The thermal expansion of the film changes the thickness of the film when the temperature changes.

EXECUTE: For this film on this glass, there is a net $\lambda/2$ phase change due to reflection and the condition for destructive interference is $2t = m(\lambda/n)$, where $n = 1.750$.

Smallest nonzero thickness is given by $t = \lambda/2n$.

At 20.0°C, $t_0 = (582.4 \text{ nm})/[(2)(1.750)] = 166.4$ nm.

At 170°C, $t = (588.5 \text{ nm})/[(2)(1.750)] = 168.1$ nm.

$t = t_0(1+\alpha\Delta T)$ so

$\alpha = (t-t_0)/(t_0\Delta T) = (1.7 \text{ nm})/[(166.4 \text{ nm})(150\text{C}°)] = 6.8\times 10^{-5} (\text{C}°)^{-1}$.

EVALUATE: When the film is heated its thickness increases, and it takes a larger wavelength in the film to equal $2t$. The value we calculated for α is the same order of magnitude as those given in Table 17.1.

35.49. **IDENTIFY** and **SET UP:** At the $m = 3$ bright fringe for the red light there must be destructive interference at this same θ for the other wavelength.

EXECUTE: For constructive interference: $d\sin\theta = m\lambda_1 \Rightarrow d\sin\theta = 3(700 \text{ nm}) = 2100 \text{ nm}$. For destructive interference: $d\sin\theta = (m + \frac{1}{2})\lambda_2 \Rightarrow \lambda_2 = \dfrac{d\sin\theta}{m + \frac{1}{2}} = \dfrac{2100 \text{ nm}}{m + \frac{1}{2}}$. So the possible wavelengths are $\lambda_2 = 600$ nm, for $m = 3$, and $\lambda_2 = 467$ nm, for $m = 4$.

EVALUATE: Both d and θ drop out of the calculation since their combination is just the path difference, which is the same for both types of light.

35.53. **IDENTIFY:** The wave from A travels a longer distance than the wave from B to reach point P, so the two waves will be out of phase when they reach P. For constructive interference, the path difference should be a whole-number multiple of the wavelength.

SET UP: To reach point P, the wave from A travels 240.0 m and the wave from B travels a distance x. The path difference for these two waves is $240.0 \text{ m} - x$. For any wave, $\lambda f = v$. Intensity maxima occur at $x = 210.0$ m, 216.0 m, and 222.0 m, and there are others.

EXECUTE: **(a)** The distance between adjacent intensity maxima is λ. The three given values of x are 6.0 m apart, so the wavelength must be 6.0 m. The frequency is $f = c/\lambda = c/(6.0 \text{ m}) = 5.0 \times 10^7$ Hz = 50 MHz.

(b) Destructive interference occurs when $240.0 \text{ m} - x = (m + \frac{1}{2})\lambda$, which gives $x = 240.0 \text{ m} - (m + \frac{1}{2})(6.0 \text{ m})$.

The largest x occurs when $m = 0$, so $x = 240.0 \text{ m} - 3.0 \text{ m} = 237.0 \text{ m}$.

EVALUATE: According to Figure 32.4 in the textbook, a wave having a wavelength of 6.0 m is in the radiowave region of the electromagnetic spectrum, which is consistent with the fact that you are using short-wave radio antennas.

35.57. **IDENTIFY** and **SET UP:** Interference occurs when two or more waves combine.

EXECUTE: All of the students now hear the tone, so no interference is occurring. Therefore only one speaker must be on, so the professor must have disconnected one of them. This makes choice (d) correct.

EVALUATE: Turning off one of the speakers would decrease the loudness of the sound, as was observed.

DIFFRACTION

36.3. IDENTIFY: The dark fringes are located at angles θ that satisfy $\sin\theta = \dfrac{m\lambda}{a}$, $m = \pm 1, \pm 2, \ldots$.

SET UP: The largest value of $|\sin\theta|$ is 1.00.

EXECUTE: (a) Solve for m that corresponds to $\sin\theta = 1$: $m = \dfrac{a}{\lambda} = \dfrac{0.0666 \times 10^{-3}\text{ m}}{585 \times 10^{-9}\text{ m}} = 113.8$. The largest value m can have is 113. $m = \pm 1, \pm 2, \ldots, \pm 113$ gives 226 dark fringes.

(b) For $m = \pm 113$, $\sin\theta = \pm 113\left(\dfrac{585 \times 10^{-9}\text{ m}}{0.0666 \times 10^{-3}\text{ m}}\right) = \pm 0.9926$ and $\theta = \pm 83.0°$.

EVALUATE: When the slit width a is decreased, there are fewer dark fringes. When $a < \lambda$ there are no dark fringes and the central maximum completely fills the screen.

36.7. IDENTIFY: We can model the hole in the concrete barrier as a single slit that will produce a single-slit diffraction pattern of the water waves on the shore.

SET UP: For single-slit diffraction, the angles at which destructive interference occurs are given by $\sin\theta_m = m\lambda/a$, where $m = 1, 2, 3, \ldots$.

EXECUTE: (a) The frequency of the water waves is $f = 75.0\text{ min}^{-1} = 1.25\text{ s}^{-1} = 1.25$ Hz, so their wavelength is $\lambda = v/f = (15.0\text{ cm/s})/(1.25\text{ Hz}) = 12.0$ cm.

At the first point for which destructive interference occurs, we have
$\tan\theta = (0.613\text{ m})/(3.20\text{ m}) \Rightarrow \theta = 10.84°$. $a\sin\theta = \lambda$ and
$$a = \lambda/\sin\theta = (12.0\text{ cm})/(\sin 10.84°) = 63.8\text{ cm}.$$

(b) Find the angles at which destructive interference occurs.

$$\sin\theta_2 = 2\lambda/a = 2(12.0\text{ cm})/(63.8\text{ cm}) \to \theta_2 = \pm 22.1°.$$

$$\sin\theta_3 = 3\lambda/a = 3(12.0\text{ cm})/(63.8\text{ cm}) \to \theta_3 = \pm 34.3°.$$

$$\sin\theta_4 = 4\lambda/a = 4(12.0\text{ cm})/(63.8\text{ cm}) \to \theta_4 = \pm 48.8°.$$

$$\sin\theta_5 = 5\lambda/a = 5(12.0\text{ cm})/(63.8\text{ cm}) \to \theta_5 = \pm 70.1°.$$

EVALUATE: These are large angles, so we cannot use the approximation that $\theta_m \approx m\lambda/a$.

36.9. IDENTIFY and SET UP: $v = f\lambda$ gives λ. The person hears no sound at angles corresponding to diffraction minima. The diffraction minima are located by $\sin\theta = m\lambda/a$, $m = \pm 1, \pm 2, \ldots$ Solve for θ.

EXECUTE: $\lambda = v/f = (344\text{ m/s})/(1250\text{ Hz}) = 0.2752$ m; $a = 1.00$ m. $m = \pm 1$, $\theta = \pm 16.0°$; $m = \pm 2$, $\theta = \pm 33.4°$; $m = \pm 3$, $\theta = \pm 55.6°$; no solution for larger m.

EVALUATE: $\lambda/a = 0.28$ so for the large wavelength sound waves diffraction by the doorway is a large effect. Diffraction would not be observable for visible light because its wavelength is much smaller and $\lambda/a \ll 1$.

36.15. **(a) IDENTIFY:** Use $\sin\theta = m\dfrac{\lambda}{a}$ with $m = 1$ to locate the angular position of the first minimum and then use $y = x\tan\theta$ to find its distance from the center of the screen.

SET UP and EXECUTE: The diffraction pattern is sketched in Figure 36.15.

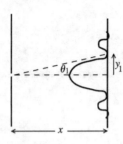

$$\sin\theta_1 = \dfrac{\lambda}{a} = \dfrac{540\times 10^{-9}\text{ m}}{0.240\times 10^{-3}\text{ m}} = 2.25\times 10^{-3}.$$

$$\theta_1 = 2.25\times 10^{-3}\text{ rad}.$$

Figure 36.15

$y_1 = x\tan\theta_1 = (3.00\text{ m})\tan(2.25\times 10^{-3}\text{ rad}) = 6.75\times 10^{-3}\text{ m} = 6.75\text{ mm}.$

(b) IDENTIFY and SET UP: Use $I_0\left(\dfrac{\sin\beta/2}{\beta/2}\right)^2$ and $\beta/2 = \dfrac{\pi a\sin\theta}{\lambda}$ to calculate the intensity at this point.

EXECUTE: Midway between the center of the central maximum and the first minimum implies

$y = \dfrac{1}{2}(6.75\text{ mm}) = 3.375\times 10^{-3}\text{ m}.$

$\tan\theta = \dfrac{y}{x} = \dfrac{3.375\times 10^{-3}\text{ m}}{3.00\text{ m}} = 1.125\times 10^{-3}; \theta = 1.125\times 10^{-3}\text{ rad}.$

The phase angle β at this point on the screen is

$\beta = \left(\dfrac{2\pi}{\lambda}\right)a\sin\theta = \dfrac{2\pi}{540\times 10^{-9}\text{ m}}(0.240\times 10^{-3}\text{ m})\sin(1.125\times 10^{-3}\text{ rad}) = \pi.$

Then $I = I_0\left(\dfrac{\sin\beta/2}{\beta/2}\right)^2 = (6.00\times 10^{-6}\text{ W/m}^2)\left(\dfrac{\sin\pi/2}{\pi/2}\right)^2.$

$I = \left(\dfrac{4}{\pi^2}\right)(6.00\times 10^{-6}\text{ W/m}^2) = 2.43\times 10^{-6}\text{ W/m}^2.$

EVALUATE: The intensity at this point midway between the center of the central maximum and the first minimum is less than half the maximum intensity. Compare this result to the corresponding one for the two-slit pattern, Exercise 35.20.

36.21. **(a) IDENTIFY and SET UP:** If the slits are very narrow then the central maximum of the diffraction pattern for each slit completely fills the screen and the intensity distribution is given solely by the two-slit interference. The maxima are given by $d\sin\theta = m\lambda$ so $\sin\theta = m\lambda/d$. Solve for θ.

EXECUTE: <u>1st order maximum:</u> $m = 1$, so $\sin\theta = \dfrac{\lambda}{d} = \dfrac{580\times 10^{-9}\text{ m}}{0.530\times 10^{-3}\text{ m}} = 1.094\times 10^{-3}; \theta = 0.0627°.$

<u>2nd order maximum:</u> $m = 2$, so $\sin\theta = \dfrac{2\lambda}{d} = 2.188\times 10^{-3}; \theta = 0.125°.$

(b) IDENTIFY and SET UP: The intensity is given by Eq. (36.12): $I = I_0\cos^2(\phi/2)\left(\dfrac{\sin\beta/2}{\beta/2}\right)^2.$ Calculate ϕ and β at each θ from part (a).

EXECUTE: $\phi = \left(\dfrac{2\pi d}{\lambda}\right)\sin\theta = \left(\dfrac{2\pi d}{\lambda}\right)\left(\dfrac{m\lambda}{d}\right) = 2\pi m,$ so $\cos^2(\phi/2) = \cos^2(m\pi) = 1.$

(Since the angular positions in part (a) correspond to interference maxima.)

$$\beta = \left(\frac{2\pi a}{\lambda}\right)\sin\theta = \left(\frac{2\pi a}{\lambda}\right)\left(\frac{m\lambda}{d}\right) = 2\pi m(a/d) = m2\pi\left(\frac{0.320 \text{ mm}}{0.530 \text{ mm}}\right) = m(3.794 \text{ rad}).$$

<u>1st order maximum:</u> $m = 1$, so $I = I_0(1)\left(\dfrac{\sin(3.794/2)\text{rad}}{(3.794/2)\text{rad}}\right)^2 = 0.249 I_0$.

<u>2nd order maximum:</u> $m = 2$, so $I = I_0(1)\left(\dfrac{\sin 3.794 \text{ rad}}{3.794 \text{ rad}}\right)^2 = 0.0256 I_0$.

EVALUATE: The first diffraction minimum is at an angle θ given by $\sin\theta = \lambda/a$ so $\theta = 0.104°$. The first order fringe is within the central maximum and the second order fringe is inside the first diffraction maximum on one side of the central maximum. The intensity here at this second fringe is much less than I_0.

36.27. IDENTIFY and SET UP: Calculate d for the grating. Use $d\sin\theta = m\lambda$ to calculate θ for the longest wavelength in the visible spectrum (750 nm) and verify that θ is small. Then use $y_m = x\dfrac{m\lambda}{a}$ to relate the linear separation of lines on the screen to the difference in wavelength.

EXECUTE: (a) $d = \left(\dfrac{1}{900}\right) \text{cm} = 1.111 \times 10^{-5} \text{ m}$.

For $\lambda = 750$ nm, $\lambda/d = 6.8 \times 10^{-2}$. The first-order lines are located at $\sin\theta = \lambda/d$; $\sin\theta$ is small enough for $\sin\theta \approx \theta$ to be an excellent approximation.

(b) $y = x\lambda/d$, where $x = 2.50$ m.

The distance on the screen between first-order bright bands for two different wavelengths is

$\Delta y = x(\Delta\lambda)/d$, so $\Delta\lambda = d(\Delta y)/x = (1.111 \times 10^{-5} \text{ m})(3.00 \times 10^{-3} \text{ m})/(2.50 \text{ m}) = 13.3$ nm.

EVALUATE: The smaller d is (greater number of lines per cm) the smaller the $\Delta\lambda$ that can be measured.

36.31. IDENTIFY: The resolving power depends on the line density and the width of the grating.
SET UP: The resolving power is given by $R = Nm = \lambda/\Delta\lambda$.
EXECUTE: (a) $R = Nm = (5000 \text{ lines/cm})(3.50 \text{ cm})(1) = 17,500$
(b) The resolving power needed to resolve the sodium doublet is
$$R = \lambda/\Delta\lambda = (589 \text{ nm})/(589.59 \text{ nm} - 589.00 \text{ nm}) = 998$$
so this grating can easily resolve the doublet.
(c) (i) $R = \lambda/\Delta\lambda$. Since $R = 17,500$ when $m = 1$, $R = 2 \times 17,500 = 35,000$ for $m = 2$. Therefore
$$\Delta\lambda = \lambda/R = (587.8 \text{ nm})/35,000 = 0.0168 \text{ nm}.$$
$$\lambda_{\min} = \lambda + \Delta\lambda = 587.8002 \text{ nm} + 0.0168 \text{ nm} = 587.8170 \text{ nm}.$$
(ii) $\lambda_{\max} = \lambda - \Delta\lambda = 587.8002 \text{ nm} - 0.0168 \text{ nm} = 587.7834$ nm

EVALUATE: (iii) Therefore the range of unresolvable wavelengths is $587.7834 \text{ nm} < \lambda < 587.8170 \text{ nm}$.

36.35. IDENTIFY and SET UP: The maxima occur at angles θ given by $2d\sin\theta = m\lambda$, where d is the spacing between adjacent atomic planes. Solve for d.
EXECUTE: Second order says $m = 2$.
$$d = \frac{m\lambda}{2\sin\theta} = \frac{2(0.0850 \times 10^{-9} \text{ m})}{2\sin 21.5°} = 2.32 \times 10^{-10} \text{ m} = 0.232 \text{ nm}.$$
EVALUATE: Our result is similar to d calculated in Example 36.5.

36.37. IDENTIFY and SET UP: The angular size of the first dark ring is given by $\sin\theta_1 = 1.22\lambda/D$. Calculate θ_1, and then the diameter of the ring on the screen is $2(4.5 \text{ m})\tan\theta_1$.

EXECUTE: $\sin\theta_1 = 1.22\left(\dfrac{620 \times 10^{-9} \text{ m}}{7.4 \times 10^{-6} \text{ m}}\right) = 0.1022$; $\theta_1 = 0.1024$ rad.

The radius of the Airy disk (central bright spot) is $r = (4.5 \text{ m})\tan\theta_1 = 0.462$ m. The diameter is $2r = 0.92 \text{ m} = 92 \text{ cm}$.

EVALUATE: $\lambda/D = 0.084$. For this small D the central diffraction maximum is broad.

36.41. IDENTIFY: The diameter D of the mirror determines the resolution.

SET UP: The resolving power is $\theta_{res} = 1.22 \dfrac{\lambda}{D}$.

EXECUTE: The same θ_{res} means that $\dfrac{\lambda_1}{D_1} = \dfrac{\lambda_2}{D_2}$. $D_2 = D_1 \dfrac{\lambda_2}{\lambda_1} = (8000 \times 10^3 \text{ m}) \left(\dfrac{550 \times 10^{-9} \text{ m}}{2.0 \times 10^{-2} \text{ m}} \right) = 220$ m.

EVALUATE: The Hubble telescope has an aperture of 2.4 m, so this would have to be an *enormous* optical telescope!

36.43. IDENTIFY and SET UP: Let y be the separation between the two points being resolved and let s be their distance from the telescope. Then the limit of resolution corresponds to $1.22 \dfrac{\lambda}{D} = \dfrac{y}{s}$.

EXECUTE: (a) Let the two points being resolved be the opposite edges of the crater, so y is the diameter of the crater. For the moon, $s = 3.8 \times 10^8$ m. $y = 1.22 \lambda s/D$.

Hubble: $D = 2.4$ m and $\lambda = 380$ nm gives the maximum resolution, so $y = 73$ m.

Arecibo: $D = 305$ m and $\lambda = 0.75$ m; $y = 1.1 \times 10^6$ m.

(b) $s = \dfrac{yD}{1.22\lambda}$. Let $y \approx 0.30$ m (the size of a license plate).

$s = (0.30 \text{ m})(2.4 \text{ m})/[(1.22)(380 \times 10^{-9} \text{ m})] = 1600$ km.

EVALUATE: D/λ is much larger for the optical telescope and it has a much larger resolution even though the diameter of the radio telescope is much larger.

36.49. IDENTIFY: In the single-slit diffraction pattern, the intensity is a maximum at the center and zero at the dark spots. At other points, it depends on the angle at which one is observing the light.

SET UP: Dark fringes occur when $\sin\theta_m = m\lambda/a$, where $m = 1, 2, 3, \ldots$, and the intensity is given by $I_0 \left(\dfrac{\sin\beta/2}{\beta/2} \right)^2$, where $\beta/2 = \dfrac{\pi a \sin\theta}{\lambda}$.

EXECUTE: (a) At the maximum possible angle, $\theta = 90°$, so
$m_{max} = (a \sin 90°)/\lambda = (0.0250 \text{ mm})/(632.8 \text{ nm}) = 39.5$.

Since m must be an integer and $\sin\theta$ must be ≤ 1, $m_{max} = 39$. The total number of dark fringes is 39 on each side of the central maximum for a total of 78.

(b) The farthest dark fringe is for $m = 39$, giving
$\sin\theta_{39} = (39)(632.8 \text{ nm})/(0.0250 \text{ mm}) \Rightarrow \theta_{39} = \pm 80.8°$.

(c) The next closer dark fringe occurs at $\sin\theta_{38} = (38)(632.8 \text{ nm})/(0.0250 \text{ mm}) \Rightarrow \theta_{38} = 74.1°$.

The angle midway these two extreme fringes is $(80.8° + 74.1°)/2 = 77.45°$, and the intensity at this angle is
$I = I_0 \left(\dfrac{\sin\beta/2}{\beta/2} \right)^2$, where $\beta/2 = \dfrac{\pi a \sin\theta}{\lambda} = \dfrac{\pi(0.0250 \text{ mm})\sin(77.45°)}{632.8 \text{ nm}} = 121.15$ rad, which gives

$I = (8.50 \text{ W/m}^2)\left[\dfrac{\sin(121.15 \text{ rad})}{121.15 \text{ rad}} \right]^2 = 5.55 \times 10^{-4}$ W/m^2.

EVALUATE: At the angle in part (c), the intensity is so low that the light would be barely perceptible.

36.51. IDENTIFY and SET UP: $\sin\theta = \lambda/a$ locates the first dark band. In the liquid the wavelength changes and this changes the angular position of the first diffraction minimum.

EXECUTE: $\sin\theta_{air} = \dfrac{\lambda_{air}}{a}$; $\sin\theta_{liquid} = \dfrac{\lambda_{liquid}}{a}$. $\lambda_{liquid} = \lambda_{air}\left(\dfrac{\sin\theta_{liquid}}{\sin\theta_{air}} \right) = \lambda_{air}\dfrac{\sin 21.6°}{\sin 38.2°} = 0.5953\lambda_{air}$.

$\lambda_{liquid} = \lambda_{air}/n$, so $n = \lambda_{air}/\lambda_{liquid} = \dfrac{\lambda_{air}}{0.5953\lambda_{air}} = 1.68$.

EVALUATE: Light travels faster in air and n must be >1.00. The smaller λ in the liquid reduces θ that located the first dark band.

36.53. **IDENTIFY:** $I = I_0 \left(\dfrac{\sin \gamma}{\gamma} \right)^2$. The maximum intensity occurs when the derivative of the intensity function with respect to γ is zero.

SET UP: $\dfrac{d \sin \gamma}{d\gamma} = \cos \gamma$. $\dfrac{d}{d\gamma}\left(\dfrac{1}{\gamma} \right) = -\dfrac{1}{\gamma^2}$.

EXECUTE: **(a)** $\dfrac{dI}{d\gamma} = I_0 \dfrac{d}{d\gamma}\left(\dfrac{\sin \gamma}{\gamma} \right)^2 = 2\left(\dfrac{\sin \gamma}{\gamma} \right)\left(\dfrac{\cos \gamma}{\gamma} - \dfrac{\sin \gamma}{\gamma^2} \right) = 0$. $\dfrac{\cos \gamma}{\gamma} - \dfrac{\sin \gamma}{\gamma^2} \Rightarrow \gamma \cos \gamma = \sin \gamma \Rightarrow \gamma = \tan \gamma$.

(b) The graph in Figure 36.53 is a plot of $f(\gamma) = \gamma - \tan \gamma$. When $f(\gamma)$ equals zero, there is an intensity maximum. Getting estimates from the graph, and then using trial and error to narrow in on the value, we find that the two smallest values of γ are $\gamma = 4.49$ rad and 7.73 rad.

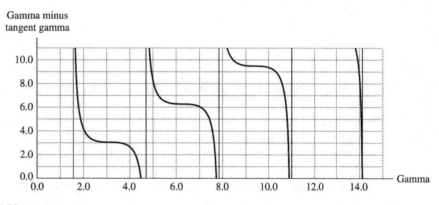

Figure 36.53

(c) The minima occur when $\gamma = \pm m\pi$, where $m = 1, 2, 3, \ldots$, so the first three positive ones are $\gamma = \pi, 2\pi,$ and 3π. From part (b), the first maximum is at $\gamma = 4.49$ rad. Midway between the first and second minima is $\gamma = \pi + \pi/2 = 3\pi/2 = 4.71$ rad. Clearly 4.49 rad $\neq 4.71$ rad, so this maximum is not midway between the adjacent minima. The second maximum is at $\gamma = 7.73$ rad. Midway between the second and third minima is $\gamma = 2\pi + \pi/2 = 5\pi/2 = 7.85$ rad. Since $7.73 \neq 7.85$, the second maximum is not midway between the adjacent minima.

(d) Using $a = 12 \lambda$, we have $\gamma = \dfrac{\pi a \sin \theta}{\lambda} = \dfrac{\pi (12\lambda)\sin \theta}{\lambda} = 12\pi \sin \theta$. Using this we get the following:

First minimum: $\gamma = \pi$, so $\pi = 12\pi \sin \theta \quad \rightarrow \quad \theta = 4.78°$.
First maximum: $\gamma = 2\pi$, so $2\pi = 12\pi \sin \theta \quad \rightarrow \quad \theta = 6.84°$.
Second minimum: $\gamma = 3\pi$, so $3\pi = 12\pi \sin \theta \quad \rightarrow \quad \theta = 9.59°$.

Midway between these minima is $\theta = (9.59° + 4.78°)/2 = 7.19°$. The first maximum is at $\theta = 6.84°$, and since $6.84° \neq 7.19°$, the first minimum is *not* midway between the adjacent minima.

EVALUATE: $\gamma = 0$ is the central maximum. The three values of γ we found are the locations of the first two secondary maxima. The first three minima are at $\gamma = 3.14$ rad, 6.28 rad, and 9.42 rad. The maxima are between adjacent minima, but not precisely midway between them.

36.55. **IDENTIFY:** Heating the plate causes it to expand, which widens the slit. The increased slit width changes the angles at which destructive interference occurs.

SET UP: First minimum is at angle θ given by $\tan \theta = \dfrac{(2.75 \times 10^{-3}/2)}{0.620}$. Therefore, θ is small and the equation $y_m = x \dfrac{m\lambda}{a}$ is accurate. The width of the central maximum is $w = \dfrac{2x\lambda}{a}$. The change in slit width is $\Delta a = a\alpha \Delta T$.

EXECUTE: $dw = 2x\lambda\left(-\dfrac{da}{a^2}\right) = -\dfrac{2x\lambda}{a^2}da = -\dfrac{w}{a}da.$ Therefore, $\Delta w = -\dfrac{w}{a}\Delta a.$ The equation for thermal expansion says $\Delta a = a\alpha\Delta T$, so $\Delta w = -w\alpha\Delta T = -(2.75 \text{ mm})(2.4\times 10^{-5}\text{ K}^{-1})(500\text{ K}) = -0.033$ mm. When the temperature of the plate increases, the width of the slit increases and the width of the central maximum decreases.

EVALUATE: The fractional change in the width of the central maximum is $\dfrac{0.033 \text{ mm}}{2.75 \text{ mm}} = 1.2\%$. This is small, but observable.

36.61. **IDENTIFY:** The liquid reduces the wavlength of the light (compared to its value in air), and the scratch causes light passing through it to undergo single-slit diffraction.

SET UP: $\sin\theta = \dfrac{\lambda}{a}$, where λ is the wavelength in the liquid. $n = \dfrac{\lambda_{air}}{\lambda}$.

EXECUTE: $\tan\theta = \dfrac{(22.4/2)\text{ cm}}{30.0 \text{ cm}}$ and $\theta = 20.47°$.

$\lambda = a\sin\theta = (1.25\times 10^{-6}\text{ m})\sin 20.47° = 4.372\times 10^{-7}$ m = 437.2 nm. $n = \dfrac{\lambda_{air}}{\lambda} = \dfrac{612 \text{ nm}}{437.2 \text{ nm}} = 1.40.$

EVALUATE: $n > 1$, as it must be, and $n = 1.40$ is reasonable for many transparent films.

36.63. **IDENTIFY:** A double-slit bright fringe is missing when it occurs at the same angle as a double-slit dark fringe.

SET UP: Single-slit diffraction dark fringes occur when $a\sin\theta = m\lambda,$ and double-slit interference bright fringes occur when $d\sin\theta = m'\lambda.$

EXECUTE: **(a)** The angle at which the first bright fringe occurs is given by
$\tan\theta_1 = (1.53 \text{ mm})/(2500 \text{ mm}) \Rightarrow \theta_1 = 0.03507°.$ $d\sin\theta_1 = \lambda$ and
$$d = \lambda/(\sin\theta_1) = (632.8 \text{ nm})/\sin(0.03507°) = 0.00103 \text{ m} = 1.03 \text{ mm}.$$

(b) The 7th double-slit interference bright fringe is just cancelled by the 1st diffraction dark fringe, so $\sin\theta_{diff} = \lambda/a$ and $\sin\theta_{interf} = 7\lambda/d.$
The angles are equal, so $\lambda/a = 7\lambda/d \rightarrow a = d/7 = (1.03 \text{ mm})/7 = 0.148$ mm.

EVALUATE: We can generalize that if $d = na$, where n is a positive integer, then every n^{th} double-slit bright fringe will be missing in the pattern.

36.65. **IDENTIFY:** The hole produces a diffraction pattern on the screen.

SET UP: The first dark ring occurs when $D\sin\theta_1 = 1.22\lambda$, and $\tan\theta_1 = r/x$.

EXECUTE: **(a)** Solving for D gives $D = (1.22\lambda)/(\sin\theta_1)$. First use $\tan\theta_1 = r/x$ to find θ_1, and then calculate D.
<u>First set</u>: $\tan\theta_1 = r/x = (5.6 \text{ cm})/(100 \text{ cm}) \rightarrow \theta_1 = 3.205°.$
$\quad D = (1.22)(562 \text{ nm})/[\sin(3.205°)] = 1.226\times 10^4$ nm.
<u>Second set</u>: $\tan\theta_1 = r/x = (8.5 \text{ cm})/(150 \text{ cm}) \rightarrow \theta_1 = 3.2433°.$
$\quad D = (1.22)(562 \text{ nm})/[\sin(3.2433°)] = 1.212\times 10^4$ nm.
<u>Third set</u>: $\tan\theta_1 = r/x = (11.6 \text{ cm})/(200 \text{ cm}) \rightarrow \theta_1 = 3.3194°.$
$\quad D = (1.22)(562 \text{ nm})/[\sin(3.3194°)] = 1.184\times 10^4$ nm.
<u>Fourth set</u>: $\tan\theta_1 = r/x = (14.1 \text{ cm})/(250 \text{ cm}) \rightarrow \theta_1 = 3.2281°.$
$\quad D = (1.22)(562 \text{ nm})/[\sin(3.2281°)] = 1.218\times 10^4$ nm.
Taking the average gives
$D_{av} = [(1.226 + 1.212 + 1.184 + 1.218)\times 10^4 \text{ nm}]/4 = 1.21\times 10^4$ nm = 12.1 μm.
(b) $\sin\theta_2 = 2.23\lambda/D = (2.23)(562 \text{ nm})/(1.21\times 10^4 \text{ nm}) \rightarrow \theta_2 = 5.945°.$
$r_2 = x_2\tan\theta_2 = (1.00 \text{ m})\tan(5.945°) = 0.104$ m = 10.4 cm.
$\sin\theta_3 = 3.24\lambda/D = (3.24)(562 \text{ nm})/(1.21\times 10^4 \text{ nm}) \rightarrow \theta_3 = 8.6551°.$
$r_3 = x_3\tan\theta_3 = (1.00 \text{ m})\tan(8.6551°) = 0.152$ m = 15.2 cm.

EVALUATE: This is a very small hole, but its diameter is still over 20 times the wavelength of the light passing through it.

36.71. **IDENTIFY** and **SET UP:** For an interference maxima, $2d\sin\theta = m\lambda_0/n$, so $\sin\theta \propto \dfrac{1}{d}$. If d is smaller, $\sin\theta$, and hence θ, is larger.

EXECUTE: At the top $\theta_{top} = 37°$, and at the bottom $\theta_{bottom} = 41°$. Since $\theta_{top} < \theta_{bottom}$, < it follows that $d_{top} > d_{bottom}$. Therefore the microspheres are closer together at the bottom than at the top, which makes choice (a) the correct one.

EVALUATE: Gravity tends to pull the spheres downward so they bunch up at the bottom, making them closer together.

37

RELATIVITY

37.5. (a) **IDENTIFY** and **SET UP:** $\Delta t_0 = 2.60 \times 10^{-8}$ s; $\Delta t = 4.20 \times 10^{-7}$ s. In the lab frame the pion is created and decays at different points, so this time is not the proper time.

EXECUTE: $\Delta t = \dfrac{\Delta t_0}{\sqrt{1-u^2/c^2}}$ says $1 - \dfrac{u^2}{c^2} = \left(\dfrac{\Delta t_0}{\Delta t}\right)^2$.

$\dfrac{u}{c} = \sqrt{1 - \left(\dfrac{\Delta t_0}{\Delta t}\right)^2} = \sqrt{1 - \left(\dfrac{2.60 \times 10^{-8} \text{ s}}{4.20 \times 10^{-7} \text{ s}}\right)^2} = 0.998$; $u = 0.998c$.

EVALUATE: $u < c$, as it must be, but u/c is close to unity and the time dilation effects are large.

(b) **IDENTIFY** and **SET UP:** The speed in the laboratory frame is $u = 0.998c$; the time measured in this frame is Δt, so the distance as measured in this frame is $d = u\Delta t$.

EXECUTE: $d = (0.998)(2.998 \times 10^8 \text{ m/s})(4.20 \times 10^{-7} \text{ s}) = 126$ m.

EVALUATE: The distance measured in the pion's frame will be different because the time measured in the pion's frame is different (shorter).

37.7. **IDENTIFY** and **SET UP:** A clock moving with respect to an observer appears to run more slowly than a clock at rest in the observer's frame. The clock in the spacecraft measurers the proper time Δt_0.

$\Delta t = 365$ days $= 8760$ hours.

EXECUTE: The clock on the moving spacecraft runs slow and shows the smaller elapsed time.

$\Delta t_0 = \Delta t \sqrt{1-u^2/c^2} = (8760 \text{ h})\sqrt{1-(4.80\times 10^6/3.00\times 10^8)^2} = 8758.88$ h. The difference in elapsed times is 8760 h $- 8758.88$ h $= 1.12$ h.

EVALUATE: 1.12 h is about 0.013% of a year. This difference would not be noticed by an astronaut, but such measurements are certainly within the capability of modern technology.

37.11. **IDENTIFY** and **SET UP:** The 2.2 μs lifetime is Δt_0 and the observer on earth measures Δt. The atmosphere is moving relative to the muon so in its frame the height of the atmosphere is l and l_0 is 10 km.

EXECUTE: (a) The greatest speed the muon can have is c, so the greatest distance it can travel in 2.2×10^{-6} s is $d = vt = (3.00 \times 10^8 \text{ m/s})(2.2 \times 10^{-6} \text{ s}) = 660$ m $= 0.66$ km.

(b) $\Delta t = \dfrac{\Delta t_0}{\sqrt{1-u^2/c^2}} = \dfrac{2.2 \times 10^{-6} \text{ s}}{\sqrt{1-(0.999)^2}} = 4.9 \times 10^{-5}$ s.

$d = vt = (0.999)(3.00 \times 10^8 \text{ m/s})(4.9 \times 10^{-5} \text{ s}) = 15$ km.

In the frame of the earth the muon can travel 15 km in the atmosphere during its lifetime.

(c) $l = l_0\sqrt{1-u^2/c^2} = (10 \text{ km})\sqrt{1-(0.999)^2} = 0.45$ km.

EVALUATE: In the frame of the muon the height of the atmosphere is less than the distance it moves during its lifetime.

37.17. IDENTIFY: The relativistic velocity addition formulas apply since the speeds are close to that of light.

SET UP: The relativistic velocity addition formula is $v'_x = \dfrac{v_x - u}{1 - \dfrac{uv_x}{c^2}}$.

EXECUTE: (a) For the pursuit ship to catch the cruiser, the distance between them must be decreasing, so the velocity of the cruiser relative to the pursuit ship must be directed toward the pursuit ship.
(b) Let the unprimed frame be Tatooine and let the primed frame be the pursuit ship. We want the velocity v' of the cruiser knowing the velocity of the primed frame u and the velocity of the cruiser v in the unprimed frame (Tatooine).

$$v'_x = \dfrac{v_x - u}{1 - \dfrac{uv_x}{c^2}} = \dfrac{0.600c - 0.800c}{1 - (0.600)(0.800)} = -0.385c.$$

The result implies that the cruiser is moving toward the pursuit ship at $0.385c$.

EVALUATE: The nonrelativistic formula would have given $-0.200c$, which is considerably different from the correct result.

37.19. IDENTIFY and SET UP: Reference frames S and S' are shown in Figure 37.19.

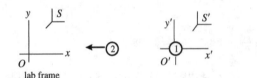

Frame S is at rest in the laboratory. Frame S' is attached to particle 1.

Figure 37.19

u is the speed of S' relative to S; this is the speed of particle 1 as measured in the laboratory. Thus $u = +0.650c$. The speed of particle 2 in S' is $0.950c$. Also, since the two particles move in opposite directions, 2 moves in the $-x'$-direction and $v'_x = -0.950c$. We want to calculate v_x, the speed of particle 2 in frame S, so use $v_x = \dfrac{v'_x + u}{1 + uv'_x/c^2}$.

EXECUTE: $v_x = \dfrac{v'_x + u}{1 + uv'_x/c^2} = \dfrac{-0.950c + 0.650c}{1 + (0.650c)(-0.950c)/c^2} = \dfrac{-0.300c}{1 - 0.6175} = -0.784c.$ The speed of the second particle, as measured in the laboratory, is $0.784c$.

EVALUATE: The incorrect Galilean expression for the relative velocity gives that the speed of the second particle in the lab frame is $0.300c$. The correct relativistic calculation gives a result more than twice this.

37.23. IDENTIFY and SET UP: Source and observer are approaching, so use $f = \sqrt{\dfrac{c+u}{c-u}} f_0$. Solve for u, the speed of the light source relative to the observer.

EXECUTE: (a) $f^2 = \left(\dfrac{c+u}{c-u}\right) f_0^2.$

$(c-u)f^2 = (c+u)f_0^2$ and $u = \dfrac{c(f^2 - f_0^2)}{f^2 + f_0^2} = c\left(\dfrac{(f/f_0)^2 - 1}{(f/f_0)^2 + 1}\right).$

$\lambda_0 = 675$ nm, $\lambda = 575$ nm.

$u = \left(\dfrac{(675 \text{ nm}/575 \text{ nm})^2 - 1}{(675 \text{ nm}/575 \text{ nm})^2 + 1}\right) c = 0.159c = (0.159)(2.998 \times 10^8 \text{ m/s}) = 4.77 \times 10^7$ m/s; definitely speeding

(b) 4.77×10^7 m/s $= (4.77 \times 10^7 \text{ m/s})(1 \text{ km}/1000 \text{ m})(3600 \text{ s}/1 \text{ h}) = 1.72 \times 10^8$ km/h. Your fine would be $\$1.72 \times 10^8$ (172 million dollars).

EVALUATE: The source and observer are approaching, so $f > f_0$ and $\lambda < \lambda_0$. Our result gives $u < c$, as it must.

37.29. **IDENTIFY:** Apply $p = \dfrac{mv}{\sqrt{1-v^2/c^2}}$ and $F = \gamma^3 ma$.

SET UP: For a particle at rest (or with $v \ll c$), $a = F/m$.

EXECUTE: (a) $p = \dfrac{mv}{\sqrt{1-v^2/c^2}} = 2mv$.

$\Rightarrow 1 = 2\sqrt{1-v^2/c^2} \Rightarrow \tfrac{1}{4} = 1 - \dfrac{v^2}{c^2} \Rightarrow v^2 = \tfrac{3}{4}c^2 \Rightarrow v = \dfrac{\sqrt{3}}{2}c = 0.866c$.

(b) $F = \gamma^3 ma = 2ma \Rightarrow \gamma^3 = 2 \Rightarrow \gamma = (2)^{1/3}$ so $\dfrac{1}{1-v^2/c^2} = 2^{2/3} \Rightarrow \dfrac{v}{c} = \sqrt{1-2^{-2/3}} = 0.608$.

EVALUATE: The momentum of a particle and the force required to give it a given acceleration both increase without bound as the speed of the particle approaches c.

37.33. **IDENTIFY and SET UP:** Use $E = mc^2 + K$ and $E^2 = (mc^2)^2 + (pc)^2$.

EXECUTE: (a) $E = mc^2 + K$, so $E = 4.00mc^2$ means $K = 3.00mc^2 = 4.50 \times 10^{-10}$ J.

(b) $E^2 = (mc^2)^2 + (pc)^2$; $E = 4.00mc^2$, so $15.0(mc^2)^2 = (pc)^2$.

$p = \sqrt{15}\,mc = 1.94 \times 10^{-18}$ kg·m/s.

(c) $E = mc^2/\sqrt{1-v^2/c^2}$.

$E = 4.00mc^2$ gives $1 - v^2/c^2 = 1/16$ and $v = \sqrt{15/16}\,c = 0.968c$.

EVALUATE: The speed is close to c since the kinetic energy is greater than the rest energy. Nonrelativistic expressions relating E, K, p, and v will be very inaccurate.

37.37. **IDENTIFY and SET UP:** The total energy is given in terms of the momentum by $E^2 = (mc^2)^2 + (pc)^2$. In terms of the total energy E, the kinetic energy K is $K = E - mc^2$. The rest energy is mc^2.

EXECUTE: (a) $E = \sqrt{(mc^2)^2 + (pc)^2} = \sqrt{\left[(6.64 \times 10^{-27})(2.998 \times 10^8)^2\right]^2 + \left[(2.10 \times 10^{-18})(2.998 \times 10^8)\right]^2}$ J.

$E = 8.67 \times 10^{-10}$ J.

(b) $mc^2 = (6.64 \times 10^{-27} \text{ kg})(2.998 \times 10^8 \text{ m/s})^2 = 5.97 \times 10^{-10}$ J.

$K = E - mc^2 = 8.67 \times 10^{-10}$ J $- 5.97 \times 10^{-10}$ J $= 2.70 \times 10^{-10}$ J.

(c) $\dfrac{K}{mc^2} = \dfrac{2.70 \times 10^{-10} \text{ J}}{5.97 \times 10^{-10} \text{ J}} = 0.452$.

EVALUATE: The incorrect nonrelativistic expressions for K and p give $K = p^2/2m = 3.3 \times 10^{-10}$ J; the correct relativistic value is less than this.

37.43. (a) **IDENTIFY and SET UP:** $\Delta t_0 = 2.60 \times 10^{-8}$ s is the proper time, measured in the pion's frame. The time measured in the lab must satisfy $d = c\Delta t$, where $u \approx c$. Calculate Δt and then use Eq. (37.6) to calculate u.

EXECUTE: $\Delta t = \dfrac{d}{c} = \dfrac{1.90 \times 10^3 \text{ m}}{2.998 \times 10^8 \text{ m/s}} = 6.3376 \times 10^{-6}$ s. $\Delta t = \dfrac{\Delta t_0}{\sqrt{1-u^2/c^2}}$ so $(1-u^2/c^2)^{1/2} = \dfrac{\Delta t_0}{\Delta t}$ and

$(1-u^2/c^2) = \left(\dfrac{\Delta t_0}{\Delta t}\right)^2$. Write $u = (1-\Delta)c$ so that $(u/c)^2 = (1-\Delta)^2 = 1 - 2\Delta + \Delta^2 \approx 1 - 2\Delta$ since Δ is small.

Using this in the above gives $1 - (1-2\Delta) = \left(\dfrac{\Delta t_0}{\Delta t}\right)^2$. $\Delta = \tfrac{1}{2}\left(\dfrac{\Delta t_0}{\Delta t}\right)^2 = \tfrac{1}{2}\left(\dfrac{2.60 \times 10^{-8} \text{ s}}{6.3376 \times 10^{-6} \text{ s}}\right)^2 = 8.42 \times 10^{-6}$.

EVALUATE: An alternative calculation is to say that the length of the tube must contract relative to the moving pion so that the pion travels that length before decaying. The contracted length must be

$l = c\Delta t_0 = (2.998 \times 10^8 \text{ m/s})(2.60 \times 10^{-8} \text{ s}) = 7.7948 \text{ m}.$ $l = l_0\sqrt{1 - u^2/c^2}$ so $1 - u^2/c^2 = \left(\dfrac{l}{l_0}\right)^2$. Then

$u = (1-\Delta)c$ gives $\Delta = \frac{1}{2}\left(\dfrac{l}{l_0}\right)^2 = \frac{1}{2}\left(\dfrac{7.7948 \text{ m}}{1.90 \times 10^3 \text{ m}}\right)^2 = 8.42 \times 10^{-6}$, which checks.

(b) IDENTIFY and SET UP: $E = \gamma mc^2$ Eq. (37.38).

EXECUTE: $\gamma = \dfrac{1}{\sqrt{1-u^2/c^2}} = \dfrac{1}{\sqrt{2\Delta}} = \dfrac{1}{\sqrt{2(8.42 \times 10^{-6})}} = 244.$

$E = (244)(139.6 \text{ MeV}) = 3.40 \times 10^4 \text{ MeV} = 34.0 \text{ GeV}.$

EVALUATE: The total energy is 244 times the rest energy.

37.45. IDENTIFY and SET UP: There must be a length contraction such that the length a becomes the same as b; $l_0 = a$, $l = b$. l_0 is the distance measured by an observer at rest relative to the spacecraft. Use $l = l_0\sqrt{1-u^2/c^2}$ and solve for u.

EXECUTE: $\dfrac{l}{l_0} = \sqrt{1-u^2/c^2}$ so $\dfrac{b}{a} = \sqrt{1-u^2/c^2}$;

$a = 1.40b$ gives $b/1.40b = \sqrt{1-u^2/c^2}$. and thus $1 - u^2/c^2 = 1/(1.40)^2$.

$u = \sqrt{1 - 1/(1.40)^2}\,c = 0.700c = 2.10 \times 10^8 \text{ m/s}.$

EVALUATE: A length on the spacecraft in the direction of the motion is shortened. A length perpendicular to the motion is unchanged.

37.47. IDENTIFY and SET UP: The proper time Δt_0 is the time that elapses in the frame of the space probe. Δt is the time that elapses in the frame of the earth. The distance traveled is 42.2 light years, as measured in the earth frame.

EXECUTE: Light travels 42.2 light years in 42.2 y, so $\Delta t = \left(\dfrac{c}{0.9930c}\right)(42.2 \text{ y}) = 42.5 \text{ y}.$

$\Delta t_0 = \Delta t\sqrt{1 - u^2/c^2} = (42.5 \text{ y})\sqrt{1 - (0.9930)^2} = 5.0 \text{ y}.$ She measures her biological age to be 19 y + 5.0 y = 24.0 y.

EVALUATE: Her age measured by someone on earth is 19 y + 42.5 y = 61.5 y.

37.51. IDENTIFY and SET UP: The clock on the plane measures the proper time Δt_0.

$\Delta t = 4.00 \text{ h} = (4.00 \text{ h})(3600 \text{ s/1 h}) = 1.44 \times 10^4 \text{ s}.$

$\Delta t = \dfrac{\Delta t_0}{\sqrt{1-u^2/c^2}}$ and $\Delta t_0 = \Delta t\sqrt{1-u^2/c^2}$.

EXECUTE: $\dfrac{u}{c}$ small so $\sqrt{1-u^2/c^2} = (1-u^2/c^2)^{1/2} \approx 1 - \frac{1}{2}\dfrac{u^2}{c^2}$; thus $\Delta t_0 = \Delta t\left(1 - \dfrac{1}{2}\dfrac{u^2}{c^2}\right).$

The difference in the clock readings is

$\Delta t - \Delta t_0 = \frac{1}{2}\dfrac{u^2}{c^2}\Delta t = \frac{1}{2}\left(\dfrac{250 \text{ m/s}}{2.998 \times 10^8 \text{ m/s}}\right)^2 (1.44 \times 10^4 \text{ s}) = 5.01 \times 10^{-9} \text{ s}.$ The clock on the plane has the shorter elapsed time.

EVALUATE: Δt_0 is always less than Δt; our results agree with this. The speed of the plane is much less than the speed of light, so the difference in the reading of the two clocks is very small.

37.55. **IDENTIFY** and **SET UP:** The energy released is $E = (\Delta m)c^2$. $\Delta m = \left(\dfrac{1}{10^4}\right)(12.0 \text{ kg})$. $P_{av} = \dfrac{E}{t}$.

The change in gravitational potential energy is $mg\Delta y$.

EXECUTE: **(a)** $E = (\Delta m)c^2 = \left(\dfrac{1}{10^4}\right)(12.0 \text{ kg})(3.00 \times 10^8 \text{ m/s})^2 = 1.08 \times 10^{14}$ J.

(b) $P_{av} = \dfrac{E}{t} = \dfrac{1.08 \times 10^{14} \text{ J}}{4.00 \times 10^{-6} \text{ s}} = 2.70 \times 10^{19}$ W.

(c) $E = \Delta U = mg\Delta y$. $m = \dfrac{E}{g\Delta y} = \dfrac{1.08 \times 10^{14} \text{ J}}{(9.80 \text{ m/s}^2)(1.00 \times 10^3 \text{ m})} = 1.10 \times 10^{10}$ kg.

EVALUATE: The mass decrease is only 1.2 grams, but the energy released is very large.

37.57. **IDENTIFY** and **SET UP:** Let S be the lab frame and let S' be the frame of the nucleus. Let the $+x$-direction be the direction the nucleus is moving. $u = 0.7500c$.

EXECUTE: **(a)** $v' = +0.9995c$. $v = \dfrac{v' + u}{1 + uv'/c^2} = \dfrac{0.9995c + 0.7500c}{1 + (0.7500)(0.9995)} = 0.999929c$.

(b) $v' = -0.9995c$. $v = \dfrac{-0.9995c + 0.7500c}{1 + (0.7500)(-0.9995)} = -0.9965c$.

(c) emitted in same direction:

(i) $K = \left(\dfrac{1}{\sqrt{1 - v^2/c^2}} - 1\right)mc^2 = (0.511 \text{ MeV})\left(\dfrac{1}{\sqrt{1 - (0.999929)^2}} - 1\right) = 42.4$ MeV.

(ii) $K' = \left(\dfrac{1}{\sqrt{1 - v^2/c^2}} - 1\right)mc^2 = (0.511 \text{ MeV})\left(\dfrac{1}{\sqrt{1 - (0.9995)^2}} - 1\right) = 15.7$ MeV.

(d) emitted in opposite direction:

(i) $K = \left(\dfrac{1}{\sqrt{1 - v^2/c^2}} - 1\right)mc^2 = (0.511 \text{ MeV})\left(\dfrac{1}{\sqrt{1 - (0.9965)^2}} - 1\right) = 5.60$ MeV.

(ii) $K' = \left(\dfrac{1}{\sqrt{1 - v^2/c^2}} - 1\right)mc^2 = (0.511 \text{ MeV})\left(\dfrac{1}{\sqrt{1 - (0.9995)^2}} - 1\right) = 15.7$ MeV.

EVALUATE: The kinetic energy in the frame of the nucleus is the same in both cases since this is the proper frame.

37.61. **IDENTIFY:** The baseball is moving toward the radar gun, so apply the Doppler effect equation $f = \sqrt{\dfrac{c + u}{c - u}} f_0$.

SET UP: The baseball had better be moving nonrelativistically, so the Doppler shift formula becomes $f \cong f_0(1 - (u/c))$. In the baseball's frame, this is the frequency with which the radar waves strike the baseball, and the baseball reradiates at f. But in the coach's frame, the reflected waves are Doppler shifted again, so the detected frequency is $f(1 - (u/c)) = f_0(1 - (u/c))^2 \approx f_0(1 - 2(u/c))$.

EXECUTE: $\Delta f = 2f_0(u/c)$ and the fractional frequency shift is $\dfrac{\Delta f}{f_0} = 2(u/c)$.

$u = \dfrac{\Delta f}{2f_0}c = \dfrac{(2.86 \times 10^{-7})}{2}(3.00 \times 10^8 \text{ m}) = 42.9 \text{ m/s} = 154 \text{ km/h} = 92.5$ mi/h.

EVALUATE: $u \ll c$, so using the approximate expression in place of $f = \sqrt{\dfrac{c + u}{c - u}} f_0$ is very accurate.

37.65. **IDENTIFY and SET UP:** The equation $\Delta t = \dfrac{\Delta t_0}{\sqrt{1-u^2/c^2}}$ relates the time interval in the laboratory (Δt) to the time interval (Δt_0) in the rest frame of the particle.

EXECUTE: **(a)** Solve the above equation for $(\Delta t)^2$. Squaring gives $(\Delta t)^2 = (\Delta t_0)^2 (1-u^2/c^2)^{-1}$. Therefore a graph of $(\Delta t)^2$ versus $(1-u^2/c^2)^{-1}$ should be a straight line with slope equal to $(\Delta t_0)^2$. Figure 37.65 shows this graph for the data in the problem. The slope of the best-fit straight line is 0.6709×10^{-15} s^2, so $\Delta t_0 = \sqrt{0.6709 \times 10^{-15} \text{ s}^2} = 2.6 \times 10^{-8}$ s = 26 ns.

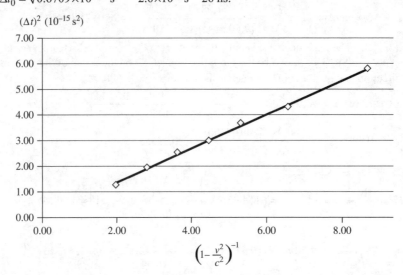

Figure 37.65

(b) Using $\Delta t = 4 \Delta t_0$, the equation $\Delta t = \dfrac{\Delta t_0}{\sqrt{1-u^2/c^2}}$ gives $4\Delta t_0 = \dfrac{\Delta t_0}{\sqrt{1-u^2/c^2}}$. Solving for u/c gives $u/c = 0.97$.

EVALUATE: At speeds near the speed of light, there is a very large difference between the lifetime measured in the laboratory frame compared to the lifetime in the rest frame of the particle.

37.67. **IDENTIFY and SET UP:** When the force on a particle is along the same line as its velocity, the force and acceleration are related by $F = \gamma^3 ma$, where $\gamma = 1/\sqrt{1-v^2/c^2}$.

EXECUTE: **(a)** Solve the equation $F = \gamma^3 ma$ for a^2.

$$a = \dfrac{F}{m\gamma^3} = \dfrac{F}{m}\left(1-\dfrac{v^2}{c^2}\right)^{3/2} \quad \rightarrow \quad a^2 = \left(\dfrac{F}{m}\right)^2 \left(1-\dfrac{v^2}{c^2}\right)^3.$$

From this result we see that a graph of a^2 versus $\left(1-\dfrac{v^2}{c^2}\right)^3$ should be a straight line with slope equal to $(F/m)^2$. Figure 37.67 shows the graph of the data in the table with the problem. It is well fit by a straight line having slope equal to 1.608×10^9 m^2/s^4. Therefore the mass is

$$(F/m)^2 = \text{slope} \quad \rightarrow \quad m = \dfrac{F}{\sqrt{\text{slope}}} = \dfrac{8.00 \times 10^{-14} \text{ N}}{\sqrt{1.608 \times 10^9 \text{ m}^2/\text{s}^4}} = 2.0 \times 10^{-18} \text{ kg}.$$

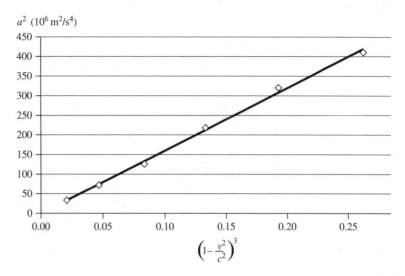

Figure 37.67

(b) In this case, $v \ll c$, so γ is essentially equal to 1. Therefore we can use the familiar form of Newton's second law, $F = ma$.
$a = F/m = (8.00\times10^{-14}\text{ N})/(2.0\times10^{-18}\text{ kg}) = 4.0\times10^{4}\text{ m/s}^2$.
EVALUATE: When v is close to c, $a = F/m$ does not give the correct result. For example, using data from the table in the problem, when $v/c = 0.85$, the acceleration is measured to be 5900 m/s². But using the familiar Newtonian formula we get $a = F/m = (8.00\times10^{-14}\text{ N})/(2.0\times10^{-18}\text{ kg}) = 4.0\times10^{4}\text{ m/s}^2 = 40{,}000$ m/s², which is *very* different from the relativistic result of 5900 m/s².

37.73. **IDENTIFY:** The rest energy is $E_0 = mc^2$.
SET UP: In our universe, $E_0 = mc^2$ and in the alternative universe the same formula would apply, except that c would be different, call it c', so $E_0' = mc'^2$. Therefore $E_0' = \left(\dfrac{E_0}{c^2}\right)c'^2$.

EXECUTE: $E_0' = \left(\dfrac{E_0}{c^2}\right)c'^2 = \left(\dfrac{8.2\times10^{-14}\text{ J}}{(3.0\times10^{8}\text{ m/s})^2}\right)(300\text{ m/s})^2 = 8.2\times10^{-26}\text{ J}$, choice (b).

EVALUATE: The rest energy of all other particles would be reduced by the same fraction in this alternate universe.

PHOTONS: LIGHT WAVES BEHAVING AS PARTICLES

38.5. **IDENTIFY** and **SET UP:** A photon has zero rest mass, so its energy is $E = pc$ and its momentum is $p = \dfrac{h}{\lambda}$.

EXECUTE: **(a)** $E = pc = (8.24 \times 10^{-28} \text{ kg} \cdot \text{m/s})(2.998 \times 10^{8} \text{ m/s}) = 2.47 \times 10^{-19} \text{ J} =$
$(2.47 \times 10^{-19} \text{ J})(1 \text{ eV}/1.602 \times 10^{-19} \text{ J}) = 1.54 \text{ eV}.$

(b) $p = \dfrac{h}{\lambda}$ so $\lambda = \dfrac{h}{p} = \dfrac{6.626 \times 10^{-34} \text{ J} \cdot \text{s}}{8.24 \times 10^{-28} \text{ kg} \cdot \text{m/s}} = 8.04 \times 10^{-7} \text{ m} = 804 \text{ nm}.$

EVALUATE: This wavelength is longer than visible wavelengths; it is in the infrared region of the electromagnetic spectrum. To check our result we could verify that the same E is given by $E = hc/\lambda$, using the λ we have calculated.

38.7. **IDENTIFY** and **SET UP:** For the photoelectric effect, the maximum kinetic energy of the photoelectrons is $\frac{1}{2} m v_{max}^2 = hf - \phi = \dfrac{hc}{\lambda} - \phi$. Take the work function ϕ from Table 38.1. Solve for v_{max}. Note that we wrote f as c/λ.

EXECUTE: $\frac{1}{2} m v_{max}^2 = \dfrac{(6.626 \times 10^{-34} \text{ J} \cdot \text{s})(2.998 \times 10^{8} \text{ m/s})}{235 \times 10^{-9} \text{ m}} - (5.1 \text{ eV})(1.602 \times 10^{-19} \text{ J}/1 \text{ eV}).$

$\frac{1}{2} m v_{max}^2 = 8.453 \times 10^{-19} \text{ J} - 8.170 \times 10^{-19} \text{ J} = 2.83 \times 10^{-20} \text{ J}.$

$v_{max} = \sqrt{\dfrac{2(2.83 \times 10^{-20} \text{ J})}{9.109 \times 10^{-31} \text{ kg}}} = 2.49 \times 10^{5} \text{ m/s}.$

EVALUATE: The work function in eV was converted to joules for use in the equation $\frac{1}{2} m v_{max}^2 = hf - \phi = \dfrac{hc}{\lambda} - \phi$. A photon with $\lambda = 235$ nm has energy greater then the work function for the surface.

38.11. **(b) IDENTIFY:** Solve part (b) first. First use $eV_0 = hf - \phi$ to find the work function ϕ.

SET UP: $eV_0 = hf - \phi$ so $\phi = hf - eV_0 = \dfrac{hc}{\lambda} - eV_0.$

EXECUTE: $\phi = \dfrac{(6.626 \times 10^{-34} \text{ J} \cdot \text{s})(2.998 \times 10^{8} \text{ m/s})}{254 \times 10^{-9} \text{ m}} - (1.602 \times 10^{-19} \text{ C})(0.181 \text{ V}).$

$\phi = 7.821 \times 10^{-19} \text{ J} - 2.900 \times 10^{-20} \text{ J} = 7.531 \times 10^{-19} \text{ J}(1 \text{ eV}/1.602 \times 10^{-19} \text{ J}) = 4.70 \text{ eV}.$

(a) IDENTIFY and **SET UP:** The threshold frequency f_{th} is the smallest frequency that still produces photoelectrons. It corresponds to $K_{max} = 0$ in the equation $\frac{1}{2} m v_{max}^2 = hf - \phi$, so $hf_{th} = \phi$.

EXECUTE: $f = \dfrac{c}{\lambda}$ says $\dfrac{hc}{\lambda_{th}} = \phi.$

$$\lambda_{\text{th}} = \frac{hc}{\phi} = \frac{(6.626\times 10^{-34}\text{ J}\cdot\text{s})(2.998\times 10^{8}\text{ m/s})}{7.531\times 10^{-19}\text{ J}} = 2.64\times 10^{-7}\text{ m} = 264\text{ nm}.$$

EVALUATE: As calculated in part (b), $\phi = 4.70$ eV. This is the value given in Table 38.1 for copper.

38.13. IDENTIFY: Apply $eV_{AC} = hf_{\max} = \dfrac{hc}{\lambda_{\min}}$.

SET UP: For a 4.00-keV electron, $eV_{AC} = 4000$ eV.

EXECUTE: $eV_{AC} = hf_{\max} = \dfrac{hc}{\lambda_{\min}} \Rightarrow \lambda_{\min} = \dfrac{hc}{eV_{AC}} = \dfrac{(6.63\times 10^{-34}\text{ J}\cdot\text{s})(3.00\times 10^{8}\text{ m/s})}{(1.60\times 10^{-19}\text{ C})(4000\text{ V})} = 3.11\times 10^{-10}$ m.

EVALUATE: This is the same answer as would be obtained if electrons of this energy were used. Electron beams are much more easily produced and accelerated than proton beams.

39.17. IDENTIFY: Apply $\lambda' - \lambda = \dfrac{h}{mc}(1-\cos\phi) = \lambda_C(1-\cos\phi)$.

SET UP: Solve for λ': $\lambda' = \lambda + \lambda_C(1-\cos\phi)$.

The largest λ' corresponds to $\phi = 180°$, so $\cos\phi = -1$.

EXECUTE: $\lambda' = \lambda + 2\lambda_C = 0.0665\times 10^{-9}$ m $+ 2(2.426\times 10^{-12}$ m$) = 7.135\times 10^{-11}$ m $= 0.0714$ nm. This wavelength occurs at a scattering angle of $\phi = 180°$.

EVALUATE: The incident photon transfers some of its energy and momentum to the electron from which it scatters. Since the photon loses energy its wavelength increases, $\lambda' > \lambda$.

38.19. IDENTIFY and SET UP: The shift in wavelength of the photon is $\lambda' - \lambda = \dfrac{h}{mc}(1-\cos\phi)$ where λ' is the wavelength after the scattering and $\dfrac{h}{mc} = \lambda_C = 2.426\times 10^{-12}$ m. The energy of a photon of wavelength λ is $E = \dfrac{hc}{\lambda} = \dfrac{1.24\times 10^{-6}\text{ eV}\cdot\text{m}}{\lambda}$. Conservation of energy applies to the collision, so the energy lost by the photon equals the energy gained by the electron.

EXECUTE: (a) $\lambda' - \lambda = \lambda_C(1-\cos\phi) = (2.426\times 10^{-12}\text{ m})(1-\cos 35.0°) = 4.39\times 10^{-13}$ m $= 4.39\times 10^{-4}$ nm.

(b) $\lambda' = \lambda + 4.39\times 10^{-4}$ nm $= 0.04250$ nm $+ 4.39\times 10^{-4}$ nm $= 0.04294$ nm.

(c) $E_\lambda = \dfrac{hc}{\lambda} = 2.918\times 10^{4}$ eV and $E_{\lambda'} = \dfrac{hc}{\lambda'} = 2.888\times 10^{4}$ eV so the photon loses 300 eV of energy.

(d) Energy conservation says the electron gains 300 eV of energy.

EVALUATE: The photon transfers energy to the electron. Since the photon loses energy, its wavelength increases.

38.23. IDENTIFY: The wavelength of the pulse tells us the momentum of the photon. The uncertainty in the momentum is determined by the uncertainty principle.

SET UP: $p = \dfrac{h}{\lambda}$ and $\Delta x \Delta p_x = \dfrac{\hbar}{2}$.

EXECUTE: $p = \dfrac{h}{\lambda} = \dfrac{6.626\times 10^{-34}\text{ J}\cdot\text{s}}{556\times 10^{-9}\text{ m}} = 1.19\times 10^{-27}$ kg$\cdot$m/s. The spatial length of the pulse is $\Delta x = c\Delta t = (2.998\times 10^{8}\text{ m/s})(9.00\times 10^{-15}\text{ s}) = 2.698\times 10^{-6}$ m. The uncertainty principle gives $\Delta x \Delta p_x = \dfrac{\hbar}{2}$. Solving for the uncertainty in the momentum, we have

$$\Delta p_x = \dfrac{\hbar}{2\Delta x} = \dfrac{1.055\times 10^{-34}\text{ J}\cdot\text{s}}{2(2.698\times 10^{-6}\text{ m})} = 1.96\times 10^{-29}\text{ kg}\cdot\text{m/s}.$$

EVALUATE: This is 1.6% of the average momentum.

38.27. **IDENTIFY** and **SET UP:** The energy added to mass m of the blood to heat it to $T_f = 100°C$ and to vaporize it is $Q = mc(T_f - T_i) + mL_v$, with $c = 4190$ J/kg·K and $L_v = 2.256 \times 10^6$ J/kg. The energy of one photon is $E = \dfrac{hc}{\lambda} = \dfrac{1.99 \times 10^{-25} \text{ J·m}}{\lambda}$.

EXECUTE: **(a)** $Q = (2.0 \times 10^{-9} \text{ kg})(4190 \text{ J/kg·K})(100°C - 33°C) + (2.0 \times 10^{-9} \text{ kg})(2.256 \times 10^6 \text{ J/kg}) = 5.07 \times 10^{-3}$ J. The pulse must deliver 5.07 mJ of energy.

(b) $P = \dfrac{\text{energy}}{t} = \dfrac{5.07 \times 10^{-3} \text{ J}}{450 \times 10^{-6} \text{ s}} = 11.3$ W.

(c) One photon has energy $E = \dfrac{hc}{\lambda} = \dfrac{1.99 \times 10^{-25} \text{ J·m}}{585 \times 10^{-9} \text{ m}} = 3.40 \times 10^{-19}$ J. The number N of photons per pulse is the energy per pulse divided by the energy of one photon:

$N = \dfrac{5.07 \times 10^{-3} \text{ J}}{3.40 \times 10^{-19} \text{ J/photon}} = 1.49 \times 10^{16}$ photons.

EVALUATE: The power output of the laser is small but it is focused on a small area, so the laser intensity is large.

38.33. **IDENTIFY** and **SET UP:** Find the average change in wavelength for one scattering and use that in $\Delta\lambda$ in $\lambda' - \lambda = \left(\dfrac{h}{mc}\right)(1 - \cos\phi)$ to calculate the average scattering angle ϕ.

EXECUTE: **(a)** The wavelength of a 1 MeV photon is

$\lambda = \dfrac{hc}{E} = \dfrac{(4.136 \times 10^{-15} \text{ eV·s})(2.998 \times 10^8 \text{ m/s})}{1 \times 10^6 \text{ eV}} = 1 \times 10^{-12}$ m.

The total change in wavelength therefore is 500×10^{-9} m $- 1 \times 10^{-12}$ m $= 500 \times 10^{-9}$ m.

If this shift is produced in 10^{26} Compton scattering events, the wavelength shift in each scattering event is

$\Delta\lambda = \dfrac{500 \times 10^{-9} \text{ m}}{1 \times 10^{26}} = 5 \times 10^{-33}$ m.

(b) Use this $\Delta\lambda$ in $\Delta\lambda = \dfrac{h}{mc}(1 - \cos\phi)$ and solve for ϕ. We anticipate that ϕ will be very small, since $\Delta\lambda$ is much less than h/mc, so we can use $\cos\phi \approx 1 - \phi^2/2$.

$\Delta\lambda = \dfrac{h}{mc}\left[1 - (1 - \phi^2/2)\right] = \dfrac{h}{2mc}\phi^2$.

$\phi = \sqrt{\dfrac{2\Delta\lambda}{(h/mc)}} = \sqrt{\dfrac{2(5 \times 10^{-33} \text{ m})}{2.426 \times 10^{-12} \text{ m}}} = 6.4 \times 10^{-11}$ rad $= (4 \times 10^{-9})°$.

ϕ in radians is much less than 1 so the approximation we used is valid.

(c) **IDENTIFY** and **SET UP:** We know the total transit time and the total number of scatterings, so we can calculate the average time between scatterings.

EXECUTE: The total time to travel from the core to the surface is $(10^6 \text{ y})(3.156 \times 10^7 \text{ s/y}) = 3.2 \times 10^{13}$ s.

There are 10^{26} scatterings during this time, so the average time between scatterings is

$t = \dfrac{3.2 \times 10^{13} \text{ s}}{10^{26}} = 3.2 \times 10^{-13}$ s.

The distance light travels in this time is $d = ct = (3.0 \times 10^8 \text{ m/s})(3.2 \times 10^{-13} \text{ s}) = 0.1$ mm.

EVALUATE: The photons are on the average scattered through a very small angle in each scattering event. The average distance a photon travels between scatterings is very small.

38.37. **IDENTIFY** and **SET UP:** Apply the photoelectric effect. $eV_0 = hf - \phi$. For a photon, $f\lambda = c$.

EXECUTE: **(a)** Using $eV_0 = hf - \phi$ and $f\lambda = c$, we get $eV_0 = hc/\lambda - \phi$. Solving for V_0 gives $V_0 = \dfrac{hc}{e} \cdot \dfrac{1}{\lambda} - \dfrac{\phi}{e}$. Therefore a graph of V_0 versus $1/\lambda$ should be a straight line with slope equal to hc/e and y-intercept equal to $-\phi/e$. Figure 38.37 shows this graph for the data given in the problem. The best-fit equation for this graph is $V_0 = (1230 \text{ V} \cdot \text{nm}) \cdot \dfrac{1}{\lambda} - 4.76 \text{ V}$. The slope is equal to $1230 \text{ V} \cdot \text{nm}$, which is equal to $1.23 \times 10^{-6} \text{ V} \cdot \text{m}$, and the y-intercept is -4.76 V.

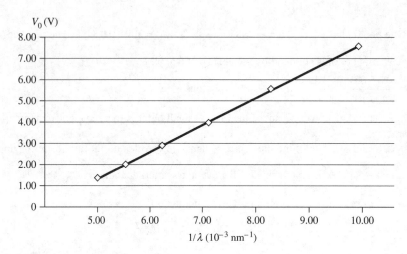

Figure 38.37

(b) Using the slope we have $hc/e =$ slope, so
$h = e(\text{slope})/c = (1.602 \times 10^{-19} \text{ C})(1.23 \times 10^{-6} \text{ V} \cdot \text{m})/(2.998 \times 10^8 \text{ m/s}) = 6.58 \times 10^{-34} \text{ J} \cdot \text{s}$.
The y-intercept is equal to $-\phi/e$, so
$\phi = -e(y\text{-intercept}) = -(1.602 \times 10^{-19} \text{ C})(-4.76 \text{ V}) = 7.63 \times 10^{-19} \text{ J} = 4.76$ eV.

(c) For the longest wavelength light, the energy of a photon is equal to the work function of the metal, so $hc/\lambda = \phi$. Solving for λ gives $\lambda = hc/\phi$. Our calculation of h was just a test of the data, so we use the accepted value for h in the calculation.
$\lambda = hc/\phi = (6.626 \times 10^{-34} \text{ J} \cdot \text{s})(2.998 \times 10^8 \text{ m/s})/(7.63 \times 10^{-19} \text{ J}) = 2.60 \times 10^{-7} \text{ m} = 260$ nm.

(d) The energy of the photon is equal to the sum of the kinetic energy of the photoelectron and the work function, so $hc/\lambda = K + \phi$. This gives
$(4.136 \times 10^{-15} \text{ eV} \cdot \text{s})(2.998 \times 10^8 \text{ m/s})/\lambda = 10.0 \text{ eV} + 4.76 \text{ eV} = 14.76$ eV, which gives
$\lambda = 8.40 \times 10^{-8}$ m $= 84.0$ nm.

EVALUATE: As we know from Table 38.1, typical metal work functions are several eV, so our results are plausible.

38.39. **IDENTIFY** and **SET UP:** We have Compton scattering, so $\lambda' - \lambda = \left(\dfrac{h}{mc}\right)(1 - \cos\phi)$, which can also be expressed as $\lambda' - \lambda = \lambda_C(1 - \cos\phi)$, where λ_C is the Compton wavelength.

EXECUTE: **(a)** Figure 38.39 shows the graph of λ' versus $1 - \cos\phi$ for the data included in the problem. The best-fit equation of the line is $\lambda' = 5.21 \text{ pm} + (2.40 \text{ pm})(1 - \cos\phi)$. The slope is 2.40 pm and the y-intercept is 5.21 pm.

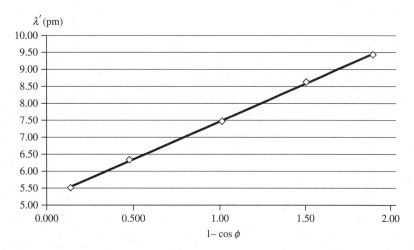

Figure 38.39

(b) Solving $\lambda' - \lambda = \lambda_C(1 - \cos\phi)$ for λ' gives $\lambda' = \lambda + \lambda_C(1 - \cos\phi)$. The graph of λ' versus $1 - \cos\phi$ should be a straight line with slope equal to λ_C and y-intercept equal to λ. From the slope, we get $\lambda_C = \text{slope} = 2.40 \text{ pm}$.

(c) From the y-intercept we get $\lambda = y\text{-intercept} = 5.21 \text{ pm}$.

EVALUATE: For backscatter, the photon wavelength would be $5.21 \text{ pm} + 2(2.40 \text{ pm}) = 10.01 \text{ pm}$.

38.41. **IDENTIFY** and **SET UP:** The specific gravity of the tumor is 1, so it has the same density as water, 1000 kg/m³. If 70 Gy are given in 35 days, the daily treatment is 2 Gy.

EXECUTE: The energy E per cell is

$$E/\text{cell} = \frac{(2 \text{ J/kg})\left(\dfrac{1000 \text{ kg}}{(100 \text{ cm})^3}\right)}{10^8 \text{ cells/cm}^3} = (2 \times 10^{-11} \text{ J/cell})(6 \times 10^{18} \text{ eV/J}) = 1.2 \times 10^8 \text{ eV/cell} = 120 \text{ MeV/cell}.$$

Choice (c) is correct.

EVALUATE: For 35 treatments the total dose would be $120 \text{ MeV} \times 35 = 4200 \text{ MeV} = 4.2 \text{ GeV}$ per cell.

PARTICLES BEHAVING AS WAVES

39.7. **IDENTIFY** and **SET UP:** A photon has zero mass and its energy and wavelength are related by $E = hc/\lambda$. An electron has mass. Its energy is related to its momentum by $E = p^2/2m$, and its wavelength is related to its momentum by $\lambda = h/p$.

EXECUTE: (a) Photon: $E = \dfrac{hc}{\lambda}$ so $\lambda = \dfrac{hc}{E} = \dfrac{(6.626\times 10^{-34}\text{ J}\cdot\text{s})(2.998\times 10^8\text{ m/s})}{(20.0\text{ eV})(1.602\times 10^{-19}\text{ J/eV})} = 62.0$ nm.

Electron: $E = p^2/(2m)$ so $p = \sqrt{2mE}$, which gives

$p = \sqrt{2(9.109\times 10^{-31}\text{ kg})(20.0\text{ eV})(1.602\times 10^{-19}\text{ J/eV})} = 2.416\times 10^{-24}$ kg·m/s. $\lambda = h/p = 0.274$ nm.

(b) Photon: $E = hc/\lambda = 7.946\times 10^{-19}$ J $= 4.96$ eV.

Electron: $\lambda = h/p$ so $p = h/\lambda = 2.650\times 10^{-27}$ kg·m/s.

$E = p^2/(2m) = 3.856\times 10^{-24}$ J $= 2.41\times 10^{-5}$ eV.

EVALUATE: (c) You should use a probe of wavelength approximately 250 nm. An electron with $\lambda = 250$ nm has much less energy than a photon with $\lambda = 250$ nm, so is less likely to damage the molecule. Note that $\lambda = h/p$ applies to all particles, those with mass and those with zero mass. $E = hf = hc/\lambda$ applies only to photons and $E = p^2/2m$ applies only to particles with mass.

39.11. **IDENTIFY:** The acceleration gives momentum to the electrons. We can use this momentum to calculate their de Broglie wavelength.
SET UP: The kinetic energy K of the electron is related to the accelerating voltage V by $K = eV$. For an electron $E = \tfrac{1}{2}mv^2 = \dfrac{p^2}{2m}$ and $\lambda = \dfrac{h}{p}$. For a photon $E = \dfrac{hc}{\lambda}$.

EXECUTE: (a) For an electron $p = \dfrac{h}{\lambda} = \dfrac{6.63\times 10^{-34}\text{ J}\cdot\text{s}}{5.00\times 10^{-9}\text{ m}} = 1.33\times 10^{-25}$ kg·m/s and

$E = \dfrac{p^2}{2m} = \dfrac{(1.33\times 10^{-25}\text{ kg}\cdot\text{m/s})^2}{2(9.11\times 10^{-31}\text{ kg})} = 9.71\times 10^{-21}$ J. $V = \dfrac{K}{e} = \dfrac{9.71\times 10^{-21}\text{ J}}{1.60\times 10^{-19}\text{ C}} = 0.0607$ V. The electrons would have kinetic energy 0.0607 eV.

(b) $E = \dfrac{hc}{\lambda} = \dfrac{1.24\times 10^{-6}\text{ eV}\cdot\text{m}}{5.00\times 10^{-9}\text{ m}} = 248$ eV.

(c) $E = 9.71\times 10^{-21}$ J so $\lambda = \dfrac{hc}{E} = \dfrac{(6.63\times 10^{-34}\text{ J}\cdot\text{s})(3.00\times 10^8\text{ m/s})}{9.71\times 10^{-21}\text{ J}} = 20.5\ \mu$m.

EVALUATE: If they have the same wavelength, the photon has vastly more energy than the electron.

39.13. **IDENTIFY:** The intensity maxima are located by $d\sin\theta = m\lambda$. Use $\lambda = \dfrac{h}{p}$ for the wavelength of the neutrons. For a particle, $p = \sqrt{2mE}$.

SET UP: For a neutron, $m = 1.675 \times 10^{-27}$ kg.

EXECUTE: For $m = 1$, $\lambda = d\sin\theta = \dfrac{h}{\sqrt{2mE}}$.

$E = \dfrac{h^2}{2md^2 \sin^2\theta} = \dfrac{(6.63\times 10^{-34}\text{ J}\cdot\text{s})^2}{2(1.675\times 10^{-27}\text{ kg})(9.10\times 10^{-11}\text{ m})^2 \sin^2(28.6°)} = 6.91\times 10^{-20}\text{ J} = 0.432\text{ eV}$.

EVALUATE: The neutrons have $\lambda = 0.0436$ nm, comparable to the atomic spacing.

39.15. **IDENTIFY:** The condition for a maximum is $d\sin\theta = m\lambda$. $\lambda = \dfrac{h}{p} = \dfrac{h}{Mv}$, so $\theta = \arcsin\left(\dfrac{mh}{dMv}\right)$.

SET UP: Here m is the order of the maximum, whereas M is the incoming particle mass.

EXECUTE: (a) $m = 1 \Rightarrow \theta_1 = \arcsin\left(\dfrac{h}{dMv}\right)$

$= \arcsin\left(\dfrac{6.63\times 10^{-34}\text{ J}\cdot\text{s}}{(1.60\times 10^{-6}\text{ m})(9.11\times 10^{-31}\text{ kg})(1.26\times 10^4\text{ m/s})}\right) = 2.07°$.

$m = 2 \Rightarrow \theta_2 = \arcsin\left(\dfrac{(2)(6.63\times 10^{-34}\text{ J}\cdot\text{s})}{(1.60\times 10^{-6}\text{ m})(9.11\times 10^{-31}\text{ kg})(1.26\times 10^4\text{ m/s})}\right) = 4.14°$.

(b) For small angles (in radians!) $y \cong D\theta$, so $y_1 \approx (50.0\text{ cm})(2.07°)\left(\dfrac{\pi\text{ radians}}{180°}\right) = 1.81$ cm,

$y_2 \approx (50.0\text{ cm})(4.14°)\left(\dfrac{\pi\text{ radians}}{180°}\right) = 3.61$ cm, and $y_2 - y_1 = 3.61\text{ cm} - 1.81\text{ cm} = 1.80$ cm.

EVALUATE: For these electrons, $\lambda = \dfrac{h}{mv} = 0.0577$ μm. λ is much less than d and the intensity maxima occur at small angles.

39.19. **IDENTIFY and SET UP:** Use the energy to calculate n for this state. Then use the Bohr equation, $L = n\hbar$, to calculate L.

EXECUTE: $E_n = -(13.6\text{ eV})/n^2$, so this state has $n = \sqrt{13.6/1.51} = 3$. In the Bohr model, $L = n\hbar$ so for this state $L = 3\hbar = 3.16\times 10^{-34}$ kg·m²/s.

EVALUATE: We will find in Section 41.1 that the modern quantum mechanical description gives a different result.

39.21. **IDENTIFY:** The force between the electron and the nucleus in Be^{3+} is $F = \dfrac{1}{4\pi\varepsilon_0}\dfrac{Ze^2}{r^2}$, where $Z = 4$ is the nuclear charge. All the equations for the hydrogen atom apply to Be^{3+} if we replace e^2 by Ze^2.

(a) **SET UP:** Modify the energy equation for hydrogen, $E_n = -\dfrac{1}{\varepsilon_0^2}\dfrac{me^4}{8n^2h^2}$ by replacing e^2 with Ze^2.

EXECUTE: $E_n = -\dfrac{1}{\varepsilon_0^2}\dfrac{me^4}{8n^2h^2}$ (hydrogen) becomes

$E_n = -\dfrac{1}{\varepsilon_0^2}\dfrac{m(Ze^2)^2}{8n^2h^2} = Z^2\left(-\dfrac{1}{\varepsilon_0^2}\dfrac{me^4}{8n^2h^2}\right) = Z^2\left(-\dfrac{13.60\text{ eV}}{n^2}\right)$ (for Be^{3+}).

The ground-level energy of Be^{3+} is $E_1 = 16\left(-\dfrac{13.60\text{ eV}}{1^2}\right) = -218$ eV.

EVALUATE: The ground-level energy of Be^{3+} is $Z^2 = 16$ times the ground-level energy of H.

(b) SET UP: The ionization energy is the energy difference between the $n \to \infty$ level energy and the $n = 1$ level energy.

EXECUTE: The $n \to \infty$ level energy is zero, so the ionization energy of Be^{3+} is 218 eV.

EVALUATE: This is 16 times the ionization energy of hydrogen.

(c) SET UP: $\dfrac{1}{\lambda} = R\left(\dfrac{1}{n_1^2} - \dfrac{1}{n_2^2}\right)$ just as for hydrogen but now R has a different value.

EXECUTE: $R_H = \dfrac{me^4}{8\varepsilon_0^2 h^3 c} = 1.097 \times 10^7 \text{ m}^{-1}$ for hydrogen becomes

$R_{Be} = Z^2 \dfrac{me^4}{8\varepsilon_0^2 h^3 c} = 16(1.097 \times 10^7 \text{ m}^{-1}) = 1.755 \times 10^8 \text{ m}^{-1}$ for Be^{3+}.

For $n = 2$ to $n = 1$, $\dfrac{1}{\lambda} = R_{Be}\left(\dfrac{1}{1^2} - \dfrac{1}{2^2}\right) = 3R_{Be}/4$.

$\lambda = 4/(3R_{Be}) = 4/(3(1.755 \times 10^8 \text{ m}^{-1})) = 7.60 \times 10^{-9}$ m = 7.60 nm.

EVALUATE: This wavelength is smaller by a factor of 16 compared to the wavelength for the corresponding transition in the hydrogen atom.

(d) SET UP: Modify the Bohr equation for hydrogen, $r_n = \varepsilon_0 \dfrac{n^2 h^2}{\pi m e^2}$, by replacing e^2 with Ze^2.

EXECUTE: $r_n = \varepsilon_0 \dfrac{n^2 h^2}{\pi m (Ze^2)}$ (Be^{3+}).

EVALUATE: For a given n the orbit radius for Be^{3+} is smaller by a factor of $Z = 4$ compared to the corresponding radius for hydrogen.

39.25. IDENTIFY and SET UP: The ionization threshold is at $E = 0$. The energy of an absorbed photon equals the energy gained by the atom and the energy of an emitted photon equals the energy lost by the atom.

EXECUTE: (a) $\Delta E = 0 - (-20 \text{ eV}) = 20$ eV.

(b) When the atom in the $n = 1$ level absorbs an 18-eV photon, the final level of the atom is $n = 4$. The possible transitions from $n = 4$ and corresponding photon energies are $n = 4 \to n = 3$, 3 eV; $n = 4 \to n = 2$, 8 eV; $n = 4 \to n = 1$, 18 eV. Once the atom has gone to the $n = 3$ level, the following transitions can occur: $n = 3 \to n = 2$, 5 eV; $n = 3 \to n = 1$, 15 eV. Once the atom has gone to the $n = 2$ level, the following transition can occur: $n = 2 \to n = 1$, 10 eV. The possible energies of emitted photons are: 3 eV, 5 eV, 8 eV, 10 eV, 15 eV, and 18 eV.

(c) There is no energy level 8 eV higher in energy than the ground state, so the photon cannot be absorbed.

(d) The photon energies for $n = 3 \to n = 2$ and for $n = 3 \to n = 1$ are 5 eV and 15 eV. The photon energy for $n = 4 \to n = 3$ is 3 eV. The work function must have a value between 3 eV and 5 eV.

EVALUATE: The atom has discrete energy levels, so the energies of emitted or absorbed photons have only certain discrete energies.

39.27. IDENTIFY and SET UP: The wavelength of the photon is related to the transition energy $E_i - E_f$ of the atom by $E_i - E_f = \dfrac{hc}{\lambda}$ where $hc = 1.240 \times 10^{-6}$ eV·m.

EXECUTE: (a) The minimum energy to ionize an atom is when the upper state in the transition has $E = 0$, so $E_1 = -17.50$ eV. For $n = 5 \to n = 1$, $\lambda = 73.86$ nm and $E_5 - E_1 = \dfrac{1.240 \times 10^{-6} \text{ eV} \cdot \text{m}}{73.86 \times 10^{-9} \text{ m}} = 16.79$ eV.

$E_5 = -17.50 \text{ eV} + 16.79 \text{ eV} = -0.71 \text{ eV}$. For $n=4 \to n=1$, $\lambda = 75.63$ nm and $E_4 = -1.10$ eV. For $n=3 \to n=1$, $\lambda = 79.76$ nm and $E_3 = -1.95$ eV. For $n=2 \to n=1$, $\lambda = 94.54$ nm and $E_2 = -4.38$ eV.

(b) $E_i - E_f = E_4 - E_2 = -1.10 \text{ eV} - (-4.38 \text{ eV}) = 3.28 \text{ eV}$ and $\lambda = \dfrac{hc}{E_i - E_f} = \dfrac{1.240 \times 10^{-6} \text{ eV} \cdot \text{m}}{3.28 \text{ eV}} = 378$ nm.

EVALUATE: The $n=4 \to n=2$ transition energy is smaller than the $n=4 \to n=1$ transition energy so the wavelength is longer. In fact, this wavelength is longer than for any transition that ends in the $n=1$ state.

39.33. **IDENTIFY and SET UP:** The number of photons emitted each second is the total energy emitted divided by the energy of one photon. The energy of one photon is given by $E = hc/\lambda$. $E = Pt$ gives the energy emitted by the laser in time t.

EXECUTE: In 1.00 s the energy emitted by the laser is $(7.50 \times 10^{-3} \text{ W})(1.00 \text{ s}) = 7.50 \times 10^{-3}$ J.

The energy of each photon is $E = \dfrac{hc}{\lambda} = \dfrac{(6.626 \times 10^{-34} \text{ J} \cdot \text{s})(2.998 \times 10^8 \text{ m/s})}{10.6 \times 10^{-6} \text{ m}} = 1.874 \times 10^{-20}$ J.

Therefore $\dfrac{7.50 \times 10^{-3} \text{ J/s}}{1.874 \times 10^{-20} \text{ J/photon}} = 4.00 \times 10^{17}$ photons/s.

EVALUATE: The number of photons emitted per second is extremely large.

39.35. **IDENTIFY:** Apply the equation $\dfrac{n_{ex}}{n_g} = e^{-(E_{ex} - E_g)/kT}$ from the section on the laser. In this case we have $\dfrac{n_{5s}}{n_{3p}} = e^{-(E_{5s} - E_{3p})/kT}$.

SET UP: $E_{5s} = 20.66$ eV and $E_{3p} = 18.70$ eV.

EXECUTE: $E_{5s} - E_{3p} = 20.66 \text{ eV} - 18.70 \text{ eV} = 1.96 \text{ eV}(1.602 \times 10^{-19} \text{ J}/1 \text{ eV}) = 3.140 \times 10^{-19}$ J.

(a) $\dfrac{n_{5s}}{n_{3p}} = e^{-(3.140 \times 10^{-19} \text{ J})/[(1.38 \times 10^{-23} \text{ J/K})(300 \text{ K})]} = e^{-75.79} = 1.2 \times 10^{-33}$.

(b) $\dfrac{n_{5s}}{n_{3p}} = e^{-(3.140 \times 10^{-19} \text{ J})/[(1.38 \times 10^{-23} \text{ J/K})(600 \text{ K})]} = e^{-37.90} = 3.5 \times 10^{-17}$.

(c) $\dfrac{n_{5s}}{n_{3p}} = e^{-(3.140 \times 10^{-19} \text{ J})/[(1.38 \times 10^{-23} \text{ J/K})(1200 \text{ K})]} = e^{-18.95} = 5.9 \times 10^{-9}$.

EVALUATE: **(d)** At each of these temperatures the number of atoms in the $5s$ excited state, the initial state for the transition that emits 632.8 nm radiation, is quite small. The ratio increases as the temperature increases.

39.37. **IDENTIFY:** Energy radiates at the rate $H = Ae\sigma T^4$.

SET UP: The surface area of a cylinder of radius r and length l is $A = 2\pi rl$.

EXECUTE: **(a)** $T = \left(\dfrac{H}{Ae\sigma}\right)^{1/4} = \left(\dfrac{100 \text{ W}}{2\pi(0.20 \times 10^{-3} \text{ m})(0.30 \text{ m})(0.26)(5.671 \times 10^{-8} \text{ W/m}^2 \cdot \text{K}^4)}\right)^{1/4}$.

$T = 2.06 \times 10^3$ K.

(b) $\lambda_m T = 2.90 \times 10^{-3}$ m $\cdot$ K; $\lambda_m = 1410$ nm.

EVALUATE: **(c)** λ_m is in the infrared. The incandescent bulb is not a very efficient source of visible light because much of the emitted radiation is in the infrared.

39.41. **IDENTIFY:** Since the stars radiate as blackbodies, they obey the Stefan-Boltzmann law and Wien's displacement law.

SET UP: The Stefan-Boltzmann law says that the intensity of the radiation is $I = \sigma T^4$, so the total radiated power is $P = \sigma A T^4$. Wien's displacement law tells us that the peak-intensity wavelength is $\lambda_m = (\text{constant})/T$.

EXECUTE: **(a)** The hot and cool stars radiate the same total power, so the Stefan-Boltzmann law gives
$\sigma A_h T_h^4 = \sigma A_c T_c^4 \Rightarrow 4\pi R_h^2 T_h^4 = 4\pi R_c^2 T_c^4 = 4\pi (3R_h)^2 T_c^4 \Rightarrow T_h^4 = 9T^4 \Rightarrow T_h = T\sqrt{3} = 1.7T$, rounded to two significant digits.

(b) Using Wien's law, we take the ratio of the wavelengths, giving

$$\frac{\lambda_m(\text{hot})}{\lambda_m(\text{cool})} = \frac{T_c}{T_h} = \frac{T}{T\sqrt{3}} = \frac{1}{\sqrt{3}} = 0.58, \text{ rounded to two significant digits.}$$

EVALUATE: Although the hot star has only 1/9 the surface area of the cool star, its absolute temperature has to be only 1.7 times as great to radiate the same amount of energy.

39.47. **IDENTIFY:** The Heisenberg uncertainty principle tells us that $\Delta x \Delta p_x \geq \hbar/2$.

SET UP: We can treat the standard deviation as a direct measure of uncertainty.

EXECUTE: Here $\Delta x \Delta p_x = (1.2 \times 10^{-10}\text{ m})(3.0 \times 10^{-25}\text{ kg}\cdot\text{m/s}) = 3.6 \times 10^{-35}\text{ J}\cdot\text{s}$, but $\hbar/2 = 5.28 \times 10^{-35}\text{ J}\cdot\text{s}$. Therefore $\Delta x \Delta p_x < \hbar/2$, so the claim is *not valid*.

EVALUATE: The uncertainty product $\Delta x \Delta p_x$ must increase by a factor of about 1.5 to become consistent with the Heisenberg uncertainty principle.

39.51. **(a) IDENTIFY and SET UP:** Apply the equation for the reduced mass, $m_r = \frac{m_1 m_2}{m_1 + m_2} = \frac{207 m_e m_p}{207 m_e + m_p}$, where m_e denotes the electron mass.

EXECUTE: $m_r = \frac{207(9.109 \times 10^{-31}\text{ kg})(1.673 \times 10^{-27}\text{ kg})}{207(9.109 \times 10^{-31}\text{ kg}) + 1.673 \times 10^{-27}\text{ kg}} = 1.69 \times 10^{-28}\text{ kg}.$

(b) IDENTIFY: In the energy equation $E_n = -\frac{1}{\varepsilon_0^2} \frac{me^4}{8n^2 h^2}$, replace $m = m_e$ by m_r: $E_n = -\frac{1}{\varepsilon_0^2} \frac{m_r e^4}{8n^2 h^2}$.

SET UP: Write as $E_n = \left(\frac{m_r}{m_H}\right)\left(-\frac{1}{\varepsilon_0^2} \frac{m_H e^4}{8n^2 h^2}\right)$, since we know that $\frac{1}{\varepsilon_0^2} \frac{m_H e^4}{8h^2} = 13.60$ eV. Here m_H denotes the reduced mass for the hydrogen atom; $m_H = (0.99946)(9.109 \times 10^{-31}\text{ kg}) = 9.104 \times 10^{-31}\text{ kg}.$

EXECUTE: $E_n = \left(\frac{m_r}{m_H}\right)\left(-\frac{13.60\text{ eV}}{n^2}\right).$

$E_1 = \frac{1.69 \times 10^{-28}\text{ kg}}{9.104 \times 10^{-31}\text{ kg}}(-13.60\text{ eV}) = 186(-13.60\text{ eV}) = -2.53\text{ keV}.$

(c) SET UP: From part (b), $E_n = \left(\frac{m_r}{m_H}\right)\left(-\frac{R_H ch}{n^2}\right)$, where $R_H = 1.097 \times 10^7$ m^{-1} is the Rydberg constant for the hydrogen atom. Use this result in $\frac{hc}{\lambda} = E_i - E_f$ to find an expression for $1/\lambda$. The initial level for the transition is the $n_i = 2$ level and the final level is the $n_f = 1$ level.

EXECUTE: $\frac{hc}{\lambda} = \frac{m_r}{m_H}\left[-\frac{R_H ch}{n_i^2} - \left(-\frac{R_H ch}{n_f^2}\right)\right].$

$\frac{1}{\lambda} = \frac{m_r}{m_H} R_H \left(\frac{1}{n_f^2} - \frac{1}{n_i^2}\right).$

$\frac{1}{\lambda} = \frac{1.69 \times 10^{-28}\text{ kg}}{9.104 \times 10^{-31}\text{ kg}}(1.097 \times 10^7\text{ m}^{-1})\left(\frac{1}{1^2} - \frac{1}{2^2}\right) = 1.527 \times 10^9\text{ m}^{-1}.$

$\lambda = 0.655$ nm.

EVALUATE: From Example 39.6, the wavelength of the radiation emitted in this transition in hydrogen is 122 nm. The wavelength for muonium is $\frac{m_H}{m_r} = 5.39 \times 10^{-3}$ times this. The reduced mass for hydrogen is very close to the electron mass because the electron mass is much less then the proton mass: $m_p/m_e = 1836$. The muon mass is $207 m_e = 1.886 \times 10^{-28}$ kg. The proton is only about 10 times more massive than the muon, so the reduced mass is somewhat smaller than the muon mass. The muon-proton atom has much more strongly bound energy levels and much shorter wavelengths in its spectrum than for hydrogen.

39.55. **(a) IDENTIFY** and **SET UP:** The photon energy is given to the electron in the atom. Some of this energy overcomes the binding energy of the atom and what is left appears as kinetic energy of the free electron. Apply $hf = E_f - E_i$, the energy given to the electron in the atom when a photon is absorbed.

EXECUTE: The energy of one photon is $\frac{hc}{\lambda} = \frac{(6.626 \times 10^{-34} \text{ J} \cdot \text{s})(2.998 \times 10^8 \text{ m/s})}{85.5 \times 10^{-9} \text{ m}}$.

$\frac{hc}{\lambda} = 2.323 \times 10^{-18}$ J$(1 \text{ eV}/1.602 \times 10^{-19}$ J$) = 14.50$ eV.

The final energy of the electron is $E_f = E_i + hf$. In the ground state of the hydrogen atom the energy of the electron is $E_i = -13.60$ eV. Thus $E_f = -13.60$ eV $+ 14.50$ eV $= 0.90$ eV.

EVALUATE: **(b)** At thermal equilibrium a few atoms will be in the $n = 2$ excited levels, which have an energy of -13.6 eV$/4 = -3.40$ eV, 10.2 eV greater than the energy of the ground state. If an electron with $E = -3.40$ eV gains 14.5 eV from the absorbed photon, it will end up with 14.5 eV $- 3.4$ eV $= 11.1$ eV of kinetic energy.

39.59. **IDENTIFY:** The energy of the peak-intensity photons must be equal to the energy difference between the $n = 1$ and the $n = 4$ states. Wien's law allows us to calculate what the temperature of the blackbody must be for it to radiate with its peak intensity at this wavelength.

SET UP: In the Bohr model, the energy of an electron in shell n is $E_n = -\frac{13.6 \text{ eV}}{n^2}$, and Wien's displacement law is $\lambda_m = \frac{2.90 \times 10^{-3} \text{ m} \cdot \text{K}}{T}$. The energy of a photon is $E = hf = hc/\lambda$.

EXECUTE: First find the energy (ΔE) that a photon would need to excite the atom. The ground state of the atom is $n = 1$ and the third excited state is $n = 4$. This energy is the *difference* between the two energy levels. Therefore $\Delta E = (-13.6 \text{ eV})\left(\frac{1}{4^2} - \frac{1}{1^2}\right) = 12.8$ eV. Now find the wavelength of the photon having this amount of energy. $hc/\lambda = 12.8$ eV and

$$\lambda = (4.136 \times 10^{-15} \text{ eV} \cdot \text{s})(3.00 \times 10^8 \text{ m/s})/(12.8 \text{ eV}) = 9.73 \times 10^{-8} \text{ m}.$$

Now use Wien's law to find the temperature. $T = (0.00290 \text{ m} \cdot \text{K})/(9.73 \times 10^{-8} \text{ m}) = 2.98 \times 10^4$ K.

EVALUATE: This temperature is well above ordinary room temperatures, which is why hydrogen atoms are not in excited states during everyday conditions.

39.65. **IDENTIFY:** The electrons behave like waves and produce a double-slit interference pattern after passing through the slits.

SET UP: The first angle at which destructive interference occurs is given by $d \sin\theta = \lambda/2$. The de Broglie wavelength of each of the electrons is $\lambda = h/mv$.

EXECUTE: **(a)** First find the wavelength of the electrons. For the first dark fringe, we have $d \sin\theta = \lambda/2$, which gives $(1.25 \text{ nm})(\sin 18.0°) = \lambda/2$, and $\lambda = 0.7725$ nm. Now solve the de Broglie wavelength equation for the speed of the electron:

$$v = \frac{h}{m\lambda} = \frac{6.626 \times 10^{-34} \text{ J} \cdot \text{s}}{(9.11 \times 10^{-31} \text{ kg})(0.7725 \times 10^{-9} \text{ m})} = 9.42 \times 10^5 \text{ m/s}$$

which is about 0.3% the speed of light, so they are *nonrelativistic*.

(b) Energy conservation gives $eV = \frac{1}{2}mv^2$ and

$$V = mv^2/2e = (9.11 \times 10^{-31} \text{ kg})(9.42 \times 10^5 \text{ m/s})^2/[2(1.60 \times 10^{-19} \text{ C})] = 2.52 \text{ V}.$$

EVALUATE: The de Broglie wavelength of the electrons is comparable to the separation of the slits.

39.67. IDENTIFY: Both the electrons and photons behave like waves and exhibit single-slit diffraction after passing through their respective slits.

SET UP: The energy of the photon is $E = hc/\lambda$ and the de Broglie wavelength of the electron is $\lambda = h/mv = h/p$. Destructive interference for a single slit first occurs when $a\sin\theta = \lambda$.

EXECUTE: (a) For the photon: $\lambda = hc/E$ and $a\sin\theta = \lambda$. Since the a and θ are the same for the photons and electrons, they must both have the same wavelength. Equating these two expressions for λ gives $a\sin\theta = hc/E$. For the electron, $\lambda = h/p = \dfrac{h}{\sqrt{2mK}}$ and $a\sin\theta = \lambda$. Equating these two expressions for λ gives $a\sin\theta = \dfrac{h}{\sqrt{2mK}}$. Equating the two expressions for $a\sin\theta$ gives $hc/E = \dfrac{h}{\sqrt{2mK}}$, which gives $E = c\sqrt{2mK} = (4.05 \times 10^{-7} \text{ J}^{1/2})\sqrt{K}$.

(b) $\dfrac{E}{K} = \dfrac{c\sqrt{2mK}}{K} = \sqrt{\dfrac{2mc^2}{K}}$. Since $v \ll c$, $mc^2 > K$, so the square root is >1. Therefore $E/K > 1$, meaning that the photon has more energy than the electron.

EVALUATE: When a photon and a particle have the same wavelength, the photon has more energy than the particle.

39.73. (a) IDENTIFY and SET UP: $\Delta x \Delta p_x \geq \hbar/2$. Estimate Δx as $\Delta x \approx 5.0 \times 10^{-15}$ m.

EXECUTE: Then the minimum allowed Δp_x is $\Delta p_x \approx \dfrac{\hbar}{2\Delta x} = \dfrac{1.055 \times 10^{-34} \text{ J} \cdot \text{s}}{2(5.0 \times 10^{-15} \text{ m})} = 1.1 \times 10^{-20}$ kg·m/s.

(b) IDENTIFY and SET UP: Assume $p \approx 1.1 \times 10^{-20}$ kg·m/s. Use $E^2 = (mc^2)^2 + (pc)^2$ to calculate E, and then $K = E - mc^2$.

EXECUTE: $E = \sqrt{(mc^2)^2 + (pc)^2}$. $mc^2 = (9.109 \times 10^{-31} \text{ kg})(2.998 \times 10^8 \text{ m/s})^2 = 8.187 \times 10^{-14}$ J.

$pc = (1.1 \times 10^{-20} \text{ kg} \cdot \text{m/s})(2.998 \times 10^8 \text{ m/s}) = 3.165 \times 10^{-12}$ J.

$E = \sqrt{(8.187 \times 10^{-14} \text{ J})^2 + (3.165 \times 10^{-12} \text{ J})^2} = 3.166 \times 10^{-12}$ J.

$K = E - mc^2 = 3.166 \times 10^{-12}$ J $- 8.187 \times 10^{-14}$ J $= 3.084 \times 10^{-12}$ J $\times (1 \text{ eV}/1.602 \times 10^{-19} \text{ J}) = 19$ MeV.

(c) IDENTIFY and SET UP: The Coulomb potential energy for a pair of point charges is given by $U = -kq_1q_2/r$. The proton has charge $+e$ and the electron has charge $-e$.

EXECUTE: $U = -\dfrac{ke^2}{r} = -\dfrac{(8.988 \times 10^9 \text{ N} \cdot \text{m}^2/\text{C}^2)(1.602 \times 10^{-19} \text{ C})^2}{5.0 \times 10^{-15} \text{ m}} = -4.6 \times 10^{-14}$ J $= -0.29$ MeV.

EVALUATE: The kinetic energy of the electron required by the uncertainty principle would be much larger than the magnitude of the negative Coulomb potential energy. The total energy of the electron would be large and positive and the electron could not be bound within the nucleus.

39.77. IDENTIFY: Assume both the x rays and electrons are at normal incidence and scatter from the surface plane of the crystal, so the maxima are located by $d\sin\theta = m\lambda$, where d is the separation between adjacent atoms in the surface plane.

SET UP: Let primed variables refer to the electrons. $\lambda' = \dfrac{h}{p'} = \dfrac{h}{\sqrt{2mE'}}$.

EXECUTE: $\sin\theta' = \dfrac{\lambda'}{\lambda}\sin\theta$, and $\lambda' = (h/p') = (h/\sqrt{2mE'})$, and so $\theta' = \arcsin\left(\dfrac{h}{\lambda\sqrt{2mE'}}\sin\theta\right)$.

39-8   Chapter 39

$$\theta' = \arcsin\left(\frac{(6.63\times10^{-34}\text{ J}\cdot\text{s})(\sin 35.8°)}{(3.00\times10^{-11}\text{ m})\sqrt{2(9.11\times10^{-31}\text{ kg})(4.50\times10^{+3}\text{ eV})(1.60\times10^{-19}\text{ J/eV})}}\right) = 20.9°.$$

EVALUATE: The x rays and electrons have different wavelengths and the $m = 1$ maxima occur at different angles.

39.81. (a) **IDENTIFY** and **SET UP:** $U = A|x|$. $F_x = -dU/dx$ relates force and potential. The slope of the function $A|x|$ is not continuous at $x = 0$ so we must consider the regions $x > 0$ and $x < 0$ separately.

EXECUTE: For $x > 0, |x| = x$ so $U = Ax$ and $F = -\dfrac{d(Ax)}{dx} = -A$. For $x < 0, |x| = -x$ so $U = -Ax$ and $F = -\dfrac{d(-Ax)}{dx} = +A$. We can write this result as $F = -A|x|/x$, valid for all x except for $x = 0$.

(b) **IDENTIFY** and **SET UP:** Use the uncertainty principle, expressed as $\Delta p \Delta x \approx h$, and as in Problem 39.80 estimate Δp by p and Δx by x. Use this to write the energy E of the particle as a function of x. Find the value of x that gives the minimum E and then find the minimum E.

EXECUTE: $E = K + U = \dfrac{p^2}{2m} + A|x|$.

$px \approx h$, so $p \approx h/x$.

Then $E \approx \dfrac{h^2}{2mx^2} + A|x|$.

For $x > 0, E = \dfrac{h^2}{2mx^2} + Ax$.

To find the value of x that gives minimum E set $\dfrac{dE}{dx} = 0$.

$0 = \dfrac{-2h^2}{2mx^3} + A$.

$x^3 = \dfrac{h^2}{mA}$ and $x = \left(\dfrac{h^2}{mA}\right)^{1/3}$.

With this x the minimum E is

$E = \dfrac{h^2}{2m}\left(\dfrac{mA}{h^2}\right)^{2/3} + A\left(\dfrac{h^2}{mA}\right)^{1/3} = \dfrac{1}{2}h^{2/3}m^{-1/3}A^{2/3} + h^{2/3}m^{-1/3}A^{2/3}$.

$E = \dfrac{3}{2}\left(\dfrac{h^2 A^2}{m}\right)^{1/3}$.

EVALUATE: The potential well is shaped like a V. The larger A is, the steeper the slope of U and the smaller the region to which the particle is confined and the greater is its energy. Note that for the x that minimizes E, $2K = U$.

39.83. **IDENTIFY** and **SET UP:** For hydrogen-like atoms (1 electron and Z protons), the energy levels are $E_n = (-13.60\text{ eV})Z^2/n^2$, with $n = 1$ for the ground state. The energy of a photon is $E = hc/\lambda$.

EXECUTE: (a) The least energy absorbed is between the ground state ($n = 1$) and the $n = 2$ state, which gives the longest wavelength. So $\Delta E_{1\to 2} = hc/\lambda$. Using the energy levels for this atom, we have

$(-13.6\text{ eV})Z^2\left(\dfrac{1}{2^2} - \dfrac{1}{1^2}\right) = \dfrac{hc}{\lambda}$ $\to$ $(10.20\text{ eV})Z^2 = hc/\lambda$. Solving Z gives

$$Z = \sqrt{\frac{hc}{(10.20 \text{ eV})\lambda}} = \sqrt{\frac{(4.136\times10^{-15} \text{ eV}\cdot\text{s})(2.998\times10^8 \text{ m/s})}{(10.20 \text{ eV})(13.56\times10^{-9} \text{ m})}} = 3.0.$$

(b) The next shortest wavelength is between the $n = 3$ and $n = 1$ states.

$$\Delta E_{1\to 3} = (-13.6 \text{ eV})(3)^2\left(\frac{1}{3^2} - \frac{1}{1^2}\right) = \frac{hc}{\lambda}.$$

Solving for λ gives

$\lambda = (4.136\times10^{-15} \text{ eV}\cdot\text{s})(2.998\times10^8 \text{ m/s})/(108.8 \text{ eV}) = 11.40$ nm.

(c) By energy conservation, $E_{\text{photon}} = E_{\text{ionization}} + K_{\text{el}}$. The ionization energy is the minimum energy to completely remove an electron from the atom, which is from the $n = 1$ state to the $n = \infty$ state. Therefore $E_{\text{ionization}} = (13.60 \text{ eV})Z^2 = (13.60 \text{ eV})(9)$. Therefore the kinetic energy of the electron is $K_{\text{el}} = E_{\text{photon}} - E_{\text{ionization}} = hc/\lambda - E_{\text{ionization}}$.

$K_{\text{el}} = (4.136\times10^{-15} \text{ eV}\cdot\text{s})(2.998\times10^8 \text{ m/s})/(6.78\times10^{-9} \text{ m}) - (13.60 \text{ eV})(9) = 60.5$ eV.

EVALUATE: The energy levels for a $Z = 3$ atom are 9 times as great as for the comparable energy levels in hydrogen, so the wavelengths of the absorbed light are much shorter than they would be for comparable transitions in hydrogen.

39.89. **IDENTIFY:** Calculate the accelerating potential V need to produce a helium ion with a wavelength of 0.1 pm to see if that potential lies within the range of 10-50 kV.

SET UP: The de Broglie wavelength of the helium ion is $\lambda = h/p$, so $p = h/\lambda$. By energy conservation, $K = eV = p^2/2m$.

EXECUTE: Combining the above equations gives

$$eV = K = p^2/2m = \frac{(h/\lambda)^2}{2m}, \text{ so } V = \frac{(h/\lambda)^2}{2me}.$$

$$V = \frac{\left[(6.626\times10^{-34} \text{ J}\cdot\text{s})/(0.1\times10^{-12} \text{ m})\right]^2}{2(7300)(9.11\times10^{-31} \text{ kg})(1.60\times10^{-19} \text{ C})} = 2.1\times10^4 \text{ V} = 21 \text{ kV}.$$

This voltage is within the 10-50 kV range, so choice (a) is correct.

EVALUATE: A large voltage is required because the desired wavelength is small.

39.91. **IDENTIFY and SET UP:** The ion loses 0.2 MeV/μm, and its energy can be determined only to within 6 keV. Call x the minimum difference in thickness that can be discerned, and realize that 0.2 MeV = 200 keV.

EXECUTE: $(0.2 \text{ MeV}/\mu\text{m})x = 6$ keV. Solving for x gives $x = (6 \text{ keV})/(200 \text{ keV}/\mu\text{m}) = 0.03$ μm, which makes choice (a) the correct one.

EVALUATE: Greater precision in determining the energy of the ion would allow one to discern smaller features.

40

QUANTUM MECHANICS I: WAVE FUNCTIONS

40.3. **IDENTIFY:** Use the wave function from Example 40.1.

SET UP: $|\Psi(x,t)|^2 = 2|A|^2\{1+\cos[(k_2-k_1)x-(\omega_2-\omega_1)t]\}$. $k_2 = 3k_1 = 3k$. $\omega = \dfrac{\hbar k^2}{2m}$, so $\omega_2 = 9\omega_1 = 9\omega$.

$|\Psi(x,t)|^2 = 2|A|^2[1+\cos(2kx-8\omega t)]$.

EXECUTE: **(a)** At $t = 2\pi/\omega$, $|\Psi(x,t)|^2 = 2|A|^2[1+\cos(2kx-16\pi)]$. $|\Psi(x,t)|^2$ is maximum for $\cos(2kx-16\pi) = 1$. This happens for $2kx - 16\pi = 0, 2\pi, \ldots$. Smallest positive x where $|\Psi(x,t)|^2$ is a maximum is $x = \dfrac{8\pi}{k}$.

(b) From the result of part (a), $v_{av} = \dfrac{8\pi/k}{2\pi/\omega} = \dfrac{4\omega}{k}$. $v_{av} = \dfrac{\omega_2-\omega_1}{k_2-k_1} = \dfrac{8\omega}{2k} = \dfrac{4\omega}{k}$.

EVALUATE: The two expressions agree.

40.11. **IDENTIFY:** An electron in the lowest energy state in this box must have the same energy as it would in the ground state of hydrogen.

SET UP: The energy of the nth level of an electron in a box is $E_n = \dfrac{nh^2}{8mL^2}$.

EXECUTE: An electron in the ground state of hydrogen has an energy of -13.6 eV, so find the width corresponding to an energy of $E_1 = 13.6$ eV. Solving for L gives

$$L = \dfrac{h}{\sqrt{8mE_1}} = \dfrac{(6.626\times10^{-34}\text{ J}\cdot\text{s})}{\sqrt{8(9.11\times10^{-31}\text{ kg})(13.6\text{ eV})(1.602\times10^{-19}\text{ J/eV})}} = 1.66\times10^{-10}\text{ m}.$$

EVALUATE: This width is of the same order of magnitude as the diameter of a Bohr atom with the electron in the K shell.

40.13. **IDENTIFY and SET UP:** The equation $E_n = \dfrac{n^2h^2}{8mL^2}$ gives the energy levels. Use this to obtain an expression for $E_2 - E_1$ and use the value given for this energy difference to solve for L.

EXECUTE: Ground state energy is $E_1 = \dfrac{h^2}{8mL^2}$; first excited state energy is $E_2 = \dfrac{4h^2}{8mL^2}$. The energy separation between these two levels is $\Delta E = E_2 - E_1 = \dfrac{3h^2}{8mL^2}$. This gives

$$L = h\sqrt{\dfrac{3}{8m\Delta E}} = L = 6.626\times10^{-34}\text{ J}\cdot\text{s}\sqrt{\dfrac{3}{8(9.109\times10^{-31}\text{ kg})(3.0\text{ eV})(1.602\times10^{-19}\text{ J/1 eV})}}$$

$= 6.1\times10^{-10}$ m $= 0.61$ nm.

EVALUATE: This energy difference is typical for an atom and L is comparable to the size of an atom.

40.17. IDENTIFY and SET UP: For the $n = 2$ first excited state the normalized wave function is given by the equation $\psi_2(x) = \sqrt{\frac{2}{L}} \sin\left(\frac{2\pi x}{L}\right)$. $|\psi_2(x)|^2 dx = \frac{2}{L}\sin^2\left(\frac{2\pi x}{L}\right) dx$. Examine $|\psi_2(x)|^2 dx$ and find where it is zero and where it is maximum.

EXECUTE: (a) $|\psi_2|^2 dx = 0$ implies $\sin\left(\frac{2\pi x}{L}\right) = 0$.

$\frac{2\pi x}{L} = m\pi$, $m = 0, 1, 2, \ldots$; $x = m(L/2)$.

For $m = 0$, $x = 0$; for $m = 1$, $x = L/2$; for $m = 2$, $x = L$.

The probability of finding the particle is zero at $x = 0$, $L/2$, and L.

(b) $|\psi_2|^2 dx$ is maximum when $\sin\left(\frac{2\pi x}{L}\right) = \pm 1$.

$\frac{2\pi x}{L} = m(\pi/2)$, $m = 1, 3, 5, \ldots$; $x = m(L/4)$.

For $m = 1$, $x = L/4$; for $m = 3$, $x = 3L/4$.

The probability of finding the particle is largest at $x = L/4$ and $3L/4$.

EVALUATE: (c) The answers to part (a) correspond to the zeros of $|\psi|^2$ shown in Figure 40.12 in the textbook and the answers to part (b) correspond to the two values of x where $|\psi|^2$ in the figure is maximum.

40.19. IDENTIFY and SET UP: $\lambda = \frac{h}{p} = \frac{h}{\sqrt{2mE}}$. The energy of the electron in level n is given By the equation $E_n = \frac{n^2 h^2}{8mL^2}$.

EXECUTE: (a) $E_1 = \frac{h^2}{8mL^2} \Rightarrow \lambda_1 = \frac{h}{\sqrt{2mh^2/8mL^2}} = 2L = 2(3.0 \times 10^{-10} \text{ m}) = 6.0 \times 10^{-10}$ m. The wavelength is twice the width of the box. $p_1 = \frac{h}{\lambda_1} = \frac{(6.63 \times 10^{-34} \text{ J} \cdot \text{s})}{6.0 \times 10^{-10} \text{ m}} = 1.1 \times 10^{-24}$ kg·m/s.

(b) $E_2 = \frac{4h^2}{8mL^2} \Rightarrow \lambda_2 = L = 3.0 \times 10^{-10}$ m. The wavelength is the same as the width of the box.

$p_2 = \frac{h}{\lambda_2} = 2p_1 = 2.2 \times 10^{-24}$ kg·m/s.

(c) $E_3 = \frac{9h^2}{8mL^2} \Rightarrow \lambda_3 = \frac{2}{3}L = 2.0 \times 10^{-10}$ m. The wavelength is two-thirds the width of the box.

$p_3 = 3p_1 = 3.3 \times 10^{-24}$ kg·m/s.

EVALUATE: In each case the wavelength is an integer multiple of $\lambda/2$. In the nth state, $p_n = np_1$.

40.23. IDENTIFY: The energy of the photon is the energy given to the electron.

SET UP: Since $U_0 = 6E_{1\text{-IDW}}$ we can use the result $E_1 = 0.625 E_{1\text{-IDW}}$ from Section 40.4. When the electron is outside the well it has potential energy U_0, so the minimum energy that must be given to the electron is $U_0 - E_1 = 5.375 E_{1\text{-IDW}}$.

EXECUTE: The maximum wavelength of the photon would be

$\lambda = \frac{hc}{U_0 - E_1} = \frac{hc}{(5.375)(h^2/8mL^2)} = \frac{8mL^2 c}{(5.375)h} = \frac{8(9.11 \times 10^{-31} \text{ kg})(1.50 \times 10^{-9} \text{ m})^2 (3.00 \times 10^8 \text{ m/s})}{(5.375)(6.63 \times 10^{-34} \text{ J} \cdot \text{s})}$

$= 1.38 \times 10^{-6}$ m.

EVALUATE: This photon is in the infrared. The wavelength of the photon decreases when the width of the well decreases.

40.27. **IDENTIFY** and **SET UP:** The probability is $T = Ge^{-2\kappa L}$, with $G = 16\dfrac{E}{U_0}\left(1 - \dfrac{E}{U_0}\right)$ and $\kappa = \dfrac{\sqrt{2m(U_0 - E)}}{\hbar}$.

$E = 32$ eV, $U_0 = 41$ eV, $L = 0.25 \times 10^{-9}$ m. Calculate T.

EXECUTE: (a) $G = 16\dfrac{E}{U_0}\left(1 - \dfrac{E}{U_0}\right) = 16\dfrac{32}{41}\left(1 - \dfrac{32}{41}\right) = 2.741$.

$\kappa = \dfrac{\sqrt{2m(U_0 - E)}}{\hbar}$.

$\kappa = \dfrac{\sqrt{2(9.109 \times 10^{-31}\text{ kg})(41\text{ eV} - 32\text{ eV})(1.602 \times 10^{-19}\text{ J/eV})}}{1.055 \times 10^{-34}\text{ J} \cdot \text{s}} = 1.536 \times 10^{10}\text{ m}^{-1}$.

$T = Ge^{-2\kappa L} = (2.741)e^{-2(1.536 \times 10^{10}\text{ m}^{-1})(0.25 \times 10^{-9}\text{ m})} = 2.741e^{-7.68} = 0.0013$.

(b) The only change is the mass m, which appears in κ.

$\kappa = \dfrac{\sqrt{2m(U_0 - E)}}{\hbar}$.

$\kappa = \dfrac{\sqrt{2(1.673 \times 10^{-27}\text{ kg})(41\text{ eV} - 32\text{ eV})(1.602 \times 10^{-19}\text{ J/eV})}}{1.055 \times 10^{-34}\text{ J} \cdot \text{s}} = 6.584 \times 10^{11}\text{ m}^{-1}$.

Then $T = Ge^{-2\kappa L} = (2.741)e^{-2(6.584 \times 10^{11}\text{ m}^{-1})(0.25 \times 10^{-9}\text{ m})} = 2.741e^{-392.2} = 10^{-143}$.

EVALUATE: The more massive proton has a much smaller probability of tunneling than the electron does.

40.31. **IDENTIFY** and **SET UP:** Use $\lambda = h/p$, where $K = p^2/2m$ and $E = K + U$.

EXECUTE: $\lambda = h/p = h/\sqrt{2mK}$, so $\lambda\sqrt{K}$ is constant. $\lambda_1\sqrt{K_1} = \lambda_2\sqrt{K_2}$; λ_1 and K_1 are for $x > L$ where $K_1 = 2U_0$ and λ_2 and K_2 are for $0 < x < L$ where $K_2 = E - U_0 = U_0$.

$\dfrac{\lambda_1}{\lambda_2} = \sqrt{\dfrac{K_2}{K_1}} = \sqrt{\dfrac{U_0}{2U_0}} = \dfrac{1}{\sqrt{2}}$.

EVALUATE: When the particle is passing over the barrier its kinetic energy is less and its wavelength is larger.

40.35. **IDENTIFY:** We can model the molecule as a harmonic oscillator. The energy of the photon is equal to the energy difference between the two levels of the oscillator.

SET UP: The energy of a photon is $E_\gamma = hf = hc/\lambda$, and the energy levels of a harmonic oscillator are given by $E_n = (n + \tfrac{1}{2})\hbar\sqrt{\dfrac{k'}{m}} = (n + \tfrac{1}{2})\hbar\omega$.

EXECUTE: (a) The photon's energy is $E_\gamma = \dfrac{hc}{\lambda} = \dfrac{(6.63 \times 10^{-34}\text{ J} \cdot \text{s})(3.00 \times 10^8\text{ m/s})}{5.8 \times 10^{-6}\text{ m}} = 0.21$ eV.

(b) The transition energy is $\Delta E = E_{n+1} - E_n = \hbar\omega = \hbar\sqrt{\dfrac{k'}{m}}$, which gives $\dfrac{2\pi\hbar c}{\lambda} = \hbar\sqrt{\dfrac{k'}{m}}$. Solving for k', we get $k' = \dfrac{4\pi^2 c^2 m}{\lambda^2} = \dfrac{4\pi^2(3.00 \times 10^8\text{ m/s})^2(5.6 \times 10^{-26}\text{ kg})}{(5.8 \times 10^{-6}\text{ m})^2} = 5{,}900$ N/m.

EVALUATE: This would be a rather strong spring in the physics lab.

40.37. **IDENTIFY:** The photon energy equals the transition energy for the atom.

SET UP: According to the energy level equation $E_n = (n + \tfrac{1}{2})\hbar\omega$, the energy released during the transition between two adjacent levels is twice the ground state energy $E_3 - E_2 = \hbar\omega = 2E_0 = 11.2$ eV.

EXECUTE: For a photon of energy E,

$E = hf \Rightarrow \lambda = \dfrac{c}{f} = \dfrac{hc}{E} = \dfrac{(6.63 \times 10^{-34}\text{ J} \cdot \text{s})(3.00 \times 10^8\text{ m/s})}{(11.2\text{ eV})(1.60 \times 10^{-19}\text{ J/eV})} = 111$ nm.

EVALUATE: This photon is in the ultraviolet.

40.39. **IDENTIFY** and **SET UP:** Use the energies given in $E_n = (n+\tfrac{1}{2})\hbar\omega$ to solve for the amplitude A and maximum speed v_{max} of the oscillator. Use these to estimate Δx and Δp_x and compute the uncertainty product $\Delta x \Delta p_x$.

EXECUTE: The total energy of a Newtonian oscillator is given by $E = \tfrac{1}{2} k' A^2$ where k' is the force constant and A is the amplitude of the oscillator. Set this equal to the energy $E = (n+\tfrac{1}{2})\hbar\omega$ of an excited level that has quantum number n, where $\omega = \sqrt{\dfrac{k'}{m}}$, and solve for A: $\tfrac{1}{2} k' A^2 = (n+\tfrac{1}{2})\hbar\omega$. $A = \sqrt{\dfrac{(2n+1)\hbar\omega}{k'}}$.

The total energy of the Newtonian oscillator can also be written as $E = \tfrac{1}{2} m v_{max}^2$. Set this equal to $E = (n+\tfrac{1}{2})\hbar\omega$ and solve for v_{max}: $\tfrac{1}{2} m v_{max}^2 = (n+\tfrac{1}{2})\hbar\omega$. $v_{max} = \sqrt{\dfrac{(2n+1)\hbar\omega}{m}}$. Thus the maximum linear momentum of the oscillator is $p_{max} = m v_{max} = \sqrt{(2n+1)\hbar m \omega}$. Now assume that $A/\sqrt{2}$ represents the uncertainty Δx in position and that $p_{max}/\sqrt{2}$ is the corresponding uncertainty Δp_x in momentum. Then the uncertainty product is

$$\Delta x \Delta p_x = \left(\dfrac{1}{\sqrt{2}}\sqrt{\dfrac{(2n+1)\hbar\omega}{k'}}\right)\left(\dfrac{1}{\sqrt{2}}\sqrt{(2n+1)\hbar m \omega}\right) = \dfrac{(2n+1)\hbar\omega}{2}\sqrt{\dfrac{m}{k'}} = \dfrac{(2n+1)\hbar\omega}{2}\left(\dfrac{1}{\omega}\right) = (2n+1)\dfrac{\hbar}{2}.$$

EVALUATE: For $n=0$ this gives $\Delta x \Delta p_x = \hbar/2$, in agreement with the result derived in Section 40.5. The uncertainty product $\Delta x \Delta p_x$ increases with n.

40.41. **IDENTIFY:** We model the atomic vibration in the crystal as a harmonic oscillator.

SET UP: The energy levels of a harmonic oscillator are given by $E_n = (n+\tfrac{1}{2})\hbar\sqrt{\dfrac{k'}{m}} = (n+\tfrac{1}{2})\hbar\omega$.

EXECUTE: **(a)** The ground state energy of a simple harmonic oscillator is

$$E_0 = \tfrac{1}{2}\hbar\omega = \tfrac{1}{2}\hbar\sqrt{\dfrac{k'}{m}} = \dfrac{(1.055\times 10^{-34}\text{ J}\cdot\text{s})}{2}\sqrt{\dfrac{12.2\text{ N/m}}{3.82\times 10^{-26}\text{ kg}}} = 9.43\times 10^{-22}\text{ J} = 5.89\times 10^{-3}\text{ eV}.$$

(b) $E_4 - E_3 = \hbar\omega = 2E_0 = 0.0118$ eV, so $\lambda = \dfrac{hc}{E} = \dfrac{(6.63\times 10^{-34}\text{ J}\cdot\text{s})(3.00\times 10^8\text{ m/s})}{1.88\times 10^{-21}\text{ J}} = 106\ \mu\text{m}$.

(c) $E_{n+1} - E_n = \hbar\omega = 2E_0 = 0.0118$ eV.

EVALUATE: These energy differences are much smaller than those due to electron transitions in the hydrogen atom.

40.43. **IDENTIFY:** We know the wave function of a particle in a box.

SET UP and EXECUTE: **(a)** $\Psi(x,t) = \dfrac{1}{\sqrt{2}}\psi_1(x)e^{-iE_1 t/\hbar} + \dfrac{1}{\sqrt{2}}\psi_3(x)e^{-iE_3 t/\hbar}$.

$\Psi^*(x,t) = \dfrac{1}{\sqrt{2}}\psi_1(x)e^{+iE_1 t/\hbar} + \dfrac{1}{\sqrt{2}}\psi_3(x)e^{+iE_3 t/\hbar}$.

$|\Psi(x,t)|^2 = \tfrac{1}{2}[\psi_1^2 + \psi_3^2 + \psi_1\psi_3(e^{i(E_3-E_1)t/\hbar} + e^{-i(E_3-E_1)t/\hbar})] = \tfrac{1}{2}\left[\psi_1^2 + \psi_3^2 + 2\psi_1\psi_3\cos\left(\dfrac{[E_3-E_1]t}{\hbar}\right)\right]$.

$\psi_1 = \sqrt{\dfrac{2}{L}}\sin\left(\dfrac{\pi x}{L}\right)$. $\psi_3 = \sqrt{\dfrac{2}{L}}\sin\left(\dfrac{3\pi x}{L}\right)$. $E_3 = \dfrac{9\pi^2\hbar^2}{2mL^2}$ and $E_1 = \dfrac{\pi^2\hbar^2}{2mL^2}$, so $E_3 - E_1 = \dfrac{4\pi^2\hbar^2}{mL^2}$.

$|\Psi(x,t)|^2 = \dfrac{1}{L}\left[\sin^2\left(\dfrac{\pi x}{L}\right) + \sin^2\left(\dfrac{3\pi x}{L}\right) + 2\sin\left(\dfrac{\pi x}{L}\right)\sin\left(\dfrac{3\pi x}{L}\right)\cos\left(\dfrac{4\pi^2 \hbar t}{mL^2}\right)\right]$. At $x = L/2$,

$\sin\left(\dfrac{\pi x}{L}\right) = \sin\left(\dfrac{\pi}{2}\right) = 1$. $\sin\left(\dfrac{3\pi x}{L}\right) = \sin\left(\dfrac{3\pi}{2}\right) = -1$. $|\Psi(x,t)|^2 = \dfrac{2}{L}\left[1 - \cos\left(\dfrac{4\pi^2 \hbar t}{mL^2}\right)\right]$.

(b) $\omega_{osc} = \dfrac{E_3 - E_1}{\hbar} = \dfrac{4\pi^2 \hbar}{mL^2}$.

EVALUATE: Note that $\Delta E = \hbar\omega$.

40.47. IDENTIFY and SET UP: The energy levels are given by the equation $E_n = \dfrac{n^2 h^2}{8mL^2}$. Calculate ΔE for the transition and set $\Delta E = hc/\lambda$, the energy of the photon.

EXECUTE: (a) Ground level, $n = 1$, $E_1 = \dfrac{h^2}{8mL^2}$. First excited level, $n = 2$, $E_2 = \dfrac{4h^2}{8mL^2}$. The transition energy is $\Delta E = E_2 - E_1 = \dfrac{3h^2}{8mL^2}$. Set the transition energy equal to the energy hc/λ of the emitted photon.

This gives $\dfrac{hc}{\lambda} = \dfrac{3h^2}{8mL^2}$. $\lambda = \dfrac{8mcL^2}{3h} = \dfrac{8(9.109 \times 10^{-31}\text{ kg})(2.998 \times 10^8 \text{ m/s})(4.18 \times 10^{-9}\text{ m})^2}{3(6.626 \times 10^{-34}\text{ J}\cdot\text{s})}$.

$\lambda = 1.92 \times 10^{-5}$ m $= 19.2$ μm.

(b) Second excited level has $n = 3$ and $E_3 = \dfrac{9h^2}{8mL^2}$. The transition energy is

$\Delta E = E_3 - E_2 = \dfrac{9h^2}{8mL^2} - \dfrac{4h^2}{8mL^2} = \dfrac{5h^2}{8mL^2}$. $\dfrac{hc}{\lambda} = \dfrac{5h^2}{8mL^2}$ so $\lambda = \dfrac{8mcL^2}{5h} = \dfrac{3}{5}(19.2\text{ μm}) = 11.5$ μm.

EVALUATE: The energy spacing between adjacent levels increases with n, and this corresponds to a shorter wavelength and more energetic photon in part (b) than in part (a).

40.49. IDENTIFY and SET UP: The normalized wave function for the $n = 2$ first excited level is

$\psi_2 = \sqrt{\dfrac{2}{L}}\sin\left(\dfrac{2\pi x}{L}\right)$. $P = |\psi(x)|^2\,dx$ is the probability that the particle will be found in the interval x to $x + dx$.

EXECUTE: (a) $x = L/4$.

$\psi(x) = \sqrt{\dfrac{2}{L}}\sin\left(\left(\dfrac{2\pi}{L}\right)\left(\dfrac{L}{4}\right)\right) = \sqrt{\dfrac{2}{L}}\sin\left(\dfrac{\pi}{2}\right) = \sqrt{\dfrac{2}{L}}$.

$P = (2/L)\,dx$.

(b) $x = L/2$.

$\psi(x) = \sqrt{\dfrac{2}{L}}\sin\left(\left(\dfrac{2\pi}{L}\right)\left(\dfrac{L}{2}\right)\right) = \sqrt{\dfrac{2}{L}}\sin(\pi) = 0$.

$P = 0$.

(c) $x = 3L/4$.

$\psi(x) = \sqrt{\dfrac{2}{L}}\sin\left(\left(\dfrac{2\pi}{L}\right)\left(\dfrac{3L}{4}\right)\right) = \sqrt{\dfrac{2}{L}}\sin\left(\dfrac{3\pi}{2}\right) = -\sqrt{\dfrac{2}{L}}$.

$P = (2/L)\,dx$.

EVALUATE: Our results are consistent with the $n = 2$ part of Figure 40.12 in the textbook. $|\psi|^2$ is zero at the center of the box and is symmetric about this point.

40.53. **IDENTIFY:** Carry out the calculations that are specified in the problem.

SET UP: For a free particle, $U(x) = 0$ so Schrödinger's equation becomes $\dfrac{d^2\psi(x)}{dx^2} = -\dfrac{2m}{\hbar^2} E\psi(x)$.

EXECUTE: **(a)** The graph is given in Figure 40.53.

(b) For $x < 0$: $\psi(x) = e^{+\kappa x}$. $\dfrac{d\psi(x)}{dx} = \kappa e^{+\kappa x}$. $\dfrac{d^2\psi(x)}{dx^2} = \kappa^2 e^{+\kappa x}$. So $\kappa^2 = -\dfrac{2m}{\hbar^2}E \Rightarrow E = -\dfrac{\hbar^2 \kappa^2}{2m}$.

(c) For $x > 0$: $\psi(x) = e^{-\kappa x}$. $\dfrac{d\psi(x)}{dx} = -\kappa e^{-\kappa x}$. $\dfrac{d^2\psi(x)}{dx^2} = \kappa^2 e^{-\kappa x}$. So again $\kappa^2 = -\dfrac{2m}{\hbar^2}E \Rightarrow E = \dfrac{-\hbar^2 \kappa^2}{2m}$.

Parts (b) and (c) show $\psi(x)$ satisfies the Schrödinger's equation, provided $E = \dfrac{-\hbar^2 \kappa^2}{2m}$.

EVALUATE: **(d)** $\dfrac{d\psi(x)}{dx}$ is discontinuous at $x = 0$. (That is, it is negative for $x > 0$ and positive for $x < 0$.) Therefore, this ψ is not an acceptable wave function; $d\psi/dx$ must be continuous everywhere, except where $U \to \infty$.

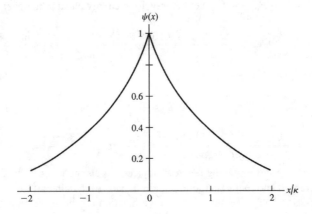

Figure 40.53

40.61. **IDENTIFY and SET UP:** The energy levels for an infinite potential well are $E_n = \dfrac{n^2 h^2}{8mL^2} = E_1 n^2$. The energy of the absorbed photon is equal to the energy difference between the levels. The energy of a photon is $E = hf = hc/\lambda$, so $\Delta E = hf = E_{n+1} - E_n$.

EXECUTE: **(a)** For the first transition, we have $hf_1 = E_1(n^2 - 1^2)$, and for the second transition we have $hf_2 = E_1[(n+1)^2 - 1^2]$. Taking the ratio of these two equations gives

$\dfrac{hf_2}{hf_1} = \dfrac{16.9}{9.0} = \dfrac{(n+1)^2 - 1}{n^2 - 1} = \dfrac{n^2 + 2n + 1 - 1}{n^2 - 1} = \dfrac{n^2 + 2n}{n^2 - 1}$.

Rearranging and collecting terms gives the quadratic equation $n^2\left(\dfrac{16.9}{9.0} - 1\right) - 2n - \dfrac{16.9}{9.0} = 0$. Using the quadratic formula and taking the positive root gives $n = 3.0$, so $n = 3$. Therefore the transitions are from the $n = 3$ and $n = 4$ levels to the $n = 1$ level.

(b) Using the $3 \to 1$ transition with $f_1 = 9.0 \times 10^{14}$ Hz, we have $hf_1 = (h^2/8mL^2)(3^2 - 1^2) = h^2/mL^2$.

$L = \sqrt{\dfrac{h}{f_1 m}} = \sqrt{\dfrac{6.626 \times 10^{-34}\ \text{J} \cdot \text{s}}{(9.0 \times 10^{14}\ \text{Hz})(9.109 \times 10^{-31}\ \text{kg})}} = 9.0 \times 10^{-10}$ m $= 0.90$ nm.

(c) The longest wavelength is for the smallest energy, and that would be for a transition between $n = 1$ and $n = 2$ levels. Comparing the $1 \to 3$ transition and the $1 \to 2$ transition, we have

$$\frac{hf_{1 \to 2}}{hf_{1 \to 3}} = \frac{E_1(2^2 - 1^2)}{E_1(3^2 - 1^2)} \quad \to \quad f_{1 \to 2} = \frac{3}{8} f_{1 \to 3} = \frac{3}{8}(9.0 \times 10^{14} \text{ Hz}).$$

$$\lambda = c/f = \frac{3.00 \times 10^8 \text{ m/s}}{\frac{3}{8}(9.0 \times 10^{14} \text{ Hz})} = 890 \text{ nm}.$$

EVALUATE: This wavelength is too long to be visible light. The wavelength of the 9.0×10^{14} Hz photon is 333 nm, which is too short to be visible, as is the 16.9×10^{14} Hz photon. So none of these photons will be visible.

40.63. **IDENTIFY** and **SET UP:** The transmission coefficient T is equal to 1 when the width L of the barrier is $L = \frac{1}{2}\lambda, \lambda, \frac{3}{2}\lambda, 2\lambda, \ldots = n\lambda/2$, where $n = 1, 2, 3, \ldots$, and where λ is the de Broglie wavelength of the electron, given by $\lambda = h/p$. The total energy of the electron is $E = U + K$, and $K = p^2/2m$.

EXECUTE: From the condition on λ, we have $\lambda_n = 2L/n$. Therefore $\lambda = h/p = 2L/n$, which gives $p = nh/2L$. The kinetic energy is $K = p^2/2m$, so $K_n = \frac{p^2}{2m} = \frac{\left(\frac{nh}{2L}\right)^2}{2m} = n^2\left(\frac{h^2}{8mL^2}\right) = n^2 K_1$. The three lowest values of K are for $n = 1, 2$, and 3.

$K_1 = h^2/8mL^2 = (6.626 \times 10^{-34} \text{ J} \cdot \text{s})^2/[8(9.11 \times 10^{-31} \text{ kg})(1.8 \times 10^{-10} \text{ m})^2] = 1.86 \times 10^{-18}$ J = 11.6 eV.
The total energy is $E = K + U$, so $E_1 = K_1 + U = 11.6$ eV + 10 eV = 22 eV.
For the $n = 2$ state, we have
$K_2 = 2^2 K_1 = 4(11.6 \text{ eV}) = 46.4$ eV, so $E_2 = 46.4$ eV + 10 eV = 56 eV.
For the $n = 3$ state, we have
$K_3 = 3^2 K_1 = 9(11.6 \text{ eV}) = 104.4$ eV, so $E_3 = 104.4$ eV + 10 eV = 114 eV, which rounds to 110 eV.
EVALUATE: We cannot use Eq. (40.42) because T is not small.

40.69. **IDENTIFY** and **SET UP:** The energy levels are $E_{m,n} = (m^2 + n^2)(\pi^2 \hbar^2)/2ML^2$. $\Delta E = \frac{hc}{\lambda}$ for the photon emitted during a transition through energy difference ΔE.

EXECUTE: Since $\Delta E \propto 1/M$ and $\Delta E = \frac{hc}{\lambda}$, it follows that $\lambda \propto 1/M$. Since $\lambda_1 > \lambda_2$, it is true that $M_1 > M_2$, which makes choice (a) correct.

EVALUATE: When M is large, the energy difference between states is small, so the energy of an emitted (or absorbed) photon is small, so its wavelength is large.

QUANTUM MECHANICS II: ATOMIC STRUCTURE

41.3. **IDENTIFY:** The energy of the photon is equal to the energy difference between the states. We can use this energy to calculate its wavelength.

SET UP: $E_{1,1,1} = \dfrac{3\pi^2 \hbar^2}{2mL^2}$. $E_{2,2,1} = \dfrac{9\pi^2 \hbar^2}{2mL^2}$. $\Delta E = \dfrac{3\pi^2 \hbar^2}{mL^2}$. $\Delta E = \dfrac{hc}{\lambda}$.

EXECUTE: $\Delta E = \dfrac{3\pi^2 (1.055 \times 10^{-34} \text{ J} \cdot \text{s})^2}{(9.109 \times 10^{-31} \text{ kg})(8.00 \times 10^{-11} \text{ m})^2} = 5.653 \times 10^{-17}$ J. $\Delta E = \dfrac{hc}{\lambda}$ gives

$\lambda = \dfrac{hc}{\Delta E} = \dfrac{(6.626 \times 10^{-34} \text{ J} \cdot \text{s})(2.998 \times 10^8 \text{ m/s})}{5.653 \times 10^{-17} \text{ J}} = 3.51 \times 10^{-9}$ m $= 3.51$ nm.

EVALUATE: This wavelength is much shorter than that of visible light.

41.5. **IDENTIFY:** A particle is in a three-dimensional box. At what planes is its probability function zero?

SET UP: $|\psi_{2,2,1}|^2 = \left(\dfrac{L}{2}\right)^3 \left(\sin^2 \dfrac{2\pi x}{L}\right)\left(\sin^2 \dfrac{2\pi y}{L}\right)\left(\sin^2 \dfrac{\pi z}{L}\right)$.

EXECUTE: $|\psi_{2,2,1}|^2 = 0$ for $\dfrac{2\pi x}{L} = 0, \pi, 2\pi, \ldots$. $x = 0$ and $x = L$ correspond to walls of the box. $x = \dfrac{L}{2}$ is the other plane where $|\psi_{2,2,1}|^2 = 0$. Similarly, $|\psi_{2,2,1}|^2 = 0$ on the plane $y = \dfrac{L}{2}$. The $\sin^2 \dfrac{\pi z}{L}$ factor is zero only on the walls of the box. Therefore, for this state $|\psi_{2,2,1}|^2 = 0$ on the following two planes other than walls of the box: $x = \dfrac{L}{2}$ and $y = \dfrac{L}{2}$.

$|\psi_{2,1,1}|^2 = \left(\dfrac{L}{2}\right)^3 \left(\sin^2 \dfrac{2\pi x}{L}\right)\left(\sin^2 \dfrac{\pi y}{L}\right)\left(\sin^2 \dfrac{\pi z}{L}\right)$ is zero only on one plane $(x = L/2)$ other than the walls of the box.

$|\psi_{1,1,1}|^2 = \left(\dfrac{L}{2}\right)^3 \left(\sin^2 \dfrac{\pi x}{L}\right)\left(\sin^2 \dfrac{\pi y}{L}\right)\left(\sin^2 \dfrac{\pi z}{L}\right)$ is zero only on the walls of the box; for this state there are zero additional planes.

EVALUATE: For comparison, (2,1,1) has two nodal planes, (2,1,1) has one nodal and (1,1,1) has no nodal planes. The number of nodal planes increases as the energy of the state increases.

41.9. **IDENTIFY and SET UP:** The magnitude of the orbital angular momentum L is related to the quantum number l by Eq. (41.22): $L = \sqrt{l(l+1)}\hbar$, $l = 0, 1, 2, \ldots$

EXECUTE: $l(l+1) = \left(\dfrac{L}{\hbar}\right)^2 = \left(\dfrac{4.716 \times 10^{-34} \text{ kg} \cdot \text{m}^2/\text{s}}{1.055 \times 10^{-34} \text{ J} \cdot \text{s}}\right)^2 = 20$.

And then $l(l+1) = 20$ gives that $l = 4$.

EVALUATE: l must be integer.

41.15. **IDENTIFY:** For the 5g state, $l = 4$, which limits the other quantum numbers.

SET UP: $m_l = 0, \pm 1, \pm 2, \ldots, \pm l$. g means $l = 4$. $\cos\theta = L_z/L$, with $L = \sqrt{l(l+1)}\,\hbar$ and $L_z = m_l \hbar$.

EXECUTE: (a) There are eighteen 5g states: $m_l = 0, \pm 1, \pm 2, \pm 3, \pm 4$, with $m_s = \pm\tfrac{1}{2}$ for each.

(b) The largest θ is for the most negative m_l. $L = 2\sqrt{5}\,\hbar$. The most negative L_z is $L_z = -4\hbar$.

$\cos\theta = \dfrac{-4\hbar}{2\sqrt{5}\,\hbar}$ and $\theta = 153.4°$.

(c) The smallest θ is for the largest positive m_l, which is $m_l = +4$. $\cos\theta = \dfrac{4\hbar}{2\sqrt{5}\,\hbar}$ and $\theta = 26.6°$.

EVALUATE: The minimum angle between $\vec{L}$ and the z-axis is for $m_l = +l$ and for that m_l, $\cos\theta = \dfrac{l}{\sqrt{l(l+1)}}$.

41.19. **IDENTIFY:** Apply $\Delta U = \mu_B B$.

SET UP: For a 3p state, $l = 1$ and $m_l = 0, \pm 1$.

EXECUTE: (a) $B = \dfrac{U}{\mu_B} = \dfrac{(2.71 \times 10^{-5}\text{ eV})}{(5.79 \times 10^{-5}\text{ eV/T})} = 0.468$ T.

(b) Three: $m_l = 0, \pm 1$.

EVALUATE: The $m_l = +1$ level will be highest in energy and the $m_l = -1$ level will be lowest. The $m_l = 0$ level is unaffected by the magnetic field.

41.21. **IDENTIFY and SET UP:** The interaction energy between an external magnetic field and the orbital angular momentum of the atom is given by $U = m_l \mu_B B$. The energy depends on m_l with the most negative m_l value having the lowest energy.

EXECUTE: (a) For the 5g level, $l = 4$ and there are $2l + 1 = 9$ different m_l states. The 5g level is split into 9 levels by the magnetic field.

(b) Each m_l level is shifted in energy an amount given by $U = m_l \mu_B B$. Adjacent levels differ in m_l by one, so $\Delta U = \mu_B B$.

$\mu_B = \dfrac{e\hbar}{2m} = \dfrac{(1.602 \times 10^{-19}\text{ C})(1.055 \times 10^{-34}\text{ J}\cdot\text{s})}{2(9.109 \times 10^{-31}\text{ kg})} = 9.277 \times 10^{-24}\text{ A}\cdot\text{m}^2$

$\Delta U = \mu_B B = (9.277 \times 10^{-24}\text{ A/m}^2)(0.600\text{ T}) = 5.566 \times 10^{-24}\text{ J}(1\text{ eV}/1.602 \times 10^{-19}\text{ J}) = 3.47 \times 10^{-5}\text{ eV}$

(c) The level of highest energy is for the largest m_l, which is $m_l = l = 4$; $U_4 = 4\mu_B B$. The level of lowest energy is for the smallest m_l, which is $m_l = -l = -4$; $U_{-4} = -4\mu_B B$. The separation between these two levels is $U_4 - U_{-4} = 8\mu_B B = 8(3.47 \times 10^{-5}\text{ eV}) = 2.78 \times 10^{-4}\text{ eV}$.

EVALUATE: The energy separations are proportional to the magnetic field. The energy of the $n = 5$ level in the absence of the external magnetic field is $(-13.6\text{ eV})/5^2 = -0.544$ eV, so the interaction energy with the magnetic field is much less than the binding energy of the state.

41.25. **IDENTIFY and SET UP:** The interaction energy is $U = -\vec{\mu}\cdot\vec{B}$, with μ_z given by

$\mu_z = -(2.00232)\left(\dfrac{e}{2m}\right)S_z$.

EXECUTE: $U = -\vec{\mu}\cdot\vec{B} = +\mu_z B$, since the magnetic field is in the negative z-direction.

$\mu_z = -(2.00232)\left(\dfrac{e}{2m}\right)S_z$, so $U = -(2.00232)\left(\dfrac{e}{2m}\right)S_z B$.

$S_z = m_s \hbar$, so $U = -2.00232\left(\dfrac{e\hbar}{2m}\right)m_s B$.

$\dfrac{e\hbar}{2m} = \mu_B = 5.788 \times 10^{-5}$ eV/T.

$U = -2.00232\, \mu_B m_s B.$

The $m_s = +\tfrac{1}{2}$ level has lower energy.

$\Delta U = U(m_s = -\tfrac{1}{2}) - U(m_s = +\tfrac{1}{2}) = -2.00232\, \mu_B B(-\tfrac{1}{2}-(+\tfrac{1}{2})) = +2.00232\, \mu_B B.$

$\Delta U = +2.00232(5.788\times 10^{-5}\ \text{eV/T})(1.45\ \text{T}) = 1.68\times 10^{-4}\ \text{eV}.$

EVALUATE: The interaction energy with the electron spin is the same order of magnitude as the interaction energy with the orbital angular momentum for states with $m_l \neq 0$. But a $1s$ state has $l = 0$ and $m_l = 0$, so there is no orbital magnetic interaction.

41.29. IDENTIFY: Write out the electron configuration for ground-state beryllium.
SET UP: Beryllium has 4 electrons.
EXECUTE: (a) $1s^2 2s^2$.

(b) $1s^2 2s^2 2p^6 3s^2$. $Z = 12$ and the element is magnesium.

(c) $1s^2 2s^2 2p^6 3s^2 3p^6 4s^2$. $Z = 20$ and the element is calcium.

EVALUATE: Beryllium, calcium, and magnesium are all in the same column of the periodic table.

41.33. IDENTIFY and SET UP: Use the exclusion principle to determine the ground-state electron configuration, as in Table 41.3. Estimate the energy by estimating Z_{eff}, taking into account the electron screening of the nucleus.

EXECUTE: (a) $Z = 7$ for nitrogen so a nitrogen atom has 7 electrons. N^{2+} has 5 electrons: $1s^2 2s^2 2p$.

(b) $Z_{\text{eff}} = 7 - 4 = 3$ for the $2p$ level.

$E_n = -\left(\dfrac{Z_{\text{eff}}^2}{n^2}\right)(13.6\ \text{eV}) = -\dfrac{3^2}{2^2}(13.6\ \text{eV}) = -30.6\ \text{eV}.$

(c) $Z = 15$ for phosphorus so a phosphorus atom has 15 electrons. P^{2+} has 13 electrons: $1s^2 2s^2 2p^6 3s^2 3p$.

(d) $Z_{\text{eff}} = 15 - 12 = 3$ for the $3p$ level.

$E_n = -\left(\dfrac{Z_{\text{eff}}^2}{n^2}\right)(13.6\ \text{eV}) = -\dfrac{3^2}{3^2}(13.6\ \text{eV}) = -13.6\ \text{eV}.$

EVALUATE: In these ions there is one electron outside filled subshells, so it is a reasonable approximation to assume full screening by these inner-subshell electrons.

41.37. IDENTIFY and SET UP: Apply $E_{K\alpha} \cong (Z-1)^2(10.2\ \text{eV})$. $E = hf$ and $c = f\lambda$.

EXECUTE: (a) $Z = 20$: $f = (2.48\times 10^{15}\ \text{Hz})(20-1)^2 = 8.95\times 10^{17}\ \text{Hz}.$

$E = hf = (4.14\times 10^{-15}\ \text{eV}\cdot\text{s})(8.95\times 10^{17}\ \text{Hz}) = 3.71\ \text{keV}.$ $\lambda = \dfrac{c}{f} = \dfrac{3.00\times 10^8\ \text{m/s}}{8.95\times 10^{17}\ \text{Hz}} = 3.35\times 10^{-10}\ \text{m}.$

(b) $Z = 27$: $f = 1.68\times 10^{18}\ \text{Hz}.$ $E = 6.96\ \text{keV}.$ $\lambda = 1.79\times 10^{-10}\ \text{m}.$

(c) $Z = 48$: $f = 5.48\times 10^{18}\ \text{Hz}$, $E = 22.7\ \text{keV}$, $\lambda = 5.47\times 10^{-11}\ \text{m}.$

EVALUATE: f and E increase and λ decreases as Z increases.

41.41. IDENTIFY: Calculate the probability of finding a particle in a given region within a cubical box.

(a) **SET UP and EXECUTE:** The box has volume L^3. The specified cubical space has volume $(L/4)^3$. Its fraction of the total volume is $\dfrac{1}{64} = 0.0156.$

(b) SET UP and EXECUTE: $P = \left(\dfrac{2}{L}\right)^3 \left[\displaystyle\int_0^{L/4} \sin^2 \dfrac{\pi x}{L} dx\right]\left[\displaystyle\int_0^{L/4} \sin^2 \dfrac{\pi y}{L} dy\right]\left[\displaystyle\int_0^{L/4} \sin^2 \dfrac{\pi z}{L} dz\right].$

From Example 41.1, each of the three integrals equals $\dfrac{L}{8} - \dfrac{L}{4\pi} = \dfrac{1}{2}\left(\dfrac{L}{2}\right)\left(\dfrac{1}{2} - \dfrac{1}{\pi}\right).$

$P = \left(\dfrac{2}{L}\right)^3 \left(\dfrac{L}{2}\right)^3 \left(\dfrac{1}{2}\right)^3 \left(\dfrac{1}{2} - \dfrac{1}{\pi}\right)^3 = 7.50 \times 10^{-4}.$

EVALUATE: Note that this is the cube of the probability of finding the particle anywhere between $x = 0$ and $x = L/4$. This probability is much less that the fraction of the total volume that this space represents. In this quantum state the probability distribution function is much larger near the center of the box than near its walls.

(c) SET UP and EXECUTE: $|\psi_{2,1,1}|^2 = \left(\dfrac{L}{2}\right)^3 \left(\sin^2 \dfrac{2\pi x}{L}\right)\left(\sin^2 \dfrac{\pi y}{L}\right)\left(\sin^2 \dfrac{\pi z}{L}\right).$

$P = \left(\dfrac{2}{L}\right)^3 \left[\displaystyle\int_0^{L/4} \sin^2 \dfrac{2\pi x}{L} dx\right]\left[\displaystyle\int_0^{L/4} \sin^2 \dfrac{\pi y}{L} dy\right]\left[\displaystyle\int_0^{L/4} \sin^2 \dfrac{\pi z}{L} dz\right].$

$\left[\displaystyle\int_0^{L/4}\sin^2\dfrac{\pi y}{L}dy\right] = \left[\displaystyle\int_0^{L/4}\sin^2\dfrac{\pi z}{L}dz\right] = \dfrac{L}{2}\left(\dfrac{1}{2}\right)\left(\dfrac{1}{2}-\dfrac{1}{\pi}\right).\quad \displaystyle\int_0^{L/4}\sin^2\dfrac{2\pi x}{L}dx = \dfrac{L}{8}.$

$P = \left(\dfrac{2}{L}\right)^3 \left(\dfrac{L}{2}\right)^2 \left(\dfrac{1}{2}\right)^2 \left(\dfrac{1}{2}-\dfrac{1}{\pi}\right)^2 \left(\dfrac{L}{8}\right) = 2.06\times 10^{-3}.$

EVALUATE: This is about a factor of three larger than the probability when the particle is in the ground state.

41.45. IDENTIFY: Find solutions to Eq. (41.5).

SET UP: $\omega_1 = \sqrt{k_1'/m},\ \omega_2 = \sqrt{k_2'/m}.$ Let $\psi_{n_x}(x)$ be a solution of Eq. (40.44) with $E_{n_x} = (n_x + \tfrac{1}{2})\hbar\omega_1,\ \psi_{n_y}(y)$ be a similar solution, and let $\psi_{n_z}(z)$ be a solution of Eq. (40.44) but with z as the independent variable instead of x, and energy $E_{n_z} = (n_z + \tfrac{1}{2})\hbar\omega_2.$

EXECUTE: **(a)** As in Problem 41.44, look for a solution of the form $\psi(x,y,z) = \psi_{n_x}(x)\psi_{n_y}(y)\psi_{n_z}(z).$

Then, $-\dfrac{\hbar^2}{2m}\dfrac{\partial^2\psi}{\partial x^2} = (E_{n_x} - \tfrac{1}{2}k_1'x^2)\psi$ with similar relations for $\dfrac{\partial^2\psi}{\partial y^2}$ and $\dfrac{\partial^2\psi}{\partial z^2}.$ Adding,

$-\dfrac{\hbar^2}{2m}\left(\dfrac{\partial^2\psi}{\partial x^2} + \dfrac{\partial^2\psi}{\partial y^2} + \dfrac{\partial^2\psi}{\partial z^2}\right) = (E_{n_x} + E_{n_y} + E_{n_z} - \tfrac{1}{2}k_1'x^2 - \tfrac{1}{2}k_1'y^2 - \tfrac{1}{2}k_2'z^2)\psi$

$= (E_{n_x} + E_{n_y} + E_{n_z} - U)\psi = (E - U)\psi$

where the energy E is $E = E_{n_x} + E_{n_y} + E_{n_z} = \hbar\left[(n_x + n_y + 1)\omega_1^2 + (n_z + \tfrac{1}{2})\omega_2^2\right]$, with $n_x, n_y,$ and n_z all nonnegative integers.

(b) The ground level corresponds to $n_x = n_y = n_z = 0$, and $E = \hbar(\omega_1 + \tfrac{1}{2}\omega_2).$ The first excited level corresponds to $n_x = n_y = 0$ and $n_z = 1$, since $\omega_1^2 > \omega_2^2$, and $E = \hbar(\omega_1 + \tfrac{3}{2}\omega_2).$

(c) There is only one set of quantum numbers for both the ground state and the first excited state.

EVALUATE: For the isotropic oscillator of Problem 41.44 there are three states for the first excited level but only one for the anisotropic oscillator.

41.53. IDENTIFY: Use Figure 41.6 in the textbook to relate θ_L to L_z and L: $\cos\theta_L = \dfrac{L_z}{L}$ so $\theta_L = \arccos\left(\dfrac{L_z}{L}\right).$

(a) SET UP: The smallest angle $(\theta_L)_{\min}$ is for the state with the largest L and the largest L_z. This is the state with $l = n-1$ and $m_l = l = n-1.$

EXECUTE: $L_z = m_l\hbar = (n-1)\hbar$.

$L = \sqrt{l(l+1)}\hbar = \sqrt{(n-1)n}\,\hbar$.

$(\theta_L)_{\min} = \arccos\left(\dfrac{(n-1)\hbar}{\sqrt{(n-1)n}\,\hbar}\right) = \arccos\left(\dfrac{(n-1)}{\sqrt{(n-1)n}}\right) = \arccos\left(\sqrt{\dfrac{n-1}{n}}\right) = \arccos(\sqrt{1-1/n})$.

EVALUATE: Note that $(\theta_L)_{\min}$ approaches $0°$ as $n \to \infty$.

(b) SET UP: The largest angle $(\theta_L)_{\max}$ is for $l = n-1$ and $m_l = -l = -(n-1)$.

EXECUTE: A similar calculation to part (a) yields $(\theta_L)_{\max} = \arccos(-\sqrt{1-1/n})$

EVALUATE: Note that $(\theta_L)_{\max}$ approaches $180°$ as $n \to \infty$.

41.55. IDENTIFY: The presence of an external magnetic field shifts the energy levels up or down, depending upon the value of m_l.

SET UP: The energy difference due to the magnetic field is $\Delta E = \mu_B B$ and the energy of a photon is $E = hc/\lambda$.

EXECUTE: For the p state, $m_l = 0$ or ± 1, and for the s state $m_l = 0$. Between any two adjacent lines, $\Delta E = \mu_B B$. Since the change in the wavelength $(\Delta\lambda)$ is very small, the energy change (ΔE) is also very small, so we can use differentials. $E = hc/\lambda$. $|dE| = \dfrac{hc}{\lambda^2}d\lambda$ and $\Delta E = \dfrac{hc\Delta\lambda}{\lambda^2}$. Since $\Delta E = \mu_B B$, we get

$\mu_B B = \dfrac{hc\Delta\lambda}{\lambda^2}$ and $B = \dfrac{hc\Delta\lambda}{\mu_B \lambda^2}$.

$B = (4.136\times 10^{-15}\text{ eV}\cdot\text{s})(3.00\times 10^8\text{ m/s})(0.0462\text{ nm})/(5.788\times 10^{-5}\text{ eV/T})(575.050\text{ nm})^2 = 3.00$ T.

EVALUATE: Even a strong magnetic field produces small changes in the energy levels, and hence in the wavelengths of the emitted light.

41.59. IDENTIFY and SET UP: m_s can take on four different values: $m_s = -\tfrac{3}{2}, -\tfrac{1}{2}, +\tfrac{1}{2}, +\tfrac{3}{2}$. Each nlm_l state can have four electrons, each with one of the four different m_s values. Apply the exclusion principle to determine the electron configurations.

EXECUTE: (a) For a filled $n = 1$ shell, the electron configuration would be $1s^4$; four electrons and $Z = 4$. For a filled $n = 2$ shell, the electron configuration would be $1s^4 2s^4 2p^{12}$; twenty electrons and $Z = 20$.

(b) Sodium has $Z = 11$; eleven electrons. The ground-state electron configuration would be $1s^4 2s^4 2p^3$.

EVALUATE: The chemical properties of each element would be very different.

41.65. IDENTIFY: The inner electrons shield much of the nuclear charge from the outer electrons, and this shielding affects the energy levels compared to the hydrogen levels. The atom behaves like hydrogen with an effective charge in the nucleus.

SET UP: The ionization energies are $E_n = \dfrac{Z_{\text{eff}}^2}{n^2}(13.6\text{ eV})$.

EXECUTE: (a) Make the conversions requested using the following conversion factor.

$E(\text{eV/atom}) = E(\text{kJ/mol})\left(\dfrac{1\text{ mol}}{6.02214\times 10^{23}\text{ atoms}}\right)\left(\dfrac{1000\text{ J}}{1\text{ kJ}}\right)\left(\dfrac{1\text{ eV}}{1.602\times 10^{-19}\text{ J}}\right) = 0.010364 E(\text{kJ/mol})$.

Li: $E = (0.010364)(520.2\text{ kJ/mol}) = 5.391$ eV
Na: $E = (0.010364)(495.8\text{ kJ/mol}) = 5.139$ eV
K: $E = (0.010364)(418.8\text{ kJ/mol}) = 4.341$ eV
Rb: $E = (0.010364)(403.0\text{ kJ/mol}) = 4.177$ eV
Cs: $E = (0.010364)(375.7\text{ kJ/mol}) = 3.894$ eV
Fr: $E = (0.010364)(380\text{ kJ/mol}) = 3.9$ eV.

(b) From the periodic chart in the Appendix, we get the following information.
Li: $Z = 3, n = 2$
Na: $Z = 11, n = 3$
K: $Z = 19, n = 4$
Rb: $Z = 37, n = 5$
Cs: $Z = 55, n = 6$
Fr: $Z = 87, n = 7$.

(c) Use $E_n = \dfrac{Z_{\text{eff}}^2}{n^2}(13.6 \text{ eV})$ and the values of n in part (b) to calculate Z_{eff}. For example, for Li we have

$5.391 \text{ eV} = \dfrac{Z_{\text{eff}}^2}{2^2}(13.6 \text{ eV}) \quad \rightarrow \quad Z_{\text{eff}} = 1.26$, and for Na we have

$5.139 \text{ eV} = \dfrac{Z_{\text{eff}}^2}{3^2}(13.6 \text{ eV}) \quad \rightarrow \quad Z_{\text{eff}} = 1.84$. Doing this for the other atoms gives

Li: $Z_{\text{eff}} = 1.26$
Na: $Z_{\text{eff}} = 1.84$
K: $Z_{\text{eff}} = 2.26$
Rb: $Z_{\text{eff}} = 2.77$
Cs: $Z_{\text{eff}} = 3.21$
Fr: $Z_{\text{eff}} = 3.8$.

EVALUATE: **(d)** We can see that Z_{eff} increases as Z increases. The outer (valence) electron has increasing probability density within the inner shells as Z increases, and therefore it "sees" more of the nuclear charge.

41.67. **IDENTIFY** and **SET UP:** The energy due to the interaction of the electron with the magnetic field is $U = -\mu_z B$. In a transition from the $m_s = -\tfrac{1}{2}$ state to the $m_s = \tfrac{1}{2}$ state, $\Delta U = 2\mu_z B$. This energy difference is the energy of the absorbed photon. The energy of the photon is $E = hf$, and $f\lambda = c$. For an electron, $S_z = \hbar/2$.

EXECUTE: **(a)** First find the frequency for each wavelength in the table with the problem. For example for the first column, $f = c/\lambda = (2.998 \times 10^8 \text{ m/s})/(0.0214 \text{ m}) = 1.40 \times 10^{10}$ Hz. Doing this for all the wavelengths gives the values in the following table. Figure 41.67 shows the graph of f versus B for this data. The slope of the best-fit straight line is 2.84×10^{10} Hz/T.

B (T)	0.51	0.74	1.03	1.52	2.02	2.48	2.97
$f(10^{10}$ Hz$)$	1.401	2.097	2.802	4.199	5.604	7.005	8.398

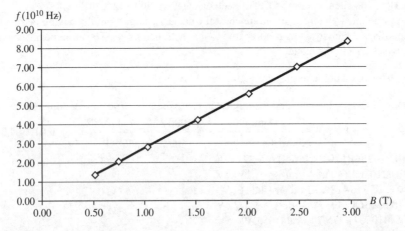

Figure 41.67

(b) The photon energy is $E = hf$, and that is the energy difference between the two levels. So $E = \Delta U = 2\mu_z B = hf$, which gives $f = \left(\dfrac{2\mu_z}{h}\right) B$. The graph of f versus B should be a straight line having slope equal to $2\mu_z/h$. Therefore the magnitude of the spin magnetic moment is
$\mu_z = \tfrac{1}{2} h(\text{slope}) = \tfrac{1}{2} (6.626 \times 10^{-34} \text{ J} \cdot \text{s})(2.84 \times 10^{10} \text{ Hz/T}) = 9.41 \times 10^{-24} \text{ J/T}$.

(c) Using $\gamma = \dfrac{|\mu_z|}{|S_z|} = \dfrac{|\mu_z|}{\hbar/2}$ gives $\gamma = \dfrac{2(9.41 \times 10^{-24} \text{ J/T})}{1.055 \times 10^{-34} \text{ J} \cdot \text{s}} = 1.784 \times 10^{11}$ Hz/T, which rounds to 1.78×10^{11} Hz/T. This gives $\dfrac{\gamma}{e/2m} = \dfrac{2\gamma m}{e} = \dfrac{2(1.784 \times 10^{11} \text{ Hz/T})(9.11 \times 10^{-31} \text{ kg})}{1.602 \times 10^{-19} \text{ C}} = 2.03$.

EVALUATE: Our result of 2.03 for the gyromagnetic ratio for electron spin is in good agreement with the currently accepted value of 2.00232.

41.71. **IDENTIFY** and **SET UP:** Particle density is the number particles divided by the volume they occupy. The distance between particles is 10 times their size, which is 20 μm.

EXECUTE: Think of each atom as being in a cubical box that is 20 μm on each side. The particle density is $(1 \text{ atom})/(20\ \mu\text{m})^3 = (1 \text{ atom})/(20 \times 10^{-4} \text{ cm})^3 = 1.25 \times 10^8$ atoms/cm$^3 \approx 10^8$ atoms/cm^3, choice (b).

EVALUATE: For rubidium, with 85 nucleons, the mass density ρ of these atoms would be $\rho = (85)(1.67 \times 10^{-24} \text{ g})(10^8 \text{ atoms/cm}^3) = 1.4 \times 10^{-14}$ g/cm^3. Ordinary rubidium has a density of 1.53 g/cm^3, so these Rydberg atoms are much farther apart than rubidium atoms under normal conditions.

MOLECULES AND CONDENSED MATTER

42.3. **IDENTIFY:** The energy of the photon is equal to the energy difference between the $l=1$ and $l=2$ states. This energy determines its wavelength.

SET UP: The reduced mass of the molecule is $m_r = \dfrac{m_H m_H}{m_H + m_H} = \tfrac{1}{2} m_H$, its moment of inertia is $I = m_r r_0^2$, the photon energy is $\Delta E = \dfrac{hc}{\lambda}$, and the energy of the state l is $E_l = l(l+1)\dfrac{\hbar^2}{2I}$.

EXECUTE: $I = m_r r_0^2 = \tfrac{1}{2}(1.67\times 10^{-27}\text{ kg})(0.074\times 10^{-9}\text{ m})^2 = 4.57\times 10^{-48}\text{ kg}\cdot\text{m}^2$. Using $E_l = l(l+1)\dfrac{\hbar^2}{2I}$, the energy levels are $E_2 = 6\dfrac{\hbar^2}{2I} = 6\dfrac{(1.055\times 10^{-34}\text{ J}\cdot\text{s})^2}{2(4.57\times 10^{-48}\text{ kg}\cdot\text{m}^2)} = 6(1.218\times 10^{-21}\text{ J}) = 7.307\times 10^{-21}\text{ J}$ and

$E_1 = 2\dfrac{\hbar^2}{2I} = 2(1.218\times 10^{-21}\text{ J}) = 2.436\times 10^{-21}\text{ J}$. $\Delta E = E_2 - E_1 = 4.87\times 10^{-21}\text{ J}$. Using $\Delta E = \dfrac{hc}{\lambda}$ gives

$\lambda = \dfrac{hc}{\Delta E} = \dfrac{(6.626\times 10^{-34}\text{ J}\cdot\text{s})(2.998\times 10^{8}\text{ m/s})}{4.871\times 10^{-21}\text{ J}} = 4.08\times 10^{-5}\text{ m} = 40.8\ \mu\text{m}$.

EVALUATE: This wavelength is much longer than that of visible light.

42.7. **IDENTIFY and SET UP:** The transition energy E and the frequency f of the absorbed photon are related by $E = hf$.

EXECUTE: The energy of the emitted photon is 1.01×10^{-5} eV, and so its frequency and wavelength are

$f = \dfrac{E}{h} = \dfrac{(1.01\times 10^{-5}\text{ eV})(1.60\times 10^{-19}\text{ J/eV})}{(6.63\times 10^{-34}\text{ J}\cdot\text{s})} = 2.44\text{ GHz} = 2440\text{ MHz}$ and

$\lambda = \dfrac{c}{f} = \dfrac{(3.00\times 10^{8}\text{ m/s})}{(2.44\times 10^{9}\text{ Hz})} = 0.123\text{ m}$.

EVALUATE: This frequency corresponds to that given for a microwave oven.

42.13. **IDENTIFY:** The vibrational energy of the molecule is related to its force constant and reduced mass, while the rotational energy depends on its moment of inertia, which in turn depends on the reduced mass.

SET UP: The vibrational energy is $E_n = (n+\tfrac{1}{2})\hbar\omega = (n+\tfrac{1}{2})\hbar\sqrt{\dfrac{k'}{m_r}}$ and the rotational energy is

$E_l = l(l+1)\dfrac{\hbar^2}{2I}$.

EXECUTE: For a vibrational transition, we have $\Delta E_v = \hbar\sqrt{\dfrac{k'}{m_r}}$, so we first need to find m_r. The energy for a rotational transition is $\Delta E_R = \dfrac{\hbar^2}{2I}[2(2+1) - 1(1+1)] = \dfrac{2\hbar^2}{I}$. Solving for I and using the fact that $I = m_r r_0^2$, we have $m_r r_0^2 = \dfrac{2\hbar^2}{\Delta E_R}$, which gives

$$m_r = \frac{2\hbar^2}{r_0^2 \Delta E_R} = \frac{2(1.055\times 10^{-34}\text{ J}\cdot\text{s})(6.583\times 10^{-16}\text{ eV}\cdot\text{s})}{(0.8860\times 10^{-9}\text{ m})^2(8.841\times 10^{-4}\text{ eV})} = 2.0014\times 10^{-28}\text{ kg}.$$

Now look at the vibrational transition to find the force constant.

$$\Delta E_v = \hbar\sqrt{\frac{k'}{m_r}} \Rightarrow k' = \frac{m_r(\Delta E_v)^2}{\hbar^2} = \frac{(2.0014\times 10^{-28}\text{ kg})(0.2560\text{ eV})^2}{(6.583\times 10^{-16}\text{ eV}\cdot\text{s})^2} = 30.27\text{ N/m}.$$

EVALUATE: This would be a rather weak spring in the laboratory.

42.19. **IDENTIFY and SET UP:** The energy ΔE deposited when a photon with wavelength λ is absorbed is $\Delta E = \dfrac{hc}{\lambda}$.

EXECUTE: $\Delta E = \dfrac{hc}{\lambda} = \dfrac{(6.63\times 10^{-34}\text{ J}\cdot\text{s})(3.00\times 10^8\text{ m/s})}{9.31\times 10^{-13}\text{ m}} = 2.14\times 10^{-13}\text{ J} = 1.34\times 10^6\text{ eV}$. So the number of electrons that can be excited to the conduction band is $n = \dfrac{1.34\times 10^6\text{ eV}}{1.12\text{ eV}} = 1.20\times 10^6$ electrons.

EVALUATE: A photon of wavelength

$$\lambda = \frac{hc}{\Delta E} = \frac{(4.13\times 10^{-15}\text{ eV}\cdot\text{s})(3.00\times 10^8\text{ m/s})}{1.12\text{ eV}} = 1.11\times 10^{-6}\text{ m} = 1110\text{ nm}$$ can excite one electron. This photon is in the infrared.

42.23. **IDENTIFY and SET UP:** The electron contribution to the molar heat capacity at constant volume of a metal is $C_V = \left(\dfrac{\pi^2 kT}{2E_F}\right)R$.

EXECUTE: (a) $C_V = \dfrac{\pi^2(1.381\times 10^{-23}\text{ J/K})(300\text{ K})}{2(5.48\text{ eV})(1.602\times 10^{-19}\text{ J/eV})}R = 0.0233R$.

(b) The electron contribution found in part (a) is $0.0233R = 0.194$ J/mol$\cdot$K. This is $0.194/25.3 = 7.67\times 10^{-3} = 0.767\%$ of the total C_V.

EVALUATE: (c) Only a small fraction of C_V is due to the electrons. Most of C_V is due to the vibrational motion of the ions.

42.25. **IDENTIFY:** The probability is given by the Fermi-Dirac distribution.

SET UP: The Fermi-Dirac distribution is $f(E) = \dfrac{1}{e^{(E-E_F)/kT} + 1}$.

EXECUTE: We calculate the value of $f(E)$, where $E = 8.520$ eV, $E_F = 8.500$ eV, $k = 1.38\times 10^{-23}$ J/K $= 8.625\times 10^{-5}$ eV/K, and $T = 20°\text{C} = 293$ K. The result is $f(E) = 0.312 = 31.2\%$.

EVALUATE: Since the energy is close to the Fermi energy, the probability is quite high that the state is occupied by an electron.

42.27. **IDENTIFY:** Use the Fermi-Dirac distribution $f(E) = \dfrac{1}{e^{(E-E_F)/kT} + 1}$. Solve for $E - E_F$.

SET UP: $e^{(E-E_F)/kT} = \dfrac{1}{f(E)} - 1$.

The problem states that $f(E) = 4.4\times 10^{-4}$ for E at the bottom of the conduction band.

EXECUTE: $e^{(E-E_F)/kT} = \dfrac{1}{4.4 \times 10^{-4}} - 1 = 2.272 \times 10^3$.

$E - E_F = kT\ln(2.272 \times 10^3) = (1.3807 \times 10^{-23} \text{ J/T})(300 \text{ K})\ln(2.272 \times 10^3) = 3.201 \times 10^{-20}$ J $= 0.20$ eV.

$E_F = E - 0.20$ eV; the Fermi level is 0.20 eV below the bottom of the conduction band.

EVALUATE: The energy gap between the Fermi level and bottom of the conduction band is large compared to kT at $T = 300$ K and as a result $f(E)$ is small.

42.29. IDENTIFY: Knowing the saturation current of a p-n junction at a given temperature, we want to find the current at that temperature for various voltages.

SET UP: $I = I_S(e^{eV/kT} - 1)$.

EXECUTE: (a) (i) For $V = 1.00$ mV, $\dfrac{eV}{kT} = \dfrac{(1.602 \times 10^{-19} \text{ C})(1.00 \times 10^{-3} \text{ V})}{(1.381 \times 10^{-23} \text{ J/K})(290 \text{ K})} = 0.0400$.

$I = (0.500 \text{ mA})(e^{0.0400} - 1) = 0.0204$ mA.

(ii) For $V = -1.00$ mV, $\dfrac{eV}{kT} = -0.0400$. $I = (0.500 \text{ mA})(e^{-0.0400} - 1) = -0.0196$ mA.

(iii) For $V = 100$ mV, $\dfrac{eV}{kT} = 4.00$. $I = (0.500 \text{ mA})(e^{4.00} - 1) = 26.8$ mA.

(iv) For $V = -100$ mV, $\dfrac{eV}{kT} = -4.00$. $I = (0.500 \text{ mA})(e^{-4.00} - 1) = -0.491$ mA.

EVALUATE: (b) For small V, between ± 1.00 mV, $R = V/I$ is approximately constant and the diode obeys Ohm's law to a good approximation. For larger V the deviation from Ohm's law is substantial.

42.31. IDENTIFY and SET UP: The voltage-current relation is given by $I = I_s(e^{eV/kT} - 1)$. Use the current for $V = +15.0$ mV to solve for the constant I_s.

EXECUTE: (a) Find I_s: $V = +15.0 \times 10^{-3}$ V gives $I = 9.25 \times 10^{-3}$ A.

$\dfrac{eV}{kT} = \dfrac{(1.602 \times 10^{-19} \text{ C})(15.0 \times 10^{-3} \text{ V})}{(1.381 \times 10^{-23} \text{ J/K})(300 \text{ K})} = 0.5800$.

$I_s = \dfrac{I}{e^{eV/kT} - 1} = \dfrac{9.25 \times 10^{-3} \text{ A}}{e^{0.5800} - 1} = 1.177 \times 10^{-2} = 11.77$ mA.

Then can calculate I for $V = 10.0$ mV: $\dfrac{eV}{kT} = \dfrac{(1.602 \times 10^{-19} \text{ C})(10.0 \times 10^{-3} \text{ V})}{(1.381 \times 10^{-23} \text{ J/K})(300 \text{ K})} = 0.3867$.

$I = I_s(e^{eV/kT} - 1) = (11.77 \text{ mA})(e^{0.3867} - 1) = 5.56$ mA.

(b) $\dfrac{eV}{kT}$ has the same magnitude as in part (a) but now V is negative so $\dfrac{eV}{kT}$ is negative.

$V = -15.0$ mV: $\dfrac{eV}{kT} = -0.5800$ and $I = I_s(e^{eV/kT} - 1) = (11.77 \text{ mA})(e^{-0.5800} - 1) = -5.18$ mA.

$V = -10.0$ mV: $\dfrac{eV}{kT} = -0.3867$ and $I = I_s(e^{eV/kT} - 1) = (11.77 \text{ mA})(e^{-0.3867} - 1) = -3.77$ mA.

EVALUATE: There is a directional asymmetry in the current, with a forward-bias voltage producing more current than a reverse-bias voltage of the same magnitude, but the voltage is small enough for the asymmetry not be pronounced.

42.35. IDENTIFY and SET UP: From Chapter 21, the electric dipole moment is $p = qd$, where the dipole consists of charges $\pm q$ separated by distance d.

EXECUTE: (a) Point charges $+e$ and $-e$ separated by distance d, so

$p = ed = (1.602 \times 10^{-19} \text{ C})(0.24 \times 10^{-9} \text{ m}) = 3.8 \times 10^{-29}$ C·m.

(b) $p = qd$ so $q = \dfrac{p}{d} = \dfrac{3.0 \times 10^{-29} \text{ C·m}}{0.24 \times 10^{-9} \text{ m}} = 1.3 \times 10^{-19}$ C.

(c) $\dfrac{q}{e} = \dfrac{1.3 \times 10^{-19} \text{ C}}{1.602 \times 10^{-19} \text{ C}} = 0.81$.

(d) $q = \dfrac{p}{d} = \dfrac{1.5 \times 10^{-30} \text{ C} \cdot \text{m}}{0.16 \times 10^{-9} \text{ m}} = 9.37 \times 10^{-21}$ C.

$\dfrac{q}{e} = \dfrac{9.37 \times 10^{-21} \text{ C}}{1.602 \times 10^{-19} \text{ C}} = 0.058$.

EVALUATE: The fractional ionic character for the bond in HI is much less than the fractional ionic character for the bond in NaCl. The bond in HI is mostly covalent and not very ionic.

42.39. **IDENTIFY:** $E_{ex} = \dfrac{L^2}{2I} = \dfrac{\hbar^2 l(l+1)}{2I}$. $E_g = 0$ ($l=0$), and there is an additional multiplicative factor of $2l+1$ because for each l state there are really $(2l+1)$ m_l-states with the same energy.

SET UP: From Example 42.2, $I = 1.449 \times 10^{-46}$ kg·m² for CO.

EXECUTE: (a) $\dfrac{n_l}{n_0} = (2l+1)e^{-\hbar^2 l(l+1)/(2IkT)}$.

(b) (i) $E_{l=1} = \dfrac{\hbar^2 (1)(1+1)}{2(1.449 \times 10^{-46} \text{ kg} \cdot \text{m}^2)} = 7.67 \times 10^{-23}$ J. $\dfrac{E_{l=1}}{kT} = \dfrac{7.67 \times 10^{-23} \text{ J}}{(1.38 \times 10^{-23} \text{ J/K})(300 \text{ K})} = 0.0185$.

$(2l+1) = 3$, so $\dfrac{n_{l=1}}{n_0} = (3)e^{-0.0185} = 2.95$.

(ii) $\dfrac{E_{l=2}}{kT} = \dfrac{\hbar^2 (2)(2+1)}{2(1.449 \times 10^{-46} \text{ kg} \cdot \text{m}^2)(1.38 \times 10^{-23} \text{ J/K})(300 \text{ K})} = 0.0556$. $(2l+1) = 5$, so

$\dfrac{n_{l=1}}{n_0} = (5)(e^{-0.0556}) = 4.73$.

(iii) $\dfrac{E_{l=10}}{kT} = \dfrac{\hbar^2 (10)(10+1)}{2(1.449 \times 10^{-46} \text{ kg} \cdot \text{m}^2)(1.38 \times 10^{-23} \text{ J/K})(300 \text{ K})} = 1.02$.

$(2l+1) = 21$, so $\dfrac{n_{l=10}}{n_0} = (21)(e^{-1.02}) = 7.57$.

(iv) $\dfrac{E_{l=20}}{kT} = \dfrac{\hbar^2 (20)(20+1)}{2(1.449 \times 10^{-46} \text{ kg} \cdot \text{m}^2)(1.38 \times 10^{-23} \text{ J/K})(300 \text{ K})} = 3.90$. $(2l+1) = 41$, so

$\dfrac{n_{l=20}}{n_0} = (41)e^{-3.90} = 0.833$.

(v) $\dfrac{E_{l=50}}{kT} = \dfrac{\hbar^2 (50)(50+1)}{2(1.449 \times 10^{-46} \text{ kg} \cdot \text{m}^2)(1.38 \times 10^{-23} \text{ J/K})(300 \text{ K})} = 23.7$. $(2l+1) = 101$, so

$\dfrac{n_{l=50}}{n_0} = (101)e^{-23.7} = 5.38 \times 10^{-9}$.

EVALUATE: (c) There is a competing effect between the $(2l+1)$ term and the decaying exponential. The $2l+1$ term dominates for small l, while the exponential term dominates for large l.

42.45. **IDENTIFY and SET UP:** Use $I = m_r r_0^2$ to calculate I. The energy levels are given by

$E_{nl} = l(l+1)\left(\dfrac{\hbar^2}{2I}\right) + (n+\tfrac{1}{2})\hbar\sqrt{\dfrac{k'}{m_r}}$. The transition energy ΔE is related to the photon wavelength by

$\Delta E = hc/\lambda$.

EXECUTE: (a) $m_r = \dfrac{m_H m_I}{m_H + m_I} = \dfrac{(1.67 \times 10^{-27} \text{ kg})(2.11 \times 10^{-25} \text{ kg})}{1.67 \times 10^{-27} \text{ kg} + 2.11 \times 10^{-25} \text{ kg}} = 1.657 \times 10^{-27}$ kg.

$I = m_r r_0^2 = (1.657 \times 10^{-27} \text{ kg})(0.160 \times 10^{-9} \text{ m})^2 = 4.24 \times 10^{-47}$ kg·m².

(b) The energy levels are $E_{nl} = l(l+1)\left(\dfrac{\hbar^2}{2I}\right) + (n + \tfrac{1}{2})\hbar\sqrt{\dfrac{k'}{m_r}}$.

$\sqrt{\dfrac{k'}{m}} = \omega = 2\pi f$ so $E_{nl} = l(l+1)\left(\dfrac{\hbar^2}{2I}\right) + (n + \tfrac{1}{2})hf$.

(i) Transition $n = 1 \to n = 0, l = 1 \to l = 0$:

$\Delta E = (2-0)\left(\dfrac{\hbar^2}{2I}\right) + (1 + \tfrac{1}{2} - \tfrac{1}{2})hf = \dfrac{\hbar^2}{I} + hf$.

$\Delta E = \dfrac{hc}{\lambda}$ so $\lambda = \dfrac{hc}{\Delta E} = \dfrac{hc}{(\hbar^2/I) + hf} = \dfrac{c}{(\hbar/2\pi I) + f}$.

$\dfrac{\hbar}{2\pi I} = \dfrac{1.055 \times 10^{-34} \text{ J·s}}{2\pi (4.24 \times 10^{-47} \text{ kg·m}^2)} = 3.960 \times 10^{11}$ Hz.

$\lambda = \dfrac{c}{(\hbar/2\pi I) + f} = \dfrac{2.998 \times 10^8 \text{ m/s}}{3.960 \times 10^{11} \text{ Hz} + 6.93 \times 10^{13} \text{ Hz}} = 4.30$ μm.

(ii) Transition $n = 1 \to n = 0, l = 2 \to l = 1$:

$\Delta E = (6-2)\left(\dfrac{\hbar^2}{2I}\right) + hf = \dfrac{2\hbar^2}{I} + hf$.

$\lambda = \dfrac{c}{2(\hbar/2\pi I) + f} = \dfrac{2.998 \times 10^8 \text{ m/s}}{2(3.960 \times 10^{11} \text{ Hz}) + 6.93 \times 10^{13} \text{ Hz}} = 4.28$ μm.

(iii) Transition $n = 2 \to n = 1, l = 2 \to l = 3$:

$\Delta E = (6-12)\left(\dfrac{\hbar^2}{2I}\right) + hf = -\dfrac{3\hbar^2}{I} + hf$.

$\lambda = \dfrac{c}{-3(\hbar/2\pi I) + f} = \dfrac{2.998 \times 10^8 \text{ m/s}}{-3(3.960 \times 10^{11} \text{ Hz}) + 6.93 \times 10^{13} \text{ Hz}} = 4.40$ μm.

EVALUATE: The vibrational energy change for the $n = 1 \to n = 0$ transition is the same as for the $n = 2 \to n = 1$ transition. The rotational energies are much smaller than the vibrational energies, so the wavelengths for all three transitions don't differ much.

42.49. **IDENTIFY and SET UP:** Use the description of the bcc lattice in Figure 42.11c in the textbook to calculate the number of atoms per unit cell and then the number of atoms per unit volume.

EXECUTE: (a) Each unit cell has one atom at its center and 8 atoms at its corners that are each shared by 8 other unit cells. So there are $1 + 8/8 = 2$ atoms per unit cell.

$\dfrac{n}{V} = \dfrac{2}{(0.35 \times 10^{-9} \text{ m})^3} = 4.66 \times 10^{-8}$ atoms/m³.

(b) $E_{F0} = \dfrac{3^{2/3} \pi^{4/3} \hbar^2}{2m}\left(\dfrac{N}{V}\right)^{2/3}$.

In this equation N/V is the number of free electrons per m³. But the problem says to assume one free electron per atom, so this is the same as n/V calculated in part (a).

$m = 9.109 \times 10^{-31}$ kg (the electron mass), so $E_{F0} = 7.563 \times 10^{-19}$ J = 4.7 eV.

EVALUATE: Our result for metallic lithium is similar to that calculated for copper in Example 42.7.

42.51. **IDENTIFY and SET UP:** The occupation probability $f(E)$ is $f(E)=\dfrac{1}{e^{(E-E_F)/kT}+1}$.

EXECUTE: (a) Figure 42.51 shows the graph of E versus $\ln\{[1/f(E)]-1\}$ for the data given in the problem. The slope of the best-fit straight line is 0.445 eV, and the y-intercept is 1.80 eV.

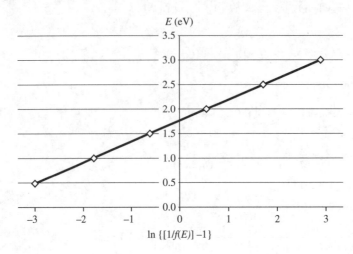

Figure 42.51

(b) Solve the $f(E)$ equation for E, giving $e^{(E-E_F)/kT}=1/f(E)-1$. Now take natural logarithms of both side of the equation, giving $(E-E_F)/kT = \ln\{[1/f(E)]-1\}$, which gives $E = kT\ln\{[1/f(E)]-1\}+E_F$. From this we see that a graph of E versus $\ln\{[1/f(E)]-1\}$ should be a straight line having a slope equal to kT and a y-intercept equal to E_F. From our graph, we get $E_F = y$-intercept = 1.80 eV.

The slope is equal to kT, so $T = $ (slope)$/k = (0.445$ eV$)/(1.38\times 10^{-23}$ J/K$) = (7.129\times 10^{-20}$ J$)/(1.38\times 10^{-23}$ J/K$) = $ 5170 K.

EVALUATE: In Example 42.7, the Fermi energy for copper was found to be 7.03 eV, so our result of 1.80 eV seems plausible.

42.57. **IDENTIFY and SET UP:** As the temperature increases, the electrons gain energy.

EXECUTE: As T increases, more electrons gain enough energy to jump into the conduction band, so for a given voltage, more current will flow at a higher temperature than at a lower temperature. Therefore choice (b) is correct.

EVALUATE: For ordinary resistors, increasing their temperature increases their resistance, so less current would flow.

NUCLEAR PHYSICS

43.3. IDENTIFY: Calculate the spin magnetic energy shift for each spin state of the 1s level. Calculate the energy splitting between these states and relate this to the frequency of the photons.
SET UP: When the spin component is parallel to the field the interaction energy is $U = -\mu_z B$. When the spin component is antiparallel to the field the interaction energy is $U = +\mu_z B$. The transition energy for a transition between these two states is $\Delta E = 2\mu_z B$, where $\mu_z = 2.7928\mu_n$. The transition energy is related to the photon frequency by $\Delta E = hf$, so $2\mu_z B = hf$.

EXECUTE: $B = \dfrac{hf}{2\mu_z} = \dfrac{(6.626 \times 10^{-34} \text{ J} \cdot \text{s})(22.7 \times 10^6 \text{ Hz})}{2(2.7928)(5.051 \times 10^{-27} \text{ J/T})} = 0.533 \text{ T}$

EVALUATE: This magnetic field is easily achievable. Photons of this frequency have wavelength $\lambda = c/f = 13.2$ m. These are radio waves.

43.7. IDENTIFY: The binding energy of the nucleus is the energy of its constituent particles minus the energy of the carbon-12 nucleus.
SET UP: In terms of the masses of the particles involved, the binding energy is
$$E_B = (6m_H + 6m_n - m_{C-12})c^2.$$
EXECUTE: **(a)** Using the values from Table 43.2, we get
$E_B = [6(1.007825 \text{ u}) + 6(1.008665 \text{ u}) - 12.000000 \text{ u}](931.5 \text{ MeV/u}) = 92.16 \text{ MeV}.$
(b) The binding energy per nucleon is $(92.16 \text{ MeV})/(12 \text{ nucleons}) = 7.680$ MeV/nucleon.
(c) The energy of the C-12 nucleus is $(12.0000 \text{ u})(931.5 \text{ MeV/u}) = 11178$ MeV. Therefore the percent of the mass that is binding energy is $\dfrac{92.16 \text{ MeV}}{11178 \text{ MeV}} = 0.8245\%.$

EVALUATE: The binding energy of 92.16 MeV binds 12 nucleons. The binding energy per nucleon, rather than just the total binding energy, is a better indicator of the strength with which a nucleus is bound.

43.9. IDENTIFY: Conservation of energy tells us that the initial energy (photon plus deuteron) is equal to the energy after the split (kinetic energy plus energy of the proton and neutron). Therefore the kinetic energy released is equal to the energy of the photon minus the binding energy of the deuteron.
SET UP: The binding energy of a deuteron is 2.224 MeV and the energy of the photon is $E = hc/\lambda$. Kinetic energy is $K = \tfrac{1}{2}mv^2$.
EXECUTE: **(a)** The energy of the photon is
$$E_{ph} = \dfrac{hc}{\lambda} = \dfrac{(6.626 \times 10^{-34} \text{ J} \cdot \text{s})(3.00 \times 10^8 \text{ m/s})}{3.50 \times 10^{-13} \text{ m}} = 5.68 \times 10^{-13} \text{ J}.$$
The binding of the deuteron is $E_B = 2.224$ MeV $= 3.56 \times 10^{-13}$ J. Therefore the kinetic energy is
$K = (5.68 - 3.56) \times 10^{-13}$ J $= 2.12 \times 10^{-13}$ J $= 1.32$ MeV.

(b) The particles share the energy equally, so each gets half. Solving the kinetic energy for v gives

$$v = \sqrt{\frac{2K}{m}} = \sqrt{\frac{2(1.06\times 10^{-13}\text{ J})}{1.6605\times 10^{-27}\text{ kg}}} = 1.13\times 10^7\text{ m/s}.$$

EVALUATE: Considerable energy has been released, because the particle speeds are in the vicinity of the speed of light.

43.15. IDENTIFY: Compare the mass of the original nucleus to the total mass of the decay products.
SET UP: Subtract the electron masses from the neutral atom mass to obtain the mass of each nucleus.
EXECUTE: If β^- decay of ^{14}C is possible, then we are considering the decay $^{14}_{6}\text{C} \rightarrow {}^{14}_{7}\text{N} + \beta^-$.
$\Delta m = M(^{14}_{6}\text{C}) - M(^{14}_{7}\text{N}) - m_e$.
$\Delta m = (14.003242\text{ u} - 6(0.000549\text{ u})) - (14.003074\text{ u} - 7(0.000549\text{ u})) - 0.0005491\text{ u}$.
$\Delta m = +1.68\times 10^{-4}$ u. So $E = (1.68\times 10^{-4}\text{ u})(931.5\text{ MeV/u}) = 0.156$ MeV $= 156$ keV.
EVALUATE: In the decay the total charge and the nucleon number are conserved.

43.17. IDENTIFY: The energy released is the energy equivalent of the difference in the masses of the original atom and the final atom produced in the capture. Apply conservation of energy to the decay products.
SET UP: 1 u is equivalent to 931.5 MeV.
EXECUTE: **(a)** As in the example, $(0.000897\text{ u})(931.5\text{ MeV/u}) = 0.836$ MeV.
(b) 0.836 MeV $- 0.122$ MeV $- 0.014$ MeV $= 0.700$ MeV.
EVALUATE: We have neglected the rest mass of the neutrino that is emitted.

43.21. IDENTIFY: From the known half-life, we can find the decay constant, the rate of decay, and the activity.
SET UP: $\lambda = \dfrac{\ln 2}{T_{1/2}}$. $T_{1/2} = 4.47\times 10^9$ yr $= 1.41\times 10^{17}$ s. The activity is $\left|\dfrac{dN}{dt}\right| = \lambda N$. The mass of one ^{238}U is approximately $238 m_p$. 1 Ci $= 3.70\times 10^{10}$ decays/s.

EXECUTE: **(a)** $\lambda = \dfrac{\ln 2}{1.41\times 10^{17}\text{ s}} = 4.92\times 10^{-18}\text{ s}^{-1}$.

(b) $N = \dfrac{|dN/dt|}{\lambda} = \dfrac{3.70\times 10^{10}\text{ Bq}}{4.92\times 10^{-18}\text{ s}^{-1}} = 7.52\times 10^{27}$ nuclei. The mass m of uranium is the number of nuclei times the mass of each one. $m = (7.52\times 10^{27})(238)(1.67\times 10^{-27}\text{ kg}) = 2.99\times 10^3$ kg.

(c) $N = \dfrac{10.0\times 10^{-3}\text{ kg}}{238 m_p} = \dfrac{10.0\times 10^{-3}\text{ kg}}{238(1.67\times 10^{-27}\text{ kg})} = 2.52\times 10^{22}$ nuclei.

$\left|\dfrac{dN}{dt}\right| = \lambda N = (4.92\times 10^{-18}\text{ s}^{-1})(2.52\times 10^{22}) = 1.24\times 10^5$ decays/s.

EVALUATE: Because ^{238}U has a very long half-life, it requires a large amount (about 3000 kg) to have an activity of a 1.0 Ci.

43.25. IDENTIFY and SET UP: Find λ from the half-life and the number N of nuclei from the mass of one nucleus and the mass of the sample. Then use Eq. (43.16) to calculate $|dN/dt|$, the number of decays per second.
EXECUTE: **(a)** $|dN/dt| = \lambda N$.

$\lambda = \dfrac{0.693}{T_{1/2}} = \dfrac{0.693}{(1.28\times 10^9\text{ y})(3.156\times 10^7\text{ s/1 y})} = 1.715\times 10^{-17}\text{ s}^{-1}$.

The mass of one ^{40}K atom is approximately 40 u, so the number of ^{40}K nuclei in the sample is

$N = \dfrac{1.63\times 10^{-9}\text{ kg}}{40\text{ u}} = \dfrac{1.63\times 10^{-9}\text{ kg}}{40(1.66054\times 10^{-27}\text{ kg})} = 2.454\times 10^{16}$.

Then $|dN/dt| = \lambda N = (1.715\times 10^{-17}\text{ s}^{-1})(2.454\times 10^{16}) = 0.421$ decays/s.

(b) $|dN/dt| = (0.421\text{ decays/s})(1\text{ Ci}/(3.70\times 10^{10}\text{ decays/s})) = 1.14\times 10^{-11}$ Ci.

EVALUATE: The very small sample still contains a very large number of nuclei. But the half-life is very large, so the decay rate is small.

43.29. IDENTIFY and SET UP: Apply $|dN/dt| = \lambda N$ with $\lambda = \ln 2/T_{1/2}$. In one half-life, one half of the nuclei decay.

EXECUTE: (a) $\left|\dfrac{dN}{dt}\right| = 7.56 \times 10^{11}$ Bq $= 7.56 \times 10^{11}$ decays/s.

$\lambda = \dfrac{0.693}{T_{1/2}} = \dfrac{0.693}{(30.8 \text{ min})(60 \text{ s/min})} = 3.75 \times 10^{-4}$ s^{-1}. $N_0 = \dfrac{1}{\lambda}\left|\dfrac{dN}{dt}\right| = \dfrac{7.56 \times 10^{11} \text{ decays/s}}{3.75 \times 10^{-4} \text{ s}^{-1}} = 2.02 \times 10^{15}$ nuclei.

(b) The number of nuclei left after one half-life is $\dfrac{N_0}{2} = 1.01 \times 10^{15}$ nuclei, and the activity is half:

$\left|\dfrac{dN}{dt}\right| = 3.78 \times 10^{11}$ decays/s.

(c) After three half-lives (92.4 minutes) there is an eighth of the original amount, so $N = 2.53 \times 10^{14}$ nuclei, and an eighth of the activity: $\left|\dfrac{dN}{dt}\right| = 9.45 \times 10^{10}$ decays/s.

EVALUATE: Since the activity is proportional to the number of radioactive nuclei that are present, the activity is halved in one half-life.

43.31. IDENTIFY: Knowing the equivalent dose in Sv, we want to find the absorbed energy.
SET UP: equivalent dose (Sv, rem) = RBE × absorbed dose (Gy, rad); 100 rad = 1 Gy.
EXECUTE: (a) RBE = 1, so 0.25 mSv corresponds to 0.25 mGy.

Energy = $(0.25 \times 10^{-3}$ J/kg$)(5.0$ kg$) = 1.2 \times 10^{-3}$ J.

(b) RBE = 1 so 0.10 mGy = 10 mrad and 10 mrem. $(0.10 \times 10^{-3}$ J/kg$)(75$ kg$) = 7.5 \times 10^{-3}$ J.

EVALUATE: (c) $\dfrac{7.5 \times 10^{-3} \text{ J}}{1.2 \times 10^{-3} \text{ J}} = 6.2$. Each chest x ray delivers only about 1/6 of the yearly background radiation energy.

43.37. IDENTIFY: Apply $|dN/dt| = \lambda N$, with $\lambda = \ln 2/T_{1/2}$, to find the number of tritium atoms that were ingested. Then use $N = N_0 e^{-\lambda t}$ to find the number of decays in one week.
SET UP: 1 rad = 0.01 J/kg. rem = RBE × rad.
EXECUTE: (a) We need to know how many decays per second occur.

$\lambda = \dfrac{0.693}{T_{1/2}} = \dfrac{0.693}{(12.3 \text{ y})(3.156 \times 10^7 \text{ s/y})} = 1.785 \times 10^{-9}$ s^{-1}. The number of tritium atoms is

$N_0 = \dfrac{1}{\lambda}\left|\dfrac{dN}{dt}\right| = \dfrac{(0.35 \text{ Ci})(3.70 \times 10^{10} \text{ Bq/Ci})}{1.79 \times 10^{-9} \text{ s}^{-1}} = 7.2540 \times 10^{18}$ nuclei. The number of remaining nuclei after

one week is $N = N_0 e^{-\lambda t} = (7.25 \times 10^{18})e^{-(1.79 \times 10^{-9} \text{ s}^{-1})(7)(24)(3600 \text{ s})} = 7.2462 \times 10^{18}$ nuclei.

$\Delta N = N_0 - N = 7.8 \times 10^{15}$ decays. So the energy absorbed is

$E_{\text{total}} = \Delta N \ E_\gamma = (7.8 \times 10^{15})(5000 \text{ eV})(1.60 \times 10^{-19}$ J/eV$) = 6.25$ J. The absorbed dose is

$\dfrac{6.25 \text{ J}}{67 \text{ kg}} = 0.0932$ J/kg $= 9.32$ rad. Since RBE = 1, then the equivalent dose is 9.32 rem.

EVALUATE: (b) In the decay, antineutrinos are also emitted. These are not absorbed by the body, and so some of the energy of the decay is lost.

43.41. (a) IDENTIFY and SET UP: Determine X by balancing the charge and the nucleon number on the two sides of the reaction equation.
EXECUTE: X must have $A = +2 + 9 - 4 = 7$ and $Z = +1 + 4 - 2 = 3$. Thus X is $^{7}_{3}$Li and the reaction is

$^{2}_{1}$H $+$ $^{9}_{4}$Be $=$ $^{7}_{3}$Li $+$ $^{4}_{2}$He.

(b) IDENTIFY and SET UP: Calculate the mass decrease and find its energy equivalent.

EXECUTE: If we use the neutral atom masses then there are the same number of electrons (five) in the reactants as in the products. Their masses cancel, so we get the same mass defect whether we use nuclear masses or neutral atom masses. The neutral atoms masses are given in Table 43.2. 1 u is equivalent to 931.5 MeV.

$^2_1$H + $^9_4$Be has mass 2.014102 u + 9.012182 u = 11.26284 u.

$^7_3$Li + $^4_2$He has mass 7.016005 u + 4.002603 u = 11.018608 u.

The mass decrease is 11.026284 u − 11.018608 u = 0.007676 u.

This corresponds to an energy release of (0.007676 u)(931.5 MeV/1 u) = 7.150 MeV.

(c) IDENTIFY and SET UP: Estimate the threshold energy by calculating the Coulomb potential energy when the $^2_1$H and $^9_4$Be nuclei just touch. Obtain the nuclear radii from $R = R_0 A^{1/3}$.

EXECUTE: The radius R_{Be} of the $^9_4$Be nucleus is $R_{Be} = (1.2 \times 10^{-15} \text{ m})(9)^{1/3} = 2.5 \times 10^{-15}$ m.

The radius R_H of the $^2_1$H nucleus is $R_H = (1.2 \times 10^{-15} \text{ m})(2)^{1/3} = 1.5 \times 10^{-15}$ m.

The nuclei touch when their center-to-center separation is

$R = R_{Be} + R_H = 4.0 \times 10^{-15}$ m.

The Coulomb potential energy of the two reactant nuclei at this separation is

$$U = \frac{1}{4\pi\varepsilon_0} \frac{q_1 q_2}{r} = \frac{1}{4\pi\varepsilon_0} \frac{e(4e)}{r}.$$

$$U = (8.988 \times 10^9 \text{ N} \cdot \text{m}^2/\text{C}^2) \frac{4(1.602 \times 10^{-19} \text{ C})^2}{(4.0 \times 10^{-15} \text{ m})(1.602 \times 10^{-19} \text{ J/eV})} = 1.4 \text{ MeV}.$$

This is an estimate of the threshold energy for this reaction.

EVALUATE: The reaction releases energy but the total initial kinetic energy of the reactants must be 1.4 MeV in order for the reacting nuclei to get close enough to each other for the reaction to occur. The nuclear force is strong but is very short-range.

43.43. **IDENTIFY and SET UP:** The energy released is the energy equivalent of the mass decrease. 1 u is equivalent to 931.5 MeV. The mass of one ^{235}U nucleus is $235 m_p$.

EXECUTE: (a) $^{235}_{92}$U + $^1_0$n → $^{144}_{56}$Ba + $^{89}_{36}$Kr + 3^1_0n. We can use atomic masses since the same number of electrons are included on each side of the reaction equation and the electron masses cancel. The mass decrease is $\Delta M = m(^{235}_{92}\text{U}) + m(^1_0\text{n}) - [m(^{144}_{56}\text{Ba}) + m(^{89}_{36}\text{Kr}) + 3m(^1_0\text{n})]$,

$\Delta M = 235.043930$ u $+ 1.0086649$ u $- 143.922953$ u $- 88.917631$ u $- 3(1.0086649$ u$)$, $\Delta M = 0.1860$ u.

The energy released is $(0.1860$ u$)(931.5$ MeV/u$) = 173.3$ MeV.

(b) The number of ^{235}U nuclei in 1.00 g is $\dfrac{1.00 \times 10^{-3} \text{ kg}}{235 m_p} = 2.55 \times 10^{21}$. The energy released per gram is

$(173.3$ MeV/nucleus$)(2.55 \times 10^{21}$ nuclei/g$) = 4.42 \times 10^{23}$ MeV/g.

EVALUATE: The energy released is 7.1×10^{10} J/kg. This is much larger than typical heats of combustion, which are about 5×10^4 J/kg.

43.49. **IDENTIFY:** The minimum energy to remove a proton or a neutron from the nucleus is equal to the energy difference between the two states of the nucleus, before and after removal.

(a) SET UP: $^{17}_8$O = $^1_0$n + $^{16}_8$O. $\Delta m = m(^1_0\text{n}) + m(^{16}_8\text{O}) - m(^{17}_8\text{O})$. The electron masses cancel when neutral atom masses are used.

EXECUTE: $\Delta m = 1.008665$ u $+ 15.994915$ u $- 16.999132$ u $= 0.004448$ u. The energy equivalent of this mass increase is $(0.004448$ u$)(931.5$ MeV/u$) = 4.14$ MeV.

(b) SET UP and EXECUTE: Following the same procedure as in part (a) gives
$\Delta M = 8M_H + 9M_n - {}^{17}_{8}M = 8(1.007825 \text{ u}) + 9(1.008665 \text{ u}) - 16.999132 \text{ u} = 0.1415 \text{ u}.$

$E_B = (0.1415 \text{ u})(931.5 \text{ MeV/u}) = 131.8 \text{ MeV}. \quad \dfrac{E_B}{A} = 7.75 \text{ MeV/nucleon}.$

EVALUATE: The neutron removal energy is about half the binding energy per nucleon.

43.55. IDENTIFY and SET UP: The amount of kinetic energy released is the energy equivalent of the mass change in the decay. $m_e = 0.0005486$ u and the atomic mass of ${}^{14}_{7}N$ is 14.003074 u. The energy equivalent of 1 u is 931.5 MeV. ${}^{14}C$ has a half-life of $T_{1/2} = 5730$ y $= 1.81 \times 10^{11}$ s. The RBE for an electron is 1.0.

EXECUTE: (a) ${}^{14}_{6}C \rightarrow e^- + {}^{14}_{7}N + \bar{\nu}_e.$

(b) The mass decrease is $\Delta M = m({}^{14}_{6}C) - [m_e + m({}^{14}_{7}N)]$. Use nuclear masses, to avoid difficulty in accounting for atomic electrons. The nuclear mass of ${}^{14}_{6}C$ is 14.003242 u $- 6m_e = 13.999950$ u. The nuclear mass of ${}^{14}_{7}N$ is 14.003074 u $- 7m_e = 13.999234$ u.

$\Delta M = 13.999950$ u $- 13.999234$ u $- 0.000549$ u $= 1.67 \times 10^{-4}$ u. The energy equivalent of ΔM is 0.156 MeV.

(c) The mass of carbon is $(0.18)(75 \text{ kg}) = 13.5$ kg. From Example 43.9, the activity due to 1 g of carbon in a living organism is 0.255 Bq. The number of decay/s due to 13.5 kg of carbon is
$(13.5 \times 10^3 \text{ g})(0.255 \text{ Bq/g}) = 3.4 \times 10^3$ decays/s.

(d) Each decay releases 0.156 MeV so 3.4×10^3 decays/s releases 530 MeV/s $= 8.5 \times 10^{-11}$ J/s.

(e) The total energy absorbed in 1 year is $(8.5 \times 10^{-11} \text{ J/s})(3.156 \times 10^7 \text{ s}) = 2.7 \times 10^{-3}$ J. The absorbed dose is $\dfrac{2.7 \times 10^{-3} \text{ J}}{75 \text{ kg}} = 3.6 \times 10^{-5}$ J/kg $= 36 \, \mu$Gy $= 3.6$ mrad. With RBE $= 1.0$, the equivalent dose is $36 \, \mu$Sv $= 3.6$ mrem.

EVALUATE: Section 43.5 says that background radiation exposure is about 1.0 mSv per year. The radiation dose calculated in this problem is much less than this.

43.57. IDENTIFY and SET UP: The mass defect is E_B/c^2.

EXECUTE: $m_{{}^{11}_{6}C} - m_{{}^{11}_{5}B} - 2m_e = 1.03 \times 10^{-3}$ u. Decay is energetically possible.

EVALUATE: The energy released in the decay is $(1.03 \times 10^{-3} \text{ u})(931.5 \text{ MeV/u}) = 0.959$ MeV.

43.61. IDENTIFY: Use $N = N_0 e^{-\lambda t}$ to relate the initial number of radioactive nuclei, N_0, to the number, N, left after time t.

SET UP: We have to be careful; after ^{87}Rb has undergone radioactive decay it is no longer a rubidium atom. Let N_{85} be the number of ^{85}Rb atoms; this number doesn't change. Let N_0 be the number of ^{87}Rb atoms on earth when the solar system was formed. Let N be the present number of ^{87}Rb atoms.

EXECUTE: The present measurements say that $0.2783 = N/(N + N_{85})$.
$(N + N_{85})(0.2783) = N$, so $N = 0.3856 N_{85}$. The percentage we are asked to calculate is $N_0/(N_0 + N_{85})$.
N and N_0 are related by $N = N_0 e^{-\lambda t}$ so $N_0 = e^{+\lambda t} N$.

Thus $\dfrac{N_0}{N_0 + N_{85}} = \dfrac{N e^{\lambda t}}{N e^{\lambda t} + N_{85}} = \dfrac{(0.3855 e^{\lambda t}) N_{85}}{(0.3856 e^{\lambda t}) N_{85} + N_{85}} = \dfrac{0.3856 e^{\lambda t}}{0.3856 e^{\lambda t} + 1}.$

$t = 4.6 \times 10^9$ y; $\lambda = \dfrac{0.693}{T_{1/2}} = \dfrac{0.693}{4.75 \times 10^{10} \text{ y}} = 1.459 \times 10^{-11}$ y^{-1}.

$e^{\lambda t} = e^{(1.459 \times 10^{-11} \text{ y}^{-1})(4.6 \times 10^9 \text{ y})} = e^{0.06711} = 1.0694.$

Thus $\dfrac{N_0}{N_0 + N_{85}} = \dfrac{(0.3856)(1.0694)}{(0.3856)(1.0694) + 1} = 29.2\%.$

EVALUATE: The half-life for ^{87}Rb is a factor of 10 larger than the age of the solar system, so only a small fraction of the ^{87}Rb nuclei initially present have decayed; the percentage of rubidium atoms that are radioactive is only a bit less now than it was when the solar system was formed.

43.63. IDENTIFY and SET UP: Find the energy emitted and the energy absorbed each second. Convert the absorbed energy to absorbed dose and to equivalent dose.

EXECUTE: (a) First find the number of decays each second:

2.6×10^{-4} Ci $\left(\dfrac{3.70 \times 10^{10} \text{ decays/s}}{1 \text{ Ci}} \right) = 9.6 \times 10^{6}$ decays/s. The average energy per decay is 1.25 MeV, and one-half of this energy is deposited in the tumor. The energy delivered to the tumor per second then is

$\frac{1}{2}(9.6 \times 10^{6} \text{ decays/s})(1.25 \times 10^{6} \text{ eV/decay})(1.602 \times 10^{-19} \text{ J/eV}) = 9.6 \times 10^{-7}$ J/s.

(b) The absorbed dose is the energy absorbed divided by the mass of the tissue:

$\dfrac{9.6 \times 10^{-7} \text{ J/s}}{0.200 \text{ kg}} = (4.8 \times 10^{-6} \text{ J/kg} \cdot \text{s})(1 \text{ rad}/(0.01 \text{ J/kg})) = 4.8 \times 10^{-4}$ rad/s.

(c) equivalent dose (REM) = RBE × absorbed dose (rad). In one second the equivalent dose is

$(0.70)(4.8 \times 10^{-4} \text{ rad}) = 3.4 \times 10^{-4}$ rem.

(d) $(200 \text{ rem})/(3.4 \times 10^{-4} \text{ rem/s}) = (5.9 \times 10^{5} \text{ s})(1 \text{ h}/3600 \text{ s}) = 164 \text{ h} = 6.9$ days.

EVALUATE: The activity of the source is small so that absorbed energy per second is small and it takes several days for an equivalent dose of 200 rem to be absorbed by the tumor. A 200-rem dose equals 2.00 Sv and this is large enough to damage the tissue of the tumor.

43.67. (a) IDENTIFY and SET UP: Use $R = R_0 A^{1/3}$ to calculate the radius R of a $^2_1$H nucleus. Calculate the Coulomb potential energy $U = \dfrac{1}{4\pi\varepsilon_0} \dfrac{q_1 q_2}{r^2}$ of the two nuclei when they just touch.

EXECUTE: The radius of $^2_1$H is $R = (1.2 \times 10^{-15} \text{ m})(2)^{1/3} = 1.51 \times 10^{-15}$ m. The barrier energy is the Coulomb potential energy of two $^2_1$H nuclei with their centers separated by twice this distance:

$U = \dfrac{1}{4\pi\varepsilon_0} \dfrac{e^2}{r} = (8.988 \times 10^9 \text{ N} \cdot \text{m}^2/\text{C}^2) \dfrac{(1.602 \times 10^{-19} \text{ C})^2}{2(1.51 \times 10^{-15} \text{ m})} = 7.64 \times 10^{-14}$ J $= 0.48$ MeV.

(b) IDENTIFY and SET UP: Find the energy equivalent of the mass decrease.

EXECUTE: $^2_1\text{H} + ^2_1\text{H} \rightarrow ^3_2\text{He} + ^1_0\text{n}$

If we use neutral atom masses there are two electrons on each side of the reaction equation, so their masses cancel. The neutral atom masses are given in Table 43.2.

$^2_1\text{H} + ^2_1\text{H}$ has mass $2(2.014102 \text{ u}) = 4.028204$ u

$^3_2\text{He} + ^1_0\text{n}$ has mass $3.016029 \text{ u} + 1.008665 \text{ u} = 4.024694$ u

The mass decrease is $4.028204 \text{ u} - 4.024694 \text{ u} = 3.510 \times 10^{-3}$ u. This corresponds to a liberated energy of $(3.510 \times 10^{-3} \text{ u})(931.5 \text{ MeV/u}) = 3.270$ MeV, or $(3.270 \times 10^6 \text{ eV})(1.602 \times 10^{-19} \text{ J/eV}) = 5.239 \times 10^{-13}$ J.

(c) IDENTIFY and SET UP: We know the energy released when two $^2_1$H nuclei fuse. Find the number of reactions obtained with one mole of $^2_1$H.

EXECUTE: Each reaction takes two $^2_1$H nuclei. Each mole of D_2 has 6.022×10^{23} molecules, so 6.022×10^{23} pairs of atoms. The energy liberated when one mole of deuterium undergoes fusion is $(6.022 \times 10^{23})(5.239 \times 10^{-13} \text{ J}) = 3.155 \times 10^{11}$ J/mol.

EVALUATE: The energy liberated per mole is more than a million times larger than from chemical combustion of one mole of hydrogen gas.

43.69. **IDENTIFY:** Apply $\left|\dfrac{dN}{dt}\right| = \lambda N_0 e^{-\lambda t}$, with $\lambda = \dfrac{\ln 2}{T_{1/2}}$.

SET UP: $\ln|dN/dt| = \ln \lambda N_0 - \lambda t$.

EXECUTE: **(a)** A least-squares fit to log of the activity *vs.* time gives a slope of magnitude $\lambda = 0.5995 \text{ h}^{-1}$, for a half-life of $\dfrac{\ln 2}{\lambda} = 1.16 \text{ h}$.

(b) The initial activity is $N_0 \lambda$, and this gives $N_0 = \dfrac{(2.00 \times 10^4 \text{ Bq})}{(0.5995 \text{ hr}^{-1})(1 \text{ hr}/3600 \text{ s})} = 1.20 \times 10^8$.

(c) $N = N_0 e^{-\lambda t} = 1.81 \times 10^6$.

EVALUATE: The activity decreases by about $\tfrac{1}{2}$ in the first hour, so the half-life is about 1 hour.

43.75. **IDENTIFY and SET UP:** A thyroid treatment administers 3.7 GBq of $^{131}_{53}\text{I}$, which has a half-life of 8.04 days. $\lambda = (\ln 2)/T_{1/2}$ and $|dN/dt| = \lambda N = \dfrac{\ln 2}{T_{1/2}} N$.

EXECUTE: Solve for N and put in the numbers.

$N = \dfrac{T_{1/2} |dN/dt|}{\ln 2} = (8.04 \text{ d})(24 \times 3600 \text{ s/d})(3.7 \times 10^9 \text{ decays/s})/(\ln 2) = 3.7 \times 10^{15}$ atoms, which is choice (d).

EVALUATE: This amount is $(3.7 \times 10^{15} \text{ atoms})/(6.02 \times 10^{23} \text{ atoms/mol}) = 6.1 \times 10^{-9}$ mol ≈ 6 nanomoles.

PARTICLE PHYSICS AND COSMOLOGY

44.3. **IDENTIFY:** The energy released is the energy equivalent of the mass decrease that occurs in the decay.
SET UP: The mass of the pion is $m_{\pi^+} = 270 m_e$ and the mass of the muon is $m_{\mu^+} = 207 m_e$. The rest energy of an electron is 0.511 MeV.
EXECUTE: **(a)** $\Delta m = m_{\pi^+} - m_{\mu^+} = 270 m_e - 207 m_e = 63 m_e \Rightarrow E = 63(0.511 \text{ MeV}) = 32 \text{ MeV}.$
EVALUATE: **(b)** A positive muon has less mass than a positive pion, so if the decay from muon to pion was to happen, you could always find a frame where energy was not conserved. This cannot occur.

44.9. **IDENTIFY and SET UP:** The angular frequency is $\omega = |q|B/m$ so $B = m\omega/|q|$. And since $\omega = 2\pi f$, this becomes $B = 2\pi m f / |q|$.

EXECUTE: **(a)** A deuteron is a deuterium nucleus ($_1^2 \text{H}$). Its charge is $q = +e$. Its mass is the mass of the neutral $_1^2 \text{H}$ atom (Table 43.2) minus the mass of the one atomic electron:
$m = 2.014102 \text{ u} - 0.0005486 \text{ u} = 2.013553 \text{ u} (1.66054 \times 10^{-27} \text{ kg/1 u}) = 3.344 \times 10^{-27} \text{ kg}.$

$$B = \frac{2\pi m f}{|q|} = \frac{2\pi(3.344 \times 10^{-27} \text{ kg})(9.00 \times 10^6 \text{ Hz})}{1.602 \times 10^{-19} \text{ C}} = 1.18 \text{ T}.$$

(b) Eq. (44.8): $K = \frac{q^2 B^2 R^2}{2m} = \frac{[(1.602 \times 10^{-19} \text{ C})(1.18 \text{ T})(0.320 \text{ m})]^2}{2(3.344 \times 10^{-27} \text{ kg})}.$

$K = 5.471 \times 10^{-13} \text{ J} = (5.471 \times 10^{-13} \text{ J})(1 \text{ eV}/1.602 \times 10^{-19} \text{ J}) = 3.42 \text{ MeV}.$

$K = \frac{1}{2} m v^2$ so $v = \sqrt{\frac{2K}{m}} = \sqrt{\frac{2(5.471 \times 10^{-13} \text{ J})}{3.344 \times 10^{-27} \text{ kg}}} = 1.81 \times 10^7 \text{ m/s}.$

EVALUATE: $v/c = 0.06,$ so it is ok to use the nonrelativistic expression for kinetic energy.

44.15. **(a) IDENTIFY and SET UP:** For a proton beam on a stationary proton target, and since E_a is much larger than the proton rest energy, we can use the equation $E_a^2 = 2mc^2 E_m$.

EXECUTE: $E_m = \frac{E_a^2}{2mc^2} = \frac{(77.4 \text{ GeV})^2}{2(0.938 \text{ GeV})} = 3200 \text{ GeV}.$

(b) IDENTIFY and SET UP: For colliding beams the total momentum is zero and the available energy E_a is the total energy for the two colliding particles.
EXECUTE: For proton-proton collisions the colliding beams each have the same energy, so the total energy of each beam is $\frac{1}{2} E_a = 38.7 \text{ GeV}.$

EVALUATE: For a stationary target less than 3% of the beam energy is available for conversion into mass. The beam energy for a colliding beam experiment is a factor of (1/83) times smaller than the required energy for a stationary target experiment.

44.21. IDENTIFY and SET UP: Find the energy equivalent of the mass decrease.
EXECUTE: The mass decrease is $m(\Sigma^+) - m(p) - m(\pi^0)$ and the energy released is
$mc^2(\Sigma^+) - mc^2(p) - mc^2(\pi^0) = 1189$ MeV $- 938.3$ MeV $- 135.0$ MeV $= 116$ MeV. (The mc^2 values for each particle were taken from Table 44.3.)
EVALUATE: The mass of the decay products is less than the mass of the original particle, so the decay is energetically allowed and energy is released.

44.23. IDENTIFY and SET UP: The lepton numbers for the particles are given in Table 44.2.
EXECUTE: **(a)** $\mu^- \to e^- + \nu_e + \bar{\nu}_\mu \Rightarrow L_\mu: +1 \neq -1, L_e: 0 \neq +1+1$, so lepton numbers are not conserved.
(b) $\tau^- \to e^- + \bar{\nu}_e + \nu_\tau \Rightarrow L_e: 0 = +1-1; L_\tau: +1 = +1$, so lepton numbers are conserved.
(c) $\pi^+ \to e^+ + \gamma$. Lepton numbers are not conserved since just one lepton is produced from zero original leptons.
(d) $n \to p + e^- + \bar{\nu}_e \Rightarrow L_e: 0 = +1-1$, so the lepton numbers are conserved.
EVALUATE: The decays where lepton numbers are conserved are among those listed in Tables 44.2 and 44.3.

44.27. IDENTIFY and SET UP: Each value for the combination is the sum of the values for each quark. Use Table 44.4.
EXECUTE: **(a)** *uds:*
$Q = \tfrac{2}{3}e - \tfrac{1}{3}e - \tfrac{1}{3}e = 0$
$B = \tfrac{1}{3} + \tfrac{1}{3} + \tfrac{1}{3} = 1$
$S = 0 + 0 - 1 = -1$
$C = 0 + 0 + 0 = 0$
(b) *c$\bar{u}$:*
The values for $\bar{u}$ are the negative for those for *u*.
$Q = \tfrac{2}{3}e - \tfrac{2}{3}e = 0$
$B = \tfrac{1}{3} - \tfrac{1}{3} = 0$
$S = 0 + 0 = 0$
$C = +1 + 0 = +1$
(c) *ddd:*
$Q = -\tfrac{1}{3}e - \tfrac{1}{3}e - \tfrac{1}{3}e = -e$
$B = \tfrac{1}{3} + \tfrac{1}{3} + \tfrac{1}{3} = +1$
$S = 0 + 0 + 0 = 0$
$C = 0 + 0 + 0 = 0$
(d) *d$\bar{c}$:*
$Q = -\tfrac{1}{3}e - \tfrac{2}{3}e = -e$
$B = \tfrac{1}{3} - \tfrac{1}{3} = 0$
$S = 0 + 0 = 0$
$C = 0 - 1 = -1$
EVALUATE: The charge, baryon number, strangeness, and charm quantum numbers of a particle are determined by the particle's quark composition.

44.31. (a) IDENTIFY and SET UP: First calculate the speed v. Then use that in Hubble's law to find r.
EXECUTE: $v = \left[\dfrac{(\lambda_0/\lambda_S)^2 - 1}{(\lambda_0/\lambda_S)^2 + 1}\right]c = \left[\dfrac{(658.5 \text{ nm}/590 \text{ nm})^2 - 1}{(658.5 \text{ nm}/590 \text{ nm})^2 + 1}\right]c = 0.1094c$
$v = (0.1094)(2.998 \times 10^8 \text{ m/s}) = 3.28 \times 10^7$ m/s. $v = rH_0$.

(b) IDENTIFY and SET UP: Use Hubble's law to calculate r.

EXECUTE: $r = \dfrac{v}{H_0} = \dfrac{3.28 \times 10^4 \text{ km/s}}{[(67.3 \text{ km/s})/\text{Mpc}](1 \text{ Mpc}/3.26 \text{ Mly})} = 1590 \text{ Mly} = 1.59 \times 10^9 \text{ ly}.$

EVALUATE: The red shift $\lambda_0/\lambda_S - 1$ for this galaxy is 0.116. It is therefore about twice as far from earth as the galaxy in Examples 44.8 and 44.9, that had a red shift of 0.053.

44.39. IDENTIFY: The energy comes from a mass decrease.

SET UP: A charged pion decays into a muon plus a neutrino. The muon in turn decays into an electron or positron plus two neutrinos.

EXECUTE: (a) $\pi^- \to \mu^- + \text{neutrino} \to e^- + \text{three neutrinos}.$

(b) If we neglect the mass of the neutrinos, the mass decrease is

$m(\pi^-) - m(e^-) = 273 m_e - m_e = 272 m_e = 2.480 \times 10^{-28}$ kg.

$E = mc^2 = 2.23 \times 10^{-11}$ J $= 139$ MeV.

(c) The total energy delivered to the tissue is $(50.0 \text{ J/kg})(10.0 \times 10^{-3} \text{ kg}) = 0.500$ J. The number of π^- mesons required is $\dfrac{0.500 \text{ J}}{2.23 \times 10^{-11} \text{ J}} = 2.24 \times 10^{10}.$

(d) The RBE for the electrons that are produced is 1.0, so the equivalent dose is

$1.0(50.0 \text{ Gy}) = 50.0 \text{ Sv} = 5.0 \times 10^3$ rem.

EVALUATE: The π are heavier than electrons and therefore behave differently as they hit the tissue.

44.41. IDENTIFY: With a stationary target, only part of the initial kinetic energy of the moving proton is available. Momentum conservation tells us that there must be nonzero momentum after the collision, which means that there must also be leftover kinetic energy. Therefore not all of the initial energy is available.

SET UP: The available energy is given by $E_a^2 = 2mc^2(E_m + mc^2)$ for two particles of equal mass when one is initially stationary. The *minimum* available energy must be equal to the rest mass energies of the products, which in this case is two protons, a K^+ and a K^-. The available energy must be at least the sum of the final rest masses.

EXECUTE: The minimum amount of available energy must be

$E_a = 2m_p + m_{K^+} + m_{K^-} = 2(938.3 \text{ MeV}) + 493.7 \text{ MeV} + 493.7 \text{ MeV} = 2864 \text{ MeV} = 2.864 \text{ GeV}.$

Solving the available energy formula for E_m gives $E_a^2 = 2mc^2(E_m + mc^2)$ and

$E_m = \dfrac{E_a^2}{2mc^2} - mc^2 = \dfrac{(2864 \text{ MeV})^2}{2(938.3 \text{ MeV})} - 938.3 \text{ MeV} = 3432.6 \text{ MeV}.$

Recalling that E_m is the *total* energy of the proton, including its rest mass energy (RME), we have

$K = E_m - \text{RME} = 3432.6 \text{ MeV} - 938.3 \text{ MeV} = 2494 \text{ MeV} = 2.494 \text{ GeV}.$

Therefore the threshold kinetic energy is $K = 2494$ MeV $= 2.494$ GeV.

EVALUATE: Considerably less energy would be needed if the experiment were done using colliding beams of protons.

44.47. IDENTIFY: Apply $\left|\dfrac{dN}{dt}\right| = \lambda N$ to find the number of decays in one year.

SET UP: Water has a molecular mass of 18.0×10^{-3} kg/mol.

EXECUTE: (a) The number of protons in a kilogram is

$(1.00 \text{ kg}) \left(\dfrac{6.022 \times 10^{23} \text{ molecules/mol}}{18.0 \times 10^{-3} \text{ kg/mol}} \right) (2 \text{ protons/molecule}) = 6.7 \times 10^{25}.$ Note that only the protons in the hydrogen atoms are considered as possible sources of proton decay. The energy per decay is $m_p c^2 = 938.3$ MeV $= 1.503 \times 10^{-10}$ J, and so the energy deposited in a year, per kilogram, is

$(6.7 \times 10^{25}) \left(\dfrac{\ln 2}{1.0 \times 10^{18} \text{ y}} \right)(1 \text{ y})(1.50 \times 10^{-10} \text{ J}) = 7.0 \times 10^{-3}$ Gy $= 0.70$ rad.

(b) For an RBE of unity, the equivalent dose is $(1)(0.70 \text{ rad}) = 0.70$ rem.

EVALUATE: The equivalent dose is much larger than that due to the natural background. It is not feasible for the proton lifetime to be as short as 1.0×10^{18} y.

44.49. IDENTIFY: The matter density is proportional to $1/R^3$.

SET UP and EXECUTE: (a) When the matter density was large enough compared to the dark energy density, the slowing due to gravitational attraction would have dominated over the cosmic repulsion due to dark energy.

(b) Matter density is proportional to $1/R^3$, so $R \propto \dfrac{1}{\rho^{1/3}}$. Therefore $\dfrac{R}{R_0} = \left(\dfrac{1/\rho_{\text{past}}}{1/\rho_{\text{now}}}\right)^{1/3} = \left(\dfrac{\rho_{\text{now}}}{\rho_{\text{past}}}\right)^{1/3}$. If ρ_m and ρ_{DE} are the present-day densities of matter of all kinds and of dark energy, we have $\rho_{DE} = 0.726\rho_{\text{crit}}$ and $\rho_m = 0.274\rho_{\text{crit}}$ at the present time. Putting this into the above equation for R/R_0 gives

$$\frac{R}{R_0} = \left(\frac{\frac{0.274}{0.726}\rho_{DE}}{2\rho_{DE}}\right)^{1/3} = 0.574.$$

EVALUATE: (c) 300 My: speeding up ($R/R_0 = 0.98$); 13.1 Gy: slowing down ($R/R_0 = 0.12$).

44.51. IDENTIFY: The kinetic energy comes from the mass difference.

SET UP and EXECUTE: $K_\Sigma = 180$ MeV. $m_\Sigma c^2 = 1197$ MeV. $m_n c^2 = 939.6$ MeV. $m_\pi c^2 = 139.6$ MeV. $E_\Sigma = K_\Sigma + m_\Sigma c^2 = 180$ MeV $+ 1197$ MeV $= 1377$ MeV. Conservation of the x-component of momentum gives $p_\Sigma = p_{nx}$. Then $p_{nx}^2 c^2 = p_\Sigma^2 c^2 = E_\Sigma^2 - (m_\Sigma c^2)^2 = (1377 \text{ MeV})^2 - (1197 \text{ MeV})^2 = 4.633 \times 10^5$ (MeV)^2. Conservation of energy gives $E_\Sigma = E_\pi + E_n$. $E_\Sigma = \sqrt{m_\pi^2 c^4 + p_\pi^2 c^2} + \sqrt{m_n^2 c^4 + p_n^2 c^2}$.

$E_\Sigma - \sqrt{m_n^2 c^4 + p_n^2 c^2} = \sqrt{m_\pi^2 c^4 + p_\pi^2 c^2}$. Square both sides:

$E_\Sigma^2 + m_n^2 c^4 + p_{nx}^2 c^2 + p_{ny}^2 c^2 - 2E_\Sigma E_n = m_\pi^2 c^4 + p_\pi^2 c^2$. $p_\pi = p_{ny}$ so

$E_\Sigma^2 + m_n^2 c^4 + p_{nx}^2 c^2 - 2E_\Sigma E_n = m_\pi^2 c^4$ and $E_n = \dfrac{E_\Sigma^2 + m_n^2 c^4 - m_\pi^2 c^4 + p_{nx}^2 c^2}{2E_\Sigma}$.

$E_n = \dfrac{(1377 \text{ MeV})^2 + (939.6 \text{ MeV})^2 - (139.6 \text{ MeV})^2 + 4.633 \times 10^5 \text{ (MeV)}^2}{2(1377 \text{ MeV})} = 1170$ MeV.

$K_n = E_n - m_n c^2 = 1170$ MeV $- 939.6$ MeV $= 230$ MeV.

$E_\pi = E_\Sigma - E_n = 1377$ MeV $- 1170$ MeV $= 207$ MeV.

$K_\pi = E_\pi - m_\pi c^2 = 207$ MeV $- 139.6$ MeV $= 67$ MeV.

$p_n^2 c^2 = E_n^2 - m_n^2 c^4 = (1170 \text{ MeV})^2 - (939.6 \text{ MeV})^2 = 4.861 \times 10^5$ (MeV)^2. The angle θ the velocity of the neutron makes with the $+x$-axis is given by $\cos\theta = \dfrac{p_{nx}}{p_n} = \sqrt{\dfrac{4.633 \times 10^5}{4.861 \times 10^5}}$ and $\theta = 12.5°$ below the $+x$-axis.

EVALUATE: The decay particles do not have equal energy because they have different masses.

44.53. IDENTIFY and SET UP: For nonrelativistic motion, the maximum kinetic energy in a cyclotron is $K_{\max} = \dfrac{q^2 R^2}{2m} B^2$. The angular frequency is $\omega = |q|B/m$.

EXECUTE: (a) The rest energy of a proton is 938 MeV, and the kinetic energies in the data table in the problem are around 1 MeV or less, so there is no need to use relativistic expressions.

(b) Figure 44.53 shows the graph of $K_{\max}$ versus B^2 for the data in the problem. The graph is clearly a straight line and has slope equal to 6.748 MeV/T^2 = 1.081×10^{-12} J/T^2. The formula for $K_{\max}$ is

$K_{max} = \dfrac{q^2 R^2}{2m} B^2$, so a graph of K_{max} versus B^2 should be a straight line with slope equal to $q^2 R^2/2m$.

Solving $q^2 R^2/2m =$ slope for R gives

$$R = \sqrt{\dfrac{2m(\text{slope})}{q^2}} = \sqrt{\dfrac{2(1.673\times 10^{-27}\ \text{kg})(1.081\times 10^{-12}\ \text{J/T}^2)}{(1.602\times 10^{-19}\ \text{C})^2}} = 0.375\ \text{m} = 37.5\ \text{cm}.$$

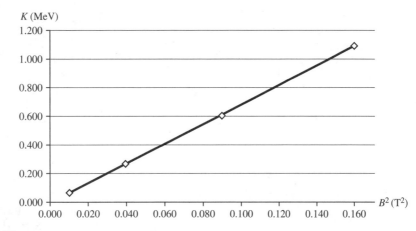

Figure 44.53

(c) Using the result from our graph, we get
$K_{max} = (\text{slope})B^2 = (6.748\ \text{MeV/T}^2)(0.25\ \text{T})^2 = 0.42\ \text{MeV}.$

(d) The angular speed is $\omega = |q|B/m = (1.602\times 10^{-19}\ \text{C})(0.40\ \text{T})/(1.67\times 10^{-27}\ \text{kg}) = 3.8\times 10^7\ \text{rad/s}$.

EVALUATE: In part (c) we can check by using $K_{max} = q^2 R^2 B^2/2m = (qRB)^2/2m$. Using $B = 0.25$ T and the standard values for the other quantities gives $K_{max} = 6.75\times 10^{-14}$ J $= 0.42$ MeV, which agrees with our result.

44.55. IDENTIFY and SET UP: Construct the diagram as specified in the problem. In part (b), use quark charges $u = +\dfrac{2}{3}, d = \dfrac{-1}{3}$, and $s = \dfrac{-1}{3}$ as a guide.

EXECUTE: (a) The diagram is given in Figure 44.55. The Ω^- particle has $Q = -1$ (as its label suggests) and $S = -3$. Its appears as a "hole" in an otherwise regular lattice in the $S - Q$ plane.

(b) The quark composition of each particle is shown in the figure.

EVALUATE: The mass difference between each S row is around 145 MeV (or so). This puts the Ω^- mass at about the right spot. As it turns out, all the other particles on this lattice had been discovered already and it was this "hole" and mass regularity that led to an accurate prediction of the properties of the Ω^-!

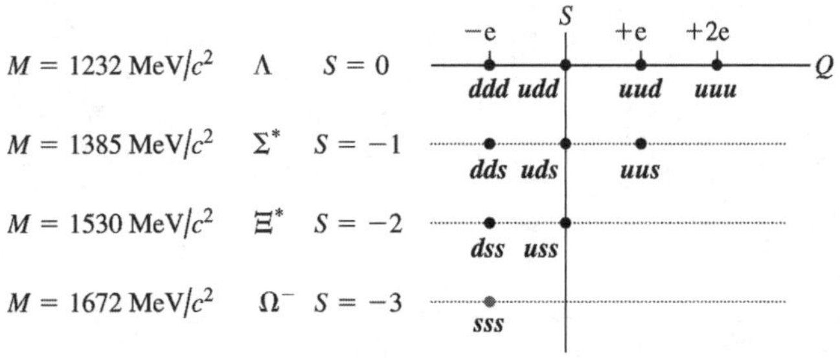

			$-e$	S	$+e$	$+2e$
$M = 1232$ MeV/c^2	Δ	$S = 0$	ddd udd		uud	uuu	Q
$M = 1385$ MeV/c^2	Σ^*	$S = -1$	dds uds		uus	
$M = 1530$ MeV/c^2	Ξ^*	$S = -2$	dss uss			
$M = 1672$ MeV/c^2	Ω^-	$S = -3$	sss			

Figure 44.55

44.59. **IDENTIFY** and **SET UP:** The absorption of photons obeys the equation $N = N_0 e^{-\mu x}$, where $\mu = 0.1$ cm^{-1}.

EXECUTE: $N/N_0 = e^{-\mu x} = e^{-(0.1 \text{ cm}^{-1})(20 \text{ cm})} = 0.14 = 14\%$. Choice (c) is correct.

EVALUATE: If 14% of the photons exit the body, 86% were absorbed within 20 cm of tissue.